# FROM ABOVE

# CRITICAL WAR STUDIES SERIES

## Series Editors

Tarak Barkawi (Department of Politics, New School for Social Research) and Shane Brighton (Department of International Relations, University of Sussex)

War transforms the social and political orders in which we live, just as it obliterates our precious certainties. Nowhere is this more obvious than in the fate of truths offered about war itself. War regularly undermines expectations, strategies and theories, and along with them the credibility of those in public life and the academy presumed to speak with authority about it. A fundamental reason for this is the frequently narrow and impoverished intellectual resources that dominate the study of war.

Critical War Studies begins with the recognition that the unsettling character of war is a profound opportunity for scholarship. Accordingly, the series welcomes submissions from across the academy as well as from reflective practitioners. It provides an open forum for critical scholarship concerned with war and armed forces and seeks to foster and develop the nascent encounter between war and contemporary approaches to society, history, politics and philosophy. It is a vehicle to reconceive the field of war studies, expand the sites where war is studied, and open the field to new voices.

DAVID KILCULLEN
The Accidental Guerrilla—Losing Small Wars in the Midst of a Big One

PATRICK PORTER
Military Orientalism—Eastern War Through Western Eyes

ANTOINE J. BOUSQUET
Scientific Way of Warfare—The Order and Chaos on the Battlefields of Modernity

AARON BELKIN
Bring Me Men—Military Masculinity and the Benign Façade of American Empire, 1898–2001

PETER ADEY, MARK WHITEHEAD, ALISON J. WILLIAMS
From Above—War, Violence and Verticality

PETER ADEY

MARK WHITEHEAD

ALISON J. WILLIAMS

# From Above

## *War, Violence and Verticality*

For the *Critical War Studies Series*
Edited by Tarak Barkawi and Shane Brighton

HURST & COMPANY, LONDON

First published in the United Kingdom in 2013 by
C. Hurst & Co. (Publishers) Ltd.,
41 Great Russell Street, London, WC1B 3PL
© Peter Adey, Mark Whitehead, and Alison J. Williams
and the Contributors, 2013
All rights reserved.
Printed in India

The right of Peter Adey, Mark Whitehead, and Alison J. Williams
and the Contributors to be identified as the authors of this
publication is asserted by them in accordance with the Copyright,
Designs and Patents Act, 1988.

A Cataloguing-in-Publication data record for this book is available
from the British Library.

ISBNs: 9781849042987 hardback
       9781849042994 paperback

**www.hurstpublishers.com**

This book is printed on paper from registered sustainable
and managed sources.

# CONTENTS

# CONTENTS

# EDITORS AND CONTRIBUTORS

**Peter Adey**, Professor of Geography, Department of Geography, Royal Holloway University of London, peter.adey@rhul.ac.uk

**Ben Anderson**, Reader, Department of Geography, Durham University, ben.anderson@durham.ac.uk

**John Armitage**, Professor of Media Arts, Winchester School of Art, University of Southampton, J.Armitage@soton.ac.uk

**Ryan Bishop**, Professor of Global Arts and Politics, and Co-director of the Winchester Global Centre for Futures in Art, Design and Media, University of Southampton, r.bishop@soton.ac.uk

**David Campbell**, Independent Scholar, Documentarian, Producer and Videographer, david@david-campbell.org

**Martin Coward**, Senior Lecturer in International Politics, School of Geography, Politics and Sociology, Newcastle University, martin.coward@ncl.ac.uk

**Jordan Crandall**, Professor and Chair of the Visual Arts Department, University of California, San Diego, actor@jordancrandall.com

**Klaus Dodds**, Professor of Geopolitics, Department of Geography, Royal Holloway University of London, K.Dodds@rhul.ac.uk

**Derek Gregory**, Peter Wall Distinguished Professor, Peter Wall Institute of Advanced Studies and Department of Geography, University of British Columbia, derek.gregory@geog.ubc.ca

# EDITORS AND CONTRIBUTORS

**Caren Kaplan**, Professor of American Studies, University of California, Davis, cjkaplan@ucdavis.edu

**Trevor Paglen**, Sound/video/installation artist, writer, and cultural geographer, Department of Geography, University of California, Berkeley, paglen.press@gmail.com

**James Robinson**, Human Geography Teaching Fellow, Department of Geography, University of Leicester, jpr23@le.ac.uk

**Paul K. Saint-Amour**, Associate Professor and Graduate Chair of English, Department of English, University of Pennsylvania, psain@english.upenn.edu

**Priya Satia**, Associate Professor of Modern British History, Department of History, Stanford University, psatia@stanford.edu

**Mark Whitehead**, Professor of Geography, Institute of Geography and Earth Sciences, Aberystwyth University, msw@aber.ac.uk

**Alison J. Williams**, Lecturer in Human Geography, School of Geography, Politics and Sociology, Newcastle University, Alison.williams1@ncl.ac.uk

# LIST OF IMAGES

## LIST OF IMAGES

# INTRODUCTION

## VISUAL CULTURE AND VERTICALITY

*Peter Adey, Mark Whitehead* and *Alison J. Williams.*

*Above All*

A grid of images pervades the screen. Partitioned into rectangular boxes with a rolling banner of text at the bottom, a news channel is broadcasting a city landscape at night. The other boxes on the screen display maps of military positions, logistics trails and aerial photography of targets. The sky suddenly erupts into colourings of flashes and fireworks; hues of red and yellow flare the camera lens. A khaki-clad reporter, noticeably wincing at the blasts, stands in the foreground explaining the beginning violence. These were the spectacular pictures of contemporary aerial warfare as operations such as Iraqi Freedom got underway. In the conflicts of today, penetrating weaponry are dropped from B-52s and F-117 bombers and launched from drone aircraft. Their perspectives are famously displayed in grainy black and white images which hone in to their distant operator's target. Their missions are rehearsed by aerial reconnaissance techniques that produce imagery of enemy installations and positions which can be seen almost holographically.[1]

When the United States Air Force (USAF) revised their slogan in 2008 they chose the simple motto: 'Above All.' Full stop. As suggested in the motto, being above is powerfully strategic. Height and verticality are values that are commonly associated with dominance and the projection of force. The USAF motto implies a totalising position difficult to equal. But as we have alluded to already, violence, security and a whole terrain of movements, technologies, practices and representations—like those portrayed on the hour and every minute during the war in Iraq—rely on height and the vertical. This book asks difficult questions of this view, as for all its spectacle and beauty, we must be careful not to celebrate it. As Stuart Elden puts it so simply, to consider the view from above can tell us a great deal about how the globe is 'divided, reformulated, fractured, and ruined'.[2] Thinking about the view from above can tell us, in short, about the remaking of the world through security, violence and war—a world of horizons bound together through what Stephen Graham has called a vertical geopolitics.[3]

The view from above all is deeply and historically contingent, and so this book is not simply an account of the present.[4] The aeroplane's position was another step in the development of a strategic aerial view or what David Matless describes as 'sky-situated knowledge'. In landscape survey, it was a simple progression from the position of the high hill or mountain; later came the balloon in Europe's and America's nineteenth-century wars, a step-up from the height of the horse-mounted cavalry or the nearest high point. The balloon's stable verticality would eventually be compared with the skyscraper, which brought with it, perhaps, the first position to permit a solid and persistent airminded way of knowing and being.[5] It transformed even the way we think about the past. In aerial archaeology—born from military aerial surveying in Palestine—the ground became a palimpsest of societal transition—an archive of prior activities that could be read into the landscape from above.[6] This book seeks to account for the seismic shifts which the changing view from the air has brought for life on the ground.

Being able to see and reach down from the aeroplane's platform has clearly permitted entirely new and violent cultural practices that have transformed the world below. Aerial knowledge has enabled the design, planning and bulldozering of one's way through the urban quagmire, from the emergence of bombing on cities in Guernica and during the Second World War.[7] Some years prior, it made legible *oriental* cities and villages to colonial air policing in the 1920s and 1930s through a form of aerial punishment, harassment (strafing and disrupting sleep) and economic repression (destroying resources) that would bring tribal leaders into line and populations into the purview of colonial government, administration and taxation. Indeed, Mark Neocleous has argued that we should think of airpower *as* police power.[8] Today the aerial view appropriates the city with similar enmity, producing visible, modern subjects capable of being identified, governed and even killed from a distance.[9] In this book we see how the techniques of government flown from the skies have brought populations into the terrain of state legibility and security so that they might become governable subjects. From map-making, aerial survey and photogrammetry,[10] the view from the air is complicit in producing, sustaining and eroding territorial sovereignty on the ground below.

The apparently precise and surgical conduct of urban counterinsurgency points to the intersection of the aerial view with scientific expertise. As a site from which atmospheric experiments were first born, *From Above* accounts for the aerial position as an imagined rational, scientific and epistemological space: it is a domain of testing, analysis, experimentation and, of course, exploration.[11]

As an interface of science, ways of seeing and militarism, there are few perspectives more culpable in their enlistment into practices of war, violence and security than the aerial one. In this collection of chapters by leading academic figures and interlocutors, we seek to explore the aerial view in a new depth, clarity and sustained analysis. Drawn in vivid detail from studies of today, to rich empirical investigations of the aerial view in the past, the chapters orbit around the politics of seeing from and being seen from the skies. The book explores how the aerial view has been radically significant for life, culture and the future of society as they have been transformed by violence, war and security. We now outline the book's three main sections and set out some of the key problematics and areas of interest the chapters contribute to.

*Science, Militarism, Distance*

The aerial view has been complicit with militarism, security and geopolitical pursuit, and the chapters in this book illustrate the complex conjunction between science and militarism, especially their truth claims.[12] That is not to say that the view from above is only imagined through those discourses and technologies concerned with killing or projecting

3

power. Of course, knowing the skies went hand in hand with inhabiting them, raising ourselves up to immerse the human body and its instruments in the winds and weathers of the sky.[13] Science meant seeing. Elevating the self to the sky could tell us more about our relationship with the earth. But as Caren Kaplan argues in her entry on the balloon prospect, intellectual pursuit was soon accompanied by the geostrategic occupation of the balloon platform in warfare. The arrival of the aerostatic balloon at the end of the nineteenth century ushered in a new perspective on the battlefield, taking over from the mount, the hill from the edge of the field of combat and the fortified tower positioned within it. In one sense, the development of aerial technologies, right up to today's drone wars fought in Afghanistan, the FATA regions of Pakistan, Somalia and Yemen, not to mention in the homeland and new domestic markets of the US–Mexico borderlands,[14] produce the view from above as a scientific, rational and calculative space. They flatten and abstract, just as they distance from the horror they might inflict. What we learn from the chapters in this book, however, is that this presumption was, and still is, far from the case.

Distance

Our notions of 'distance' have been technologically and culturally transformed, especially through militarism.[15] As Dodds shows in his chapter, the aerial view would enable new measures of touch or geopolitical projection at a distance, while it created numerous problems of coordination. In the context of British civil servants in London seeking to reinforce Britain's territorial reach into the Antarctic Peninsula, aerial survey became the method of choice in the production of new maps of the Falkland Islands Dependencies coming under Argentinean and Chilean pressure. Sovereignty claims were simultaneously metaphorically *and* literally projected on to the representational space of the charts and maps produced by aerial survey. Aerial knowledge of the Dependencies would help exert or 'project' Britain's sovereignty over them.

In some contexts, distance might seem then to justify the intervention of the aerial view. In other instances it does not simply justify but permits its execution. For example, it is the precise abstraction of the view from above that appears to make dropping bombs on people's heads— just from a very long way away—that bit more appetising and conscionable.[16] Distance permits all manner of sins in the history of airpower, and this book takes further strides in deepening our understanding of this relationship.[17]

And yet, in today's extension and intensification of the practices of aerial war, conducted from air-conditioned air force bases in Nevada and Arizona, thousands of miles away from the drones they are actually flying,[18] it does not necessarily follow that this articulation of distance is set to continue. Rather, as Derek Gregory shows, the feeling of distance is complicated by these processes, and in some instances it is intensified not lessened. In Gregory's historical contextualisation of the 'how' of drone wars, we learn how these practices have rarely occurred without public outcry, especially in the context of the first substantive use of drones during the Vietnam War.[19] What's more, today's drone wars appear to be actually transforming our senses of distance from above. The machines that now provide reconnaissance and wage violence, controlled from so far away, do something very different with distance because they are really intended to compress it, to make the 'kill chain' shorter. Might this actually make killing harder?[20] Does this bring a pilot (and an audience) that bit closer to a target? As we see in Satia's chapter later in the book, questions of bombing are not a million miles away, amazingly, from questions of love and even lust in Saint-Amour's. Love not for technology, not only for the thrill of being above, but love as an explicit 'justification for wreaking unearthly violence'. Or as John Armitage finds in theorist Paul Virilio, love for the 'blurring of destruction'. This book makes significant moves to understand the view from above within the pathos and passions of the societies that have produced and consumed it, perspectives that art, literature and other forms of expression have been more used to exploring.[21]

Grasp

Is the view from above all that simple to grasp? Underpinning the aerial view might be far less rationality and calculative certainty than we think, but actually a high degree of chance, luck and confusion.[22] Many of the chapters in the book do not shy away from but embrace the often messy and muddy waters of the view from above. For Saint-Amour, the putative distance of the aerial view is made up as much by its abstraction and the many conventions implicit within techniques such as aerial photogrammetry (which disciplined the image into the production of a map), as it was by the distortions created by those techniques of drawing and rhetorically drawing the 'accuracy' of the image.[23] Moreover, as the use of these images moved from a war-time context of the First World War to a time of peace-building and city-planning, the seams, distortions and 'events' of the image became information not to purify away as noise, but highly valuable site-specific detail.

Thus, as most of our contributors show, and specifically for Coward in the coalition's 'Shock and Awe' displays above Baghdad in 2003 (the scene we opened this introduction with), the view from above has actually been very difficult to comprehend and make sense of, despite the political rhetoric that tells us otherwise in the portrayal of 'surgical' and 'precise' and 'targeted' interventions.[24] The overwhelming vastness of the sky and the land below, or the incomprehension of displays of military might, become too much to take in. The endless space of the Antarctic, for Dodds, was culturally and scientifically impossible to grasp. The Falkland Islands Dependencies Aerial Surveying Expedition (FIDASE) ultimately provided one of the first means to 'know' the geographical landscape and what lay beneath the ice in the British attempts to map the 300,000 square miles or so of the Antarctic Peninsula region. But even then, too much light or darkness could dazzle the camera exposures or confuse the imagery's distinction between land and sea, just as Shock and Awe could daze and astonish its audiences with displays of awesome explosions.[25] A similar danger lurks in Coward's analysis. Could the data rendered actionable by the Revolution in Military Affairs easily become next to useless given the speed and proliferation of information?[26] Even the repetition of the trope of the network in this brand of 'network-centric' warfare can become seemingly endless. How far should the network be drawn? When should it be closed? As Coward and Gregory suggest, the *precision* implied in targeting is that no weapon or enemy will escape the boundaries of its points and lines.[27] The network, as opposed to territory, is supposed to be politically and legally that much cleaner, a rhetoric several of the chapters in this volume put to test.

Ground

The view from above might appear weightless, ungrounded and light. We will see that this is very far from the truth. Chapters in the volume

6

expand on the aerial view's supports, not by fleeting ephemeral forces or abstract networks, but by static, hard and rigid things, most notably *grounded* infrastructure. That is, the contributors ask through what chaotic and tumultuous conditions is the aerial view constructed and produced? How might this actually frustrate and tether the aerial gaze to a grittier sort of reality than perhaps the one we are normally presented? Thus for Kaplan, in the exploits of aerostatic balloonists and reconnaissance observers beginning in 1794, the experience is rough and frightening. Balloonists are exposed to the elements, over which they have little control. Even over the balloon itself, pilots have little motive force in the face of turbulent weathers.[28] In the FIDASE surveys, Dodds explores the labour, energy and materials such a programme of mapping entailed. We learn of helicopters and moving experts; heavy and fixed machinery, sometimes requiring transport; permanent bases and camps; runway strips. The colonial project's exertion of sovereign authority, he explains, was so often frustrated by high winds and extreme cold (–30 degrees C), with the heaters in the aircraft proving insufficient; blanket-thick cloud cover and choppy sea conditions preventing the amphibious aircraft landing or taking off. Hardly the clean and efficient scenes of the 'killing zone' that we are led to expect. Thus while it is easy to see the view from above as avoiding 'boots on the ground'—a perspective commonly espoused by or 'under' the Obama administration as part of its support for the use of drones in targeted killing—this volume illustrates the substantive costs, as well as the personnel, resources and physical infrastructure—perhaps the drone's tail end of supply, logistics and the exploitation of informants—required on the surface.[29]

*Aerial Aesthetics, Distortion and the View from Below*

The view from above has always been dialectically entwined with the struggle to be concealed from below. At each stage of the gradual human ascent to mastery of the skies, aerial counter-strategies have been sought in the subterranean spaces of concealment and the camoufleur's tricks of the eye.[30] The authors in this section explore the dialectical interplay between ground and sky through a series of conceptually informed empirical studies. Particular emphasis is placed on how the view from above is prone to forms of natural distortion and intentional obfuscation. At the same time, the different chapters illustrate how the aesthetic perspectives that are generated from the air are not only prone to deceptive measures, but are also subject to constant technological refinement and improvement. Ultimately, this section places particular emphasis on the importance of studying the experiences and aversions of those under the drones and keyhole satellites—to take the view from below seriously.[31] As the different chapters show, exploring the view from underneath the aerial gaze proves to be a source of despair, but also empowerment and hope.

Distortion and Obfuscation

Whether it is from the lookout tower or the satellite, the view from above has traditionally been associated with the empowerment of the elevated viewer. In his classic account of early modern government, James Scott recognised that *seeing like a state*, whether militarily, cartographically or administratively, was akin to an airborne and synoptic superiority.[32] From this perspective, we will see how the view from above is connected to enhanced forms of legibility, in and through which nature, peoples and settlements can be dislocated from their everyday complexities and seen (and governed) in the abstract (processes Gregory describes in his chapter). The chapters in this section explore the complex struggles that surround the construction of vertical forms of legibility.[33] Verticality, it seems, does not equal automatic and sure-footed sight. On the one hand, James Robinson charts the emergence of civil camouflage in interwar Britain. In this chapter we see how the machinery of the state combined with the work of artists, architects and engineers in an attempt to perfect the arts of distortion and obfuscation as forms of defence against the all-seeing bomber. On the other hand, Ryan Bishop's chapter explores the emergence of the US Department of Defense's programme 'Transparent Earth'. 'Transparent Earth' is a response to the natural obfuscation that the earth's surface generates to the view from above. As satellite surveillance has been enhanced, weapon systems, nuclear facilities and so-called deep underground military bases (DUMBs) have all sought illegibility within the subterranean realm. Drawing on the latest military-sponsored research, Transparent Earth reflects an attempt to generate a new era of aerial legibility, of deep penetration where there is no cover.[34]

In this section the practices of distortion and obfuscation are presented as more than merely the counter-strategies of those below against those who see from above.[35] In his analysis of the relationship between the US National Reconnaissance Office and the space programme, Trevor Paglen uncovers the covert strategies that have surrounded the 'black space race'. In an attempt to hide from view the strategic development of orchestrated, global satellite surveillance systems in space, Paglen tells of how the classification of shuttle payloads has prevented the public actually knowing what is going up in order to look down. To see the intelligence-gathering high-altitude spy planes or satellites is as if looking through a telescope from the wrong end. At the same time, the use of the NASA logo on covert space missions has enabled the US government to 'hide-in-plain sight' its other race into space. But the practices of obfuscation and distortion do not stop there. Through the development of complex cover stories about the nature of space shuttle launches, and even about the purported loss of satellites, the National Reconnaissance Office has been able confuse even an attentive public about the nature and extent of its space-based surveillance infrastructure. Furthermore, the development of stealth satellite technology has enabled once highly legible (even if officially secret) satellites to escape from view.[36] To these ends, this section appears to complete the historical arc from the development of ground-based camouflage to the camouflaging of space-based surveillance technology.

Aesthetics

The connections between the aerial perspective and evolving aesthetic sensibilities are well established. From the Italian Futurists' engagement with the aesthetics of speed and violence associated with the aeroplane, to the environmental movement's utilisation of Apollo 8's iconic *Earth Rise* photographs as the basis for a new form of ecological aesthetic, the view from above has facilitated a range of new aesthetic potentials.[37] In this section authors excavate some of the complex connections that exist between the aerial perspective and aesthetics. In his analysis of the Transparent Earth project, Ryan Bishop draws attention to the sense of aesthetic limitation that the view from above generates. Bishop contrasts the horizontal perspective with its associations with '[l]ight, distant vision, uncluttered horizon' and the vertical view from above, which must end at the surface of the ground. In many ways Transparent Earth is part of the ongoing dialectic between surface/grounded and the aerial/

elevated, as the power of the view from above is undermined by defensive measures of coverage on the ground, only to be usurped by advances in the aerial perspective.

What is aesthetically interesting about the Transparent Earth project, however, is that it seeks to overcome the aesthetic limitations of the view from above through a *synaesthesia*: the deployment of multiple sensory technologies that might have once provided the Clausewitzian 'fog' or 'friction' of war offer a different kind of clarity.[38] Once favoured by avant-garde artists as a creative mode of multisensory perception, the synaesthesia of Transparent Earth is a military experiment that seeks to see through sound. The fact that '[s]ound will let us see where vision stops' enables aerial aesthetics to move from the surface and geographical readings of the ground below to much deeper, hermeneutic and geological perspectives on the terrestrial sphere.[39] This is particularly critical as we see subterranean spaces used for interesting and important forms of political activism, and a far greater attunement to the geological and stratigraphic shaping of geopolitics.[40] While synaesthesia may not be a uniquely aerial phenomenon, it is clear that expanded military mastery of the view from above is generating a new, and expanded, context for the creative experimental combination of multiple aesthetic perspectives.

John Armitage's chapter, 'Flying into the Unknown', exposes a further terrain upon which war, aesthetics and aerial perspectives appear to coalesce. Drawing on Paul Virilio's conception of the 'aesthetics of disappearance', Armitage explores the links between contemporary cinema and aerial warfare.[41] Through the varied writings of Virilio, Armitage charts the co-evolution of the cinematic aesthetic and *aerialised war*.[42] This relationship starts with the camera-kites and cinematography that military reconnaissance aircraft increasingly came to deploy,[43] but culminates in a more systematic amalgamation between cinematic vision and aerial conflict. Through their common concern with 'indirect forms of partial vision', which are forged not from a static but from mobile perspectives, Armitage uncovers the coupling of film and war which constitutes a kind of cinematic-military complex. This cine-military perspective enables the book's attention on visual culture, to see how technological advances in warfare and film-making are part of the same system, with cinema ultimately contributing to the perceptual arsenal of military society, and cinema certainly beholden to the lens of militarism.[44]

# INTRODUCTION

From Below

Analyses of the aerial perspective naturally tend to prioritise the view and experiences of those who occupy the elevated perspective. Yet the view from above has obvious implications for the actions and practices of those who are being viewed and targeted below. In his analysis of the development of civil camouflage in interwar Britain for the protection of 'vital' buildings and infrastructure, Robinson describes how airmindedness is both an aerially constituted and terrestrial condition. In his analysis of the government's civil camouflage research unit, Robinson argues that, '[I]t was a technology and a practice which embraced the aeroplane and which encouraged its practitioners—the camoufleur—to "think", "act" and "see" aeronautically'. Resisting the view from above would involve shifting perspectives and subject positions in elaborate and creative ways that would serve to contest some of the dominant writings on airmindedness.[45]

A key theme that is touched upon by authors in this section is how being seen from above changes the mentalities and practices of those below.[46] But living below the increasingly extensive infrastructures of aerial surveillance and violence involves more than merely the creation of aeronautic neurosis, but the empowerment of terrestrial denizens. While Ryan Bishop describes a gradual retreat to the subterranean, Robinson and Paglen outline other paths of terrestrial resistance. Robinson illustrates how only by understanding what seeing from above was like could it be undermined. The camoufleurs routinely tested what their designs looked like from the air and its various vertical perspectives—from 'birds eye' to 'oblique'. They examined the atmospheric distortions caused by weather and changing light. Enhanced scientific understanding of the differences between the horizontal, vertical and oblique perspectives enabled terrestrial dwellers to think more clearly about how their earthly dwellings and everyday routines were perceived from above. Experiments explored the best techniques through which aerial targets could mimic nature to use glare, shadow and dazzle. If the book therefore indicates a gradual improvement in the ability of those below to think and see like those above, we also see in Paglen's account of amateur spy satellite trackers how, if properly trained and motivated, those below can begin to see and track those up above. The view from above is flipped. In the incredibly sophisticated and coordinated practices of satellite trackers, as shown in other contexts of the reporting of CIA

11

rendition jets during the many torture-by-proxy flights of the 'war on terror' intelligence machine, we begin to see a process whereby 'ordinary' citizen-publics expose the visual vulnerabilities of those above to surveillance from below.[47] *From Above* is thus as much about the view from the air as it is about transforming the way we think about the political possibilities of harnessing the view from below—perhaps to scrutinise or resist those activities from on high.

*From Close to Remote*

Scale and distance are, of course, key geographical components in any analysis of the ways in which the aerial view from above is conducted, while easy to take for granted within much wider narratives of the shortening or compression of time and space.[48] The chapters in this section are drawn together through the physical distances and changing scales of encounters that are affected by air- and space-borne technologies. But we must not be lulled into thinking that these spatialities occur along a linear spectrum; the movement from close to remote can and does occur in much more complex ways across a number of surfaces, through a multiplicity of spaces and at a plethora of times.[49] Thus to some extent the nomenclature employed here, 'From close to remote', is misleading, as the chapters in this section do much more than describe this linear trajectory; in fact, we might if anything argue that the trend is in reverse, especially in light of—and somewhat oxymoronically—the cinematic smart-bomb videos deployed since the first Gulf War, which brought a new kind of intimacy of war to the home, a subject Judith Butler has explored at length. It is not as if war-reporting and photography was anything new: William Howard Russell's photographic writing had given realism to the Crimea, for example, and war-photographers had captured the US Civil War in detail. Instead aerial and satellite communicated imagery meant perspectives could be transmitted and consumed in almost real time and direct to the armchair of Western viewers.[50] At the heart of the four chapters in this section are a number of cross-cutting themes that serve to link their diverse conceptual and empirical analyses together: networks, the technologies of the gaze, but first truth.

# INTRODUCTION

## Truth

Is the view from above a truthful one? Cutting through the noise, the frustration and the deceit of the ground 'elevation secures the higher truth', writes Derek Gregory elsewhere,[51] such questions reflecting the close association between the aerial and scientific epistemologies discussed in section one. Satia's discussion of the RAF's air-borne 'air policing' of the Iraq Mandate during the interwar period provides extensive detail about the ways in which RAF aircraft offered the ability for the British military to rise above the complex and orientalised deserts of Iraq, and the sublime desert landscapes and peoples the British Arabists romanticised and mythologised. To occupy a space removed could enable a true picture of the topography and occupancy to be drawn.[52] Only by abstracting themselves from the sandstorms and mysterious Arab cultures and landscape could the British truly make sense of things and begin to understand how to go about exercising control over the territory even if, in their hearts, many never left the embrace of the ground. In a similar way, but through a very different example, Campbell's analysis of satellite imagery sees how the view from above is drawn as undeniable testimony and truth about the ground below. He draws upon then US Secretary of State Colin Powell's 2003 presentation of evidence to the UN on the existence of Iraqi weapons of mass destruction. Satellite imagery of supposed decontamination vehicles and chemical weapons leaving Al-Musayyib famously provided the accumulation of facts and the bedrock of an argument.[53] If the view from above can reveal things we might want to keep hidden, then the chapters reveal the powerful uses to which satellite imagery, drones and aerial perspectives can be put. This is particularly cogent when we see international agencies and private activist groups deploying the view from above for humanitarian purposes, such as Amnesty's 'Eyes on Darfur' and 'Drones for Human Rights' who have worked to bring exposure to issues from ethnic cleansing in Sudan, inappropriate and pernicious police and military tactics, to illegal environmental contamination.

However, in their critical analyses all four of the authors in this section express the limitations of such truth-claims, illustrating the extent to which the aerial view and the technologies that enable it are engaged in architectures of myth-making. Satia discusses the extent to which the RAF's pilots in Iraq described the region in the enigmatic and magical terms of a desert sublime, and how they themselves were portrayed as

knights of the air, chivalrously flying their aircraft to control the ground
below through their ennobled omnipotence. Crandall's detailed exegesis
of the drone excavates other perceived mythologies, illustrated most
succinctly in his analysis of the 'gorgon stare' hardware that enables a
drone to record activities through up to nine cameras simultaneously,
creating a postmodern, multi-headed all-seeing technological beast.
Campbell's critique of the dependence on satellite images is critical of
the conviction that what these images portray is the truth.[54]

Networks

The concept of the network appears to differing extents through all the
chapters in this section. Perhaps most obvious is its treatment in Anderson's chapter, which focuses extensively of the notion of network-centric
warfare and the identification of the target in a network of interlinked
sites, decisions and possibilities. In critiquing the Shock and Awe doctrine of the US Air Force (and its associations with Warden's effects-
based warfare), Anderson draws our attention to the ways in which air
warfare's historical and continuing fascination with the targeting of
morale exists as part of a networked approach to combat.[55] These 'chains
of association', as he terms them, can be seen to spread out to incorporate a target population, air war strategists, government ministers and
aerospace industry contractors, as well as military personnel and the
machines they are responsible for. Crandall's chapter is also deeply concerned with teasing out the implicit and explicit networks that work to
enable drones to operate. His scale of focus switches between examining
the effects of drone crashes to uncovering the complexities of the technological systems that enable their aircraft to fly, yet his argument consistently draws upon the notion of the network as a key descriptor. In
this case we understand the network as an assemblage of human and
non-human actors all working together to enable successful drone
operations, or failing as in the case of the many drone crashes Crandall
uses to illustrate his arguments.[56] Even within her historical examples,
networks can also be perceived in Satia's chapter. For her, the RAF's
air-policing policy in Iraq existed as part of a wider network of distanci-
ated power and was enabled by a localised collection of ground-based
RAF personnel and Iraqi civilians, who formed the focus of the RAF's
operations. Similarly, Campbell's chapter elucidates the scale of network

required to enable satellite images to be utilised. He describes the historical and continuing need for teams of photo analysts whose role it is to view all of the images returned by 'their' satellites and to 'decode' them in order to make them understandable to a non-expert public. Thus despite the aircraft's apparent freedom from the lines and tethers of its nineteenth-century observation balloon counterpart—anchored to the ground by rope—this volume shows how the view from above is heavily nested within geographically extensive yet tightly coupled networks of coordination, communication and association.

Technologies of the Gaze

What all of these chapters have at their core is a concern to uncover the ways in which humans have and continue to use air- and space-borne technologies to achieve an Apollonian gaze, to borrow Cosgrove's term.[57] The chapters illustrate the levels of technology that are now at the disposal of Western militaries in their continuing desire to map territory and achieve a persistent all-seeing downward stare. Perhaps what one could take from this book is the sense that to look up at the sky today will probably involve staring—unknowingly—in the direction of numerous aerial vehicles and satellites whose eyes and sensors are pointed right back at us. However imperfectly this is achieved, Crandall and Campbell's detailed chronology of the development of drone and satellite and visual technologies illustrates how technological developments have sharpened the focus and expanded the possible in terms of capturing images of the earth from air and space. Both of these chapters elucidate the growing gap between the more-than-human imaging technologies built into these gazing devices and the need to maintain a cohort of highly trained humans to sort, analyse and interpret this data into a usable form.

Above all, the view from above is difficult to place, primarily because it is far from a flat or superficial take. The view from above is a thick field of depth—a hologram we evoked at the start of this chapter—it is less than an image, but a form made by the arrangement of many layers, some invisible. It is thick and dense despite the fact that it has often tried so hard to remove visual depth to an abstraction of surface points. Consider the legal principle *Cujus est Solum* that framed some of the early debates surrounding the establishment of aviation law and state sovereignty over the skies.[58] Those who advocated it proposed a seemingly limitless volume of territorial rights, up from the surface of the earth projected skywards. Almost emulating this short-lived claim in the evolution of air-law, the aerial view performs a deep and penetrating look. And more often than not, it is not only through sight that it sees. Looking, or targeting or tracking, it is more than a visual register or the perpendicular lines of the cross-hair, but a manner of sensing visibilities and invisibilities.[59] So we must think more of sites rather than just sight. The aerial view is embedded in different spaces and contexts. Furthermore, we have discussed how the view from above is distributed within systems and networks, embedded in practices and organisations and embodied in passionate subjects. Nor is the view, and the systems that support it, necessarily successful, complete or hegemonic. The bomb does not always find its target; the aerial view is never quite at the resolution or fidelity that its masters might hope for.

Depth, then, does not imply clarity or simplicity. Rather, the domains in which the aerial view is conceived, devised, perfected, controlled but also fractured, can boil and whirl, shift and slip away from view. The view from above has a density, a substance and a volatility that the rich and detailed explorations arranged in this book seek to trace out.

# SECTION ONE

# SCIENCE, MILITARISM AND DISTANCE

1

# THE BALLOON PROSPECT

## AEROSTATIC OBSERVATION AND THE EMERGENCE OF MILITARISED AEROMOBILITY

*Caren Kaplan*

There is no war, then, without representation ...[1]
Paul Virilio, *War and Cinema: The Logistics of Perception* (1989) [1984].

It is hard to imagine a time when war did not rain down from the skies, when satellite images did not display targets for attack, and when the freedom of flight did not signal the power of life and death over what or whoever could be sighted below. After one hundred years of airpower, even in the face of evidence that today's wars are also very much fought on the ground, the belief that force from the air is the core of a nation state's military might remains pervasive. But airpower is a set of operations much more uneven than allowed for by its official accounts. Its histories stretch back much further than are usually acknowledged, and its iconic markers of emergence and achievement are multifaceted and

19

open for diverse interpretation. Certainly, the introduction of the airplane into the strategy and tactics of war created an enormous shift in world culture in innumerable ways, many of which remain to be studied and understood. But some of the structuring elements of airpower emerged long before aviation, including visual practices and modes of thinking that constitute a cultural through line in modernity. Historicising this through line without over-generalising similarities between what can also be seen as incommensurable aspects or making too-neat claims for a worldview contributes to the study of militarisation in modernity, in particular the militarised aeromobility that generates a dynamic, complex visual culture.

It is incontrovertible that as airpower came into dominance by the end of the First World War, it played a major role in generating what Peter Adey has termed 'aerial life', a culture that promotes national identity and security through activities related to aviation.[2] As Saulo Cwerner and others have argued, the rapid movement that characterises modern aerial life makes and remakes places and spaces, producing 'aeromobilities' as subjects and objects circulate in an increasingly globalised world.[3] This aeromobility of 'people, goods, and ideas' has been supported by state and industrial investments in scientific and commercial innovation and management.[4] It is possible to view airpower as the military segment of a more general aeromobility that marks late modern society, but it can also be argued that aeromobility is intrinsically militarised. That is, any asserted division between military and civilian aerial life masks the foundational practices of preparing for and waging war that support the security and boundary concerns of the state and its related institutions.[5]

Views from the air are part of the intrinsically militarised foundation of aeromobility in directly overt as well as in less obvious ways. Direct line of sight from above led first to the observation of enemy installations and positions, as well as, increasingly, to the delivery of weapons to specific locations on the ground.[6] Despite high hopes, the record of aerial bombardment in the First World War was checkered: the somewhat chaotic dropping of bombs from planes resembled more closely the process of heaving a weapon of destruction overboard and hoping for the best.[7] The terror of that kind of inaccuracy for populations on the ground was soon integrated into a practice that deliberately targeted civilians on the supposedly 'humane' grounds that their devastated morale would bring a war to a quick close (thereby saving the lives and property that would be lost in a more prolonged conflict).[8] After the First World War, the preferred populations and terrain for the European advocates of the developing doctrine of airpower were exclusively colonised.[9] Throughout the so-called 'interwar' period, the newly formed air forces of the major powers bombed and strafed people in territories they controlled who exhibited tendencies to rebel directly or indirectly. By the advent of the Second World War, the aviation industries and the air branches of the militaries were ready to put airpower into full operation with greatly increased tactical accuracy and strategic sophistication.[10] As aerial reconnaissance and photographic interpretation became more advanced, a heightened interest in precision bombing required even more investment in visual technologies.[11] Thus as the 1930s waned and war in Europe and the Pacific loomed, the 'eyes in the skies' became an integral part of weapons delivery as well as observation.[12]

This narrative of airpower as a specific doctrine and operation of militarised aeromobility is compelling, if horrible in its effects. But it privileges the airplane, especially the bomber, to a degree that begins to obscure the complexities of aeromobility in modernity, lending to aerial viewing an omniscient power that becomes naturalised by its own teleology. What the military does or does not do is a matter of vital concern to everyone in today's world. Yet what is less understood is how military ways of seeing and doing are similar if not foundational to areas of culture that seem unrelated to the project of state security or waging war.[13] Parsing this murky or seemingly tangential connection provides new ways of understanding militarisation not so much as a separate, unpopular part of modernity, but as a constitutive aspect of democratic nation states.[14] Acknowledging that militarisation is intrinsic to the political and economic operations of modern nation states also requires that we analyse cultural instances not so much as corrupted or touched by the evil that is war-making, but as part of a civilisation that enacts power in any way that it can, through objects as well as subjects of security and nationalism.[15] Aerial views are assembled through these operations of power and also generate power relations by creating differences through representational practices.[16] These differences—between military and non-military, soldiers and civilians, citizens and non-citizens, those who target and those who are targeted, for example—become operational as culture over time, circulating as ways of seeing, modes of making and reproducing art, industrial design and technologies, composing the ground of everyday life.

*Modernity's Aerial 'Watching Machine'*

Thus, alongside the 'war machine', there has always existed an ocular ... 'watching machine' capable of providing soldiers, and particularly commanders, with a visual perspective on the military action under way. From the original watchtower through the anchored balloon to the reconnaissance aircraft and remote-sensing satellites, one and the same function has been indefinitely repeated, the eye's function being the function of a weapon.[17]

When do views from above become consolidated in modernity as a form of scopic mastery, as components in a militarised visual culture machine? Does the 'watching machine', as Virilio puts it, originate in the Renaissance inventions of linear perspective or bird's-eye views? Or, does it stretch back much further to the geometry and cartography of classical Greece? Is the universalised 'god's eye' view of a world displayed in its entirety intrinsic to Judeo-Christian Western culture, or can it be linked to Confucian art or early Islamic philosophy and mathematics? While diverse fields and disciplines make specific claims for a generalised view 'from above', the link between aerial perspective and military aims has been advanced energetically by both cultural geographers and theorists of contemporary electronic or digital visual culture who are concerned with the modern period.[18] In this body of work aerial imagery is read as representing pure power, as a transcendental visual practice that brings together sight and knowledge, always 'one and the same function'.[19] As Denis Cosgrove has argued, this 'Apollonian gaze' gathers and unifies diverse elements through vision, real or imagined, thereby producing a 'mastering' view from a 'single perspective'.[20] The process of combining multiple perspectives into one, singular, panoptic 'view from the heavens' has powered various representations of not only terrain and individual communities but the Western, modern state as a political institution. From the French Revolution onwards, cultural representations of this ocular mastery became incorporated frequently into official state declarations, coins, and paper money.[21] Symbolised as the single eye, often at the apex or centre of a triangle, the 'watching machine' observed the nation state's population from on high, a 'supreme controller', transforming optical surveillance 'into a common practice' of reason.[22]

Bentham's late eighteenth-century design for a rationalised prison, the panopticon, could be considered to be the prototype of modernity's 'watching machine'. In this emblematic structure, the single eye is embodied by the representative of the state (the supervisor) who occupies a central, elevated position in order to achieve a total, circular view of inmates enclosed in segmented but viewable cells. Extrapolating from this specifically penal application, Foucault described a more general 'panopticism' in the nineteenth century that organised space through the creation of binary subjects who occupied different zones, with the least powerful coerced through 'differential distribution' and 'constant surveillance'.[23] This compelling description of an architectural structuring of social control, a 'state of conscious and permanent visibility that assures the automatic functioning of power', holds undeniable analytical force.[24] But as Kevin Haggerty has argued, the panopticon has become reified as a concept through ahistorical over-extension to any and all instances, thereby skewing the study of surveillance in favour of a 'select subset of attributes'.[25] Foucault himself cautioned against losing the historical specificity of his argument, writing that the 'procedures of power that are at work in modern societies are much more numerous, diverse, and rich' than can be accounted for by 'the principle of visibility' alone.[26] The incorporation of a political ideal of a 'transparent society' in France during the revolutionary period combined with panoptic social projects to produce sometimes conflicting aims and effects. As Martin Jay has commented, 'visual primacy was by no means without its complications'.[27]

Does this widespread cultural tradition of privileging a pervasive visuality inherently combine extreme military force with state power and technological innovation to transform all viewing, at least potentially, into a machine not only of controlling watchfulness but also of destruction? As Rey Chow argues, drawing on Heidegger's famous formulation, if modernity is marked by the propensity for scientific research to configure the world as a picture, in 'the age of bombing' that picture must now be grasped as a target, and 'to conceive of the world as a target is to conceive of it as an object to be destroyed'.[28] Heidegger's notion of the 'world picture' has been as widely generalised as the panopticon. What is useful for our discussion here is to remember that, following Kant and Descartes, the interiorisation of perception as a ground for knowledge led to the belief in modernity that all we can truly know is visualised within the mind. This 'picturing' or representation becomes a primary function of subjectivity, key to processes of identification and power, and has become extended to geographical pursuits such as mapping and other forms of visual information.[29]

The emergence of visual logics of control, or dominance through the visualisation of the 'whole picture' in the 'age of bombing', has been most persuasively demonstrated in relation to media and informatics.[30] As Paul Virilio has argued across numerous works, aviation and visual media such as cinema must be understood as weaponised technologies, arising in a specific time period from industrial innovations that privilege speed and scopic reach. Forecasting the visual culture now associated most closely with the first and second Gulf Wars as well as the expanded conflicts in Afghanistan, North Africa and the Middle East, Virilio has insisted that wars are contests of perception as well as destruction. Tracing the evolution of a visual culture of militarised state power from the first observation balloon at the Battle of Fleurus in 1794 through the height of airpower in the Second World War to the development of unmanned drones and the emergence of digital information systems, Virilio has declared that 'direct vision' is 'now a thing of the past'.[31] Accordingly, he claims that battlefields today resemble film sets, while the interpretation of images on screens has replaced the direct line of sight that used to be required for accurate observation and the effective firing of weapons. Thus the crux of Virilio's argument holds that a war of 'pictures and sounds' has replaced the war of 'objects (projectiles and missiles)'; 'the world disappears in war, and war as a phenomenon disappears from the eyes of the world'.[32]

In a time period when barely a day passes without news of the military's increasing reliance on unmanned drones for reconnaissance and targeting in wars around the globe, it is not difficult to concede Virilio's larger point on the logistics of perception.[33] A panoptic culture of scopic mastery emerged throughout the nineteenth century that generated fear as well as celebration of loss of autonomy and control. These concerns now resonate through our contemporary understanding of the powers of networked technologies. Accordingly, Virilio and the theorists of perception and technology in modernity remind us of the prevalence of sight as both a metaphor and a direct enactment of individual and state power in an era of increasing mobility and speed. But vision has meant different things to different people across time and space. The challenge for a critique of the visual culture of militarised aeromobility is to avoid, as Giuliana Bruno has warned, reducing all spectatorship to a 'fixed, unified geometry of a transcendental, disembodied gaze'.[34] Unfixing the gaze does not require denigrating or discounting vision or visual culture.[35] Rather, the project is to think through the ways in which militaries use vision as a means of making divisions between military and non-military, targets and non-targets, or kinds of terrain, for example, and then to note the ways in which these divisions are made and remade by observers of all kinds throughout culture in general.

*Lines of Sight: Aerostatic Observation as a 'New Instrument of War'*

One of the discoveries which has had the most astonishing effects, and strikes the imagination by the position it gives to man, by raising him on the wings of the wind, is the aerostatic machine, which … has since become for your Committee a new instrument of war, which our enemies have recognised as the pioneer to victory. The Convention will learn with interest that many *savans* [*sic*] have devoted ten months of zealous study to perfecting the art of aerostation, and to render it of easy use in camps, fortresses, and even on the theatre of war … Soon all our armies will have complete aerostatic companies …[36]

Militarised aeromobility first became realised through aerostatic flight on 26 June 1794. On that day the French revolutionary army faced a coalition of Austrian and Dutch forces in Fleurus, Belgium. Although General Jean-Baptiste Jourdan's 75,000 troops outnumbered the prince of Saxe-Coburg's 52,000, the French advantage in this decisive battle is often attributed not to numbers of soldiers but to the use of a completely new 'instrument of war', the observation balloon. Tethered to the ground by cables requiring the efforts of at least twenty handlers, the hydrogen-powered *L'Entreprenant* stayed aloft throughout the battle. Its two-person crew hauled tactical queries up by the tethering cables while sending replies and directives down by the same method. In addition to the reportedly spectacular advantage of viewing the entire battlefield from the air, the startling appearance of the balloon itself reportedly amazed and terrified the enemy troops, significantly affecting morale.[37]

From the moment that balloons were flown successfully more than ten years before the Battle of Fleurus, they were closely linked to fantasies of waging war from the air. Joseph Montgolfier was fully aware of the military potential of the hot-air balloon he had invented with his brother, Étienne. Referring to the conflict between France and England over the island of Gibraltar, he wrote: 'I possess ... a super-human means of introducing our soldiers into this impregnable fortress ... By making the bag large enough, it will be possible to introduce into Gibraltar an entire army, which, borne by the wind, will enter right above the heads of the English.'[38] This was not an isolated notion. One of the Montgolfier brothers' first human passengers, André Giraud de Vilette, immediately grasped the possibilities of military reconnaissance, writing soon after his historic flight: 'I was convinced that this apparatus, costing but little, could be made useful to the army for discovering the positions of its enemy, his movements, his advances, and his dispositions, and that information could be conveyed to the troops operating the machine.'[39] The spectacular flights that followed as physicist Jacques Alexandre César Charles launched his hydrogen-powered balloons to demonstrate their superiority to the Montgolfiers' hot-air method brought out almost half the population of Paris to watch each launch.[40] Serving as a US diplomat in Paris at the time, Benjamin Franklin reported back to his superiors that the extraordinary new aerostatic technology could be used not only for reconnaissance and communications but also, echoing Montgolfier, for transporting troops behind enemy lines.[41]

*L'Entreprenant* was created when the first aerostatic corps was formed in 1793 by two scientists, Jean Marie Joseph Coutelle and Nicolas Jacques Conté. Charged with developing innovative solutions to the new French republic's many military challenges by the Committee of Public Safety and its Commission Scientifique, these two chemists solved several problems for battlefield conditions including a method to produce hydrogen fuel on-site and methods to aid the rapid inflation and deflation of the envelope. With state support, Coutelle and Conté set up an aerostatic unit in Meudon outside of Paris, and by the early spring of 1794 they were confident enough to offer their services to General Jourdan. Their primary balloon, *L'Entreprenant*, performed well overall and was used in several sorties and battles including the iconic event at Fleurus. On the basis of these successes, a second aerostatic company was formed, and Coutelle was named the battalion commander and director of the training programme at Meudon.[42]

Throughout these 'small wars', the advocates of war balloons argued that visual observation would provide a decisive advantage for national forces. Of his view from *L'Entreprenant* at several battles in the spring of 1794, Coutelle wrote a measured evaluation: 'The rocking is troublesome, and increases with the force of the wind, and sometimes prevents the use of glasses; but I must remark that one can see the movements of infantry, cavalry, and artillery with the naked eye: and at Mauberge, Mayence, and Mannheim I could count the pieces on the redoubts and ramparts without any extraneous assistance.'[43] A British major-general who had been up in military observation balloons three times by 1803 argued vigorously for their expanded use, writing: 'There never was any doubt in my mind on the subject; you see from them everything you wish to see.'[44] Despite these reports that observation was immeasurably aided by aerostation, the 'old guard' in the French armies was not persuaded by the new technology.[45] Although Napoleon grudgingly sent war balloons to Egypt, he was not an enthusiastic supporter.[46] Without commitment from the top leadership, the balloon school at Meudon was closed in 1799 and by 1804 the use of war balloons for observation by the French military was over.[47] The aerostatic corps was created and disbanded in the short period of a decade. Yet public support and interest in ballooning was so intense that advocates continued to press for their military use throughout the nineteenth century.

As this brief history indicates, early advocates of the use of balloons in battle were themselves embattled, arguing for what they deemed was a revolutionary, modern technology that would supersede ground forces in the midst of militaries that had hardly changed their tactics for centuries. The 'new instrument of war', a complex combination of 'aeronautics, observation, and communication', was not comprehensible as a modern practice to all military commanders. Until the industrial era and even afterwards, military tactics continued to draw on the lessons learned from field and siege warfare. Martin van Creveld argues that field warfare can be said to have remained fairly stable across centuries and civilisations, relying upon 'endless alterations and combinations' of existing weaponry and tactics.[48] According to van Creveld, sixteenth- and seventeenth-century commanders, such as Machiavelli, Maurice of Nassau and Gustavus Adolphus, consciously sought to profit from the lessons of earlier armies, even to the point of directly imitating weapons, formations and tactics.[49] In many ways, fortifications had a similar history in that they varied according to specific needs, but their underlying strategy remained the same for centuries. Forts were built either to create continuous barriers in strategic areas or to guard a specific position. Until the introduction of gunpowder to Europe, military architecture relied on the 'height, thickness and solidity' of walls to maintain effective defences against weapons and field tactics that could be predicted to operate in much the same way across centuries.[50]

There are various explanations for the introduction of gunpowder from China into Europe sometime in the thirteenth century, but most military historians agree that by the seventeenth century a transformation had finally occurred in European military operations.[51] The increasing role of firearms and their continual improvement in function and reliability throughout the Renaissance period had changed the shape and size of armies as well as their operations. Longer-range artillery and the heavier firepower created conditions that altered the design of forts and city walls.[52] As fortifications grew larger they became much more expensive to build and operate and, inevitably, increasingly complicated to defend. To avoid frontal assaults by heavy cannon-fire, fortresses were built lower and lower to the ground to the point of being half-buried.[53] At the same time, as printing and communications improved, commanders were no longer required to fight at the head of their troops in order to keep track of the battle. With larger fighting forces and more terrain involved, along with a lessened need to be positioned at the very front of the fighting force, commanders began to situate themselves most ideally on a distant hill.[54] Overlooking the battlefield from a distance, often aided by telescopes, fielding written and oral communications from scouts on horseback, the modern commander developed a reliance on a field of vision provided by an elevated position.

Lines of sight had been important in the conceptualisation of early modern defence as well as offence. The most forward-thinking military architecture began to advocate the construction of fortresses that integrated elements of the local terrain rather than following generalised or ideal plans.[55] The resulting irregular lengths of walls, odd shapes and strange angles not only followed the contours of a particular piece of ground but also generated improved lines of sight such that the enemy could not attempt to scale any section of the fort without being seen. As improvements on medieval fortifications, the early modern forts were designed with geometric rigor in response to increasingly elaborate studies of terrain and military tactics. But by the end of the eighteenth century, these heavily manned, extended fortifications studded across wide areas of territory strained the finances of the powerful nation states to a severe degree. The massive watch-towers became useless; too vulnerable to artillery fire. Across nineteenth-century Europe, demolition of the old fortifications changed the way wars were fought in the countryside and cleared the way for new construction in the major cities.

Thus the arrival of the balloon at the end of eighteenth century augured a dramatic change in warfare that drew upon and heightened visual practices as the traditional fortress was rendered increasingly obsolete. Aerostatic flight offered to combine the valuable view from the defended watch-tower with the elevated observational post at the edge of the battlefield. Reconnaissance, always the privileged preserve of the most mobile unit—the cavalry with its highly trained scouts—would begin to move off the ground and into the air. Although most militaries did not develop aerostatic corps until the US Civil War, when some limited use of balloons for observation was introduced,[56] the war balloon fulfilled several of Clausewitz's celebrated elements of military strategy as elaborated in his treatise, *Principles of War* (published posthumously in 1832).[57] The view from the observation balloon provided the commander with information concerning the entire scene of battle including the size and organisation of the forces, their exact position, and specifics of the terrain. The aerostatic view also provided precise geometric calculations and could assist in assessing the routes for vital supplies for both sides of the conflict. The observation balloons could also influence troop morale through intimidation or distraction. Based on these proven advantages, advocates in all of the major militaries argued that, with improvements in the timing of launches and descents as well as innovations in navigation, aerostatic units could benefit the overall war mission. Equally persistent and vociferous objections to the use of war balloons pointed out that fuel sources remained a problem of expense and supply, while accurate navigation remained impossible. Moreover, the balloons themselves presented ideal targets for long-range artillery. Early airpower advocates were thwarted at every turn until the advent of heavier-than-air machines (and even long into the twentieth century, and despite a growing desire and capacity for visual information of the kind made possible by aerostatic flight, resistance to air forces from other military branches remained constant).[58] Militarised aeromobility proceeded by other means.

# FROM ABOVE: WAR, VIOLENCE AND VERTICALITY

*The Balloon Prospect*

> Forms of war and forms of life are … always intimately correlated.[59]

The 'balloon craze' of the late eighteenth and early nineteenth centuries has been linked to almost primal or transhistorical desires to achieve flight or to attain an extreme elevated viewpoint. As Beaumont Newhall has written, 'The face of the earth from the air fascinated man before he was able to fly.'[60] Or, as Jean-Marc Besse puts it, the view from the air is one of Western culture's 'oldest imaginative resources'.[61] These generalisations almost always precede the assertion that either aerostatic or powered flight made possible the realisation of this ancient desire, thereby revolutionising visual culture; thus Hallion notes that a 'transformation of perspective' had occurred when humans 'gazed down on the earth', 'forever altering their perception of the world'.[62] But it is, perhaps, Newhall's late 1960s classic, *Airborne Camera*, that best exemplifies the notion that human flight expanded worldviews literally and figuratively.[63] In *Airborne Camera*, Newhall, the first curator of photography for the Museum of Modern Art and a key figure in the movement to recognise photographic work as art, makes several points about the exceptional and transformative aspects of aerostatic observation for modern visual culture. First, Newhall points out that, until aerostation, depictions of views from above were completely imaginary, while after the first lighter-than-air ascents such unprecedented views became 'real'. Second, the first aeronauts were often overcome by the sublime view—the panoramic sweep above, around and below them had never before been experienced even from mountaintops. Third, the appearance of the earth from above was often very different to what had been expected. Newhall's fourth point is that sketches and images based on these 'real' views were rarely accepted as landscape art and were more likely to be regarded as scientific or informational material. Nevertheless, the 'unfamiliar' view generated its own aesthetic—it could be both 'stimulating' and 'beautiful'.[64]

First- and second-person accounts of aerostatic flight support Newhall's conclusions. Books, reviews, newspaper accounts and lectures on the topic of aerostation contributed to a wave of intense interest in Europe and North America during the late eighteenth and early nineteenth centuries. Ballooning represented a modern, scientific, even republican, attitude to newly empowered middle-class enthusiasts. Hallion describes the affiliation with signs and symbols of this new technology as pervasive: 'Balloon motifs decorated furniture and clothing; images of the aeronauts adorned pendants and jewelry; elegant crystal chandeliers and clocks took on balloon shapes; and dinnerware reflected the personalities, events, and paraphernalia of the new balloon era.'[65] In France, iconic scenes from the Battle of Fleurus decorated fans and porcelain objects. Prints and paintings depicted balloon launches, scenes from historic flights and even tragic accidents in which aeronauts lost their lives. The globular shape of most balloons became emblematic of an increasingly mobile, experimental culture that embraced risk-taking, adventure and scientific endeavour. The 'stimulating' and 'beautiful' views from these lighter-than-air globes were a key part of the discursive apparatus of aerostation. Scenes of 'grandeur and beauty', 'rapture' and 'new perspectives' were breathlessly reported by successive aeronauts to a public eager to learn more.[66]

Although the almost immediate militarisation of aerostatic observation on the part of military avant-gardes seems to support the argument that scopic mastery was the only mode of seeing made possible by this new mode of transport, the records of viewing from above suggest a less totalising set of visual practices. As an instrument of war, the balloon did seem to mobilise desires for a panoptic watch-tower in the sky, an über-commander, elevated far above the battle with unlimited views. In this line of argument, aerostatic flight inaugurated the kind of totalising viewpoint that resulted in a distanced objectification of human diversity and an increasing propensity for aerial warfare. But aeromobility, even militarised, did not cohere immediately in this manner. Views from balloons fostered ideas about landscapes that contributed to national and military projects, but viewing practices were uneven and disorganised at first.

There were at least four representational operations at work in early accounts of aerostatic viewing: changes in scale, access to the previously inaccessible, sharp delineation and equalisation or flattening. These elements of visual aeromobility became part of efforts to organise the aerial view into something more generic, intelligible and, indeed, representable. Through balloon views, Western metropolitans were able to see locations or objects that had been scattered or too far distant as linked or part of a previously unknown whole. They also became newly cognisant of compression in time via this expansion of space—that is, the rapidity and mobility offered by the balloon extended the imaginary reach for the individual as well as the nation.[67] It is not far-fetched, then, to argue that the vistas described in written accounts and in sketches and prints produced by balloonists expanded a visual field in culture, inspiring different ways of depicting rural landscape, cities and the world, in general, if not a transformed spatialisation.

This new spatialisation participated in the picturesque, defined by Bruno as a 'new type of spatiality' in which 'spectacle was displayed through motion … by an observer moving through space'.[68] Picturesque prospects were hardly benign, tied to emerging national identities linked to terrain that positioned colonial landscapes as resources to be literally or figuratively mined over and against the domestic, cultivated geographies of 'home'.[69] But from the very first, aerial viewers experienced many of the emotions and sensations elaborated by Newhall. They were bowled over, stunned, delighted, frightened, disoriented, confused, lost, calmed and, above all, keen to *see* something, anything, from their perch hundreds if not thousands of feet above the earth. The panoramic vista, so extensive and vast, was hardly empty. Aerostatic viewers were extremely interested in clouds and weather as well as the views above and below.[70] The balloon 'prospect' was immersive, mobile, spatialising and embodied.[71]

This dynamic prospect that encompassed both the act of observing the view from above as well as its representation was relatively new. As Barbara Stafford points out, the terms *landscape* and *prospect* were used interchangeably in the eighteenth century, but *prospect* 'typically connoted a more extensive view'.[72] The Latin origin of 'prospect' is linked to Justinian-era legal contexts, referring to either one side of a house (and the minimum distance between buildings that was required to guarantee a specific view) or 'the vista that could be had from the house itself'.[73] By the sixteenth century the term took on the meaning of an 'extended view' or 'vista'. Building on the Renaissance perceptual system of linear perspective, the practice of recreating a view as if seen from a fixed position became the norm in Western European art, generating both picturesque and realist depictions of nature. While landscape art had ranked fairly low in the valuation established by the European academies, during the eighteenth century the upsurge in more democratic social practices led to subjects and methods that did not require royal patronage, academic training or arcane knowledge. Thus historical paintings and monumental art were joined by views of more ordinary and accessible subjects sometimes produced in less formal modes such as the 'sketch'. Picturesque vistas, or prospects, emphasised principles of harmonious beauty and intentional composition in the service of a constructed 'naturalism' that was at the same time extremely idealised and often imbued with nationalist sentiment.[74] The new subjects of democracy were offered views of their cities, towns, seashores, rivers, lakes, hillsides and moors in these more accessible, less academic styles. Tourism to favoured 'picturesque' sites increased along with sales and collections of either printed or painted souvenir views. As M. Christine Boyer points out, seeing the world as landscape or illustrated picture was a 'visual and mental experience', one that 'moved from vista to vista, viewing objects from different perspectives'.[75]

Along with rural scenes, the balloon prospect incorporated elements of bird's-eye views of urban sites. These views from above were pictorial; that is they were drawn, painted or carved from what is now called an 'imaginary' or 'impossible' vantage point. Thus Louis Marin argues that bird's-eye views of European cities, such as Jacopo de' Barbari's woodcut of Venice dating from 1500 or the print of Oxford dating from 1575 (included in Braun's *Civitates Orbis*), compose total and finite images; he refers to them as utopic views.[76] But as Besse points out, bird's-eye views in the early modern period transform cities into 'possible visual experiences' that can then be reproduced and disseminated in atlases, geography books, and so on.[77] They are, then, virtual views; as Besse writes, 'the images are the locations in which *virtual* experiences are illustrated and portrayed'.[78] Putting it another way, Newhall points out that these early bird's-eye views were 'earthbound': they portrayed the land as imagined from a great height but they could not realistically convey the view from 'points suspended in space'.[79] Utopic, virtual, impossible, unrealisable; the bird's-eye views seem to mark a desire for an expanded vista that could contain all the parts of the growing cities of Europe including borders and topographic specificities.

The desire for a total view did not necessarily produce a holistic image. Before the 1780s (and even well afterwards), prints that portrayed bird's-eye views were assembled from sketches made of many different points across a city or specific site. The resulting montage strove to create the *effect* of a fixed single perspective, but the condition of its production introduced innumerable viewpoints and uneven distances. Thus, counter to most interpretations (while engaging Marin's terms), Bruno describes the bird's-eye view not as totalising so much as a view from both 'nowhere' and 'now here'—that is, as a view that brings space and time together in an imaginatively 'dislocated' and 'mobilising' present that does not exclude past and future.[80] Without a true focal point, the observer could 'wander around' in a 'permeable place of encounters between the map and the landscape'.[81] Thus the bird's-eye view from 'nowhere' and 'now here' prefigured the balloon prospect's unstable moves between precisely locative, panoptic and unintelligible or startling views and ways of seeing.

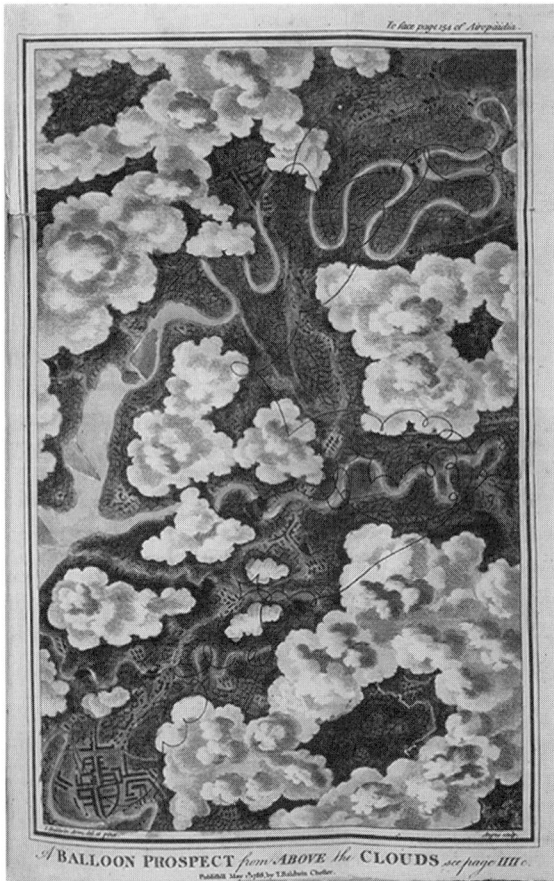

Image 1: A Balloon Prospect from Above the Clouds (Baldwin, 1786).

An interesting example of this dynamic, evolving mode of representation is contained in Thomas Baldwin's 'A Balloon Prospect from Above the Clouds.' In 1785 Baldwin borrowed a balloon from a much more famous figure, aeronaut Vincenzo Lunardi,[82] to make a solo flight from Chester castle in Cheshire, England, taking detailed notes and making drawings of the countryside along the way. These records of flight were published under the title *Airopaidia* in the following year.[83] Baldwin has been credited with the first published aerial views based on sketches undertaken by a person in flight. 'A Balloon Prospect from Above the Clouds' depicts the Cheshire countryside, the River Mersey and the town of Warrington as seen from the height of approximately 7,000 feet. It is a vertical view with an accompanying 'explanatory print' that shows the balloon's route and important landmarks. 'A Balloon Prospect' is bounded by the dimensions of the page in a book—framed much more tightly than the 'real' view available from the balloon. As a 'picture', then, it emphasises a variety of elements to be seen as well as an entirely new view of terrain. We see the winding River Mersey, variations of vegetation and habitation, clouds, and the shadows of the clouds. Contemporary reviewers found some fault with Baldwin's written style but 'The Balloon Prospect' earned praise as a reflection of the novel aerial scenes that 'constitute the true sublime and beautiful' and that 'raise the most careless observer to a high degree, not only of pleasure, but of rapture and enthusiasm'.[84]

The balloon prospect in 1785, literally and figuratively, reinforces not so much a scopic mastery as Stafford's assertion that aerostatic views are unsettled and unsettling, unstable. As she writes, from the perspective of a balloon: 'Objects and the amorphous medium that bathes them are ever in motion, sliding away from beneath the eye, evading the pursuing mind.'[85] Thus Baldwin writes of the effect of peering through clouds, the earth appearing and disappearing; a view of the ground could be discovered by a 'glance of the eye' before 'repeatedly escaping from sight'.[86] The early nineteenth-century aeronaut Thomas Monck Mason described the disconcerting effects of aerostatic flight, especially the sensation of remaining still while the visual field appears to be in motion:

Insensible of motion from any direct impression on himself, and beholding the fast-retreating forms, the rapidly diminishing size of all those objects which so lately were by his side, an idea, almost amounting to conviction, involuntarily seizes upon his mind, that the earth with all its inhabitants had, by some unaccountable effort of nature, been suddenly precipitated from its hold, and was in the act of slipping away from beneath his feet into the murky recesses of some unfathomable abyss below. Everything in fact but himself seems to have been suddenly endowed with motion, and in the confusion of the moment, the novelty of his situation, and the rapidity of his ascent, he almost feels as if, the usual community of sentiment between his mind and body having been dissolved, the former alone retained the consciousness of motion, whereof the latter had by some extraordinary interference been suddenly and unaccountably deprived.[87]

Viewing the earth from a balloon moving upwards or descending downwards did not so much shift the view in precise registers as portrayed in twentieth-century work such as Kees Boeke's *Cosmic View: the Universe in Forty Jumps* or the Eames' *Powers of Ten* as mix things up.[88] Thus Mason wrote:

There projected upon a plane at right angles to his line of vision, the whole adjacent surface of the earth lies stretched beneath him, affording an heterogeneous display of matters at once the most interesting and incongruous. Distances which he used to regard as important, contracted to a span; objects once imposing to him from their dimensions, dwindled into insignificance; localities which he never beheld or expected to behold at one and the same view, standing side by side in friendly juxtaposition; all the most striking productions of art, the most interesting varieties of nature, town and country, sea and land, mountains and plains, mixed up together in the one scene, appear before him as if suddenly called into existence by the magic virtues of some great enchanter's wand.[89]

The contraction of distance and the diminution of large objects that early aerial observers reported along with the mixture of incongruous elements and the sensation of movement is part of the related thrill of viewing previously unknowable or inaccessible sights: 'The summits of mountains, the tops of buildings, the upper surfaces of woods, those parts, in short, of all objects which by their natural or artificial positions have hitherto been excluded from his view, are now almost the only ones that come within the scope of his observations.'[90] And along with these deviations from the rules of linear perspective and scale, the aerial observer was also surprised by the clarity and sharpness of delineation when clouds permitted an unobstructed view of the earth. A contemporaneous reviewer of *Airopaidia* was enthused to describe this very effect:

The circular opening in the clouds, through which the earth's surface was presented to the eye, discovered a smooth level plain; a sort of shining carpet, enriched with an endless variety of figures depicted without shadow, as on a map; what was really shadow forming a separate color, and not being considered at the time as shadow. All was coloring; no outline; yet each appearance curiously defined by a striking contrast of simple colors, which served to distinguish the respective boundaries with more exact precision, and inconceivable elegance. Red waters, yellow roads, enclosures yellow and light green, woods and hedges dark green, were the only objects clearly distinguishable, and their coloring was extremely vivid: the sun's rays, reflected from the sea and other waters (which appeared all red) dazzled the sight.[91]

In her study of balloonists' accounts, Barbara Stafford argues that the sharpness of delineation that is achieved through distance is made possible by the extreme alterations of the sense of time and space that aerostatic mobility made possible.[92] Mason refers to the 'remarkable sharpness of outline' that is 'strengthened' by distance; thus, objects 'instead of appearing obscured and rendered more indistinct by their remotion [sic] from the point of sight, seem on the contrary to augment in clearness and decision, and absolutely gain in intensity what they lose in the magnitude of their proportions'.[93] Mid-nineteenth-century aeronaut James Glaisher also underscored the remarkable 'distinctness superior to that on earth': 'the line of sight is through a purer and less dense medium, everything seems clearer, though smaller'.[94]

The final new mode of perception engendered by aerostatic flight was only noted by aerial observers who ventured into the greater heights. An 'equalisation' of the field of vision accompanied the immersive experience as the earth below appeared to flatten into an 'uninterrupted surface'.[95] This flat and 'shadowless' view, more abstractly geometric, eliminated once and for all the Renaissance linear perspective: 'no horizon line, no vanishing point, no human scale, and few nuances of light and shadow'.[96] If clouds did not prevent the aeronaut from viewing the earth below, all of the elements of aerial viewing combined for radical shifts in scale, altered perception, intense clarity or delineation even from a great height, and a 'universal disfiguration' leading to a flattening effect. As Mason wrote: 'the whole face of nature … appear[s] to have undergone a process of equalization; the houses and the trees, the mountains and the very clouds by which they are capped, have long since been consigned to the one level; all the natural irregularities of its surface completely obliterated, and the character of the model entirely superseded by that of the plan'.[97]

From several miles above the earth, these early aeronauts entered an increasingly objectless field of vision, one that amazed and fascinated the scientist as well as the artist or the adventure traveller. It is here, at the furthest point reached by the balloonists of the late eighteenth and early nineteenth centuries that we can ponder what followed. Both a humanist cosmopolitanism *and* a scopic totalising regime are generated as possible viewing practices by the aerial observer floating free in the earth's upper atmosphere. Rather than choose between a naive humanism or a demonised war-mongering, let us consider instead the counter-intuitive possibilities of intimacy in distance and of equalisation in distinction. Aerial viewers strove to discern, to make differences meaningful, through the process of aerostatic flight. If Baldwin could identify the blue slate rooftops of Chester as a specific characteristic of the British north, beginning a visual navigational practice that would become codified and written in manuals for pilots from the First World War on, his perception of the 'charming' view was also full of observations of confusion and slippage. Similarly, Mason, flying several decades later, reported contradictory aspects of observation; known laws of nature were 'annihilated' even as viewing was sharper and enhanced in ways he had never before experienced.

It has become a truism in critiques of airpower to charge that views from far above the earth dehumanise subjects as targets, leading to slaughter by bombing raids and even genocide.[98] To say that such attitudes are a truism is not to suggest that they are incorrect or not deeply felt. But the argument that representation from a distance blunts subjectivity and, in particular, empathy or identification, has a discursive history, one that might indicate important variations or diverse possibilities for the politics of perception and visual culture. In particular, the argument that distance is death-dealing circulated in debates about the uses and abuses of photography after the First World War. For example, Ernst Jünger argued that the photographic apparatus could dull the pain and shock of war and, by extension, industrial society by introducing a perceptual distance, both literally and figuratively.[99] Aerial reconnaissance photography seemed to literalise this philosophy, proposing to some commentators a 'detached and disembodied gaze' of abstracted landscapes, normalising the horror of war and industrial capitalism.[100] By extension, aerial photography wages war through representational operations.[101]

This desensitised, almost robotic killing-machine view contrasts vividly with the embodied, affective accounts of early aerostatic flight and its views from above. The first aeronauts commented frequently on the new or unexpected sensations they experienced in flight—the point is that flight and aerial sight were full of feeling and sensation rather than devoid of such. Derek McCormack has described the process of becoming aerostatic or lighter-than-air as threefold, moving through envelopment, inflation and buoyancy.[102] This emphasis on different ways of being in the air, what McCormack refers to as aerostatic *spacing*,[103] underscores the processual relationships between persons and things that are made possible by the embodied experience of flight. It points to Jen Southern's historically postmodern inquiry into aerial perspectives that are not panoptic, an aerial view that 'can mediate a closer relationship between the individual, specific locations and connections to the global'.[104] This attention to potential views that are not panoptic or to operations that include and mediate other senses besides sight necessitate what Southern calls a 'greater connection to the ground' through the embodied practice of aerostation.[105]

I want to be careful, however, not to pose early aerostatic flight as a rustic or romantic alternative to fully technologised powered flight. All of the ingredients for a future weaponised visual logics of control were available in the early days of aerostation. Thus we can grasp that airpower is present; imagined, advocated for, practised and innovated from the start. Aeromobility is militarised from its inception and not only in its most obvious incorporation as an 'instrument of war'. The exigencies of governing over greater distances, of transporting people and goods further and further, of seeing as much as possible to navigate or plan for these activities—industrial society's global and imperial aims favoured the promise of aerial modes. The control of the vast spaces of modernity required speed as well as enhanced properties of vision—both more easily achievable by air as technological innovation advanced and supported these needs.

Militarised aeromobility was not, perhaps, inevitable. But with the rise of aerostation in countries almost always already at war, nation states that arose through the process of making their differences visible and obvious in innumerable ways to the point of fighting to the death to maintain them, it would be difficult to avoid. Ironically, the untethered balloon flights, so impractical for the kind of warfare waged at the time, initiated experiments in perception that were ideal for future military use. Aerial observers became mobilised to the modern projects of cartography, planning and surveillance through falling in love with flying and, therefore, becoming enraptured by the sights that could be seen.[106] As those sights became increasingly organised into more precise registers, and as powered flight (and therefore more accurate navigation) became a real possibility, reconnaissance became more and more viable. Geography aided reconnaissance, and both were supported by aerial observation. The recognition of specific terrain, the myriad activities of distinguishing and differentiating increasingly detailed elements of the view, created disciplines and modes of visual analytics that came into full use during the First World War and after.

The balloon prospect contributed to militarised aeromobility in innumerable ways. After all, in addition to giving us the first printed image of a view based on a sketch made by a human being in a balloon, Baldwin also published in 1784 a folding plan titled *Proposals at large, for the construction of a grand naval air-balloon furnished with an apparatus corresponding to that of sails oars and rudder, to be occasionally applied.* No matter how 'intimate' and 'proximate'[107] the view of the new subject of aerostation, the comprehension of that view seems to have almost always been accompanied by commercial, industrial or military concerns. Aerostation brought access to the 'core of a living universe'[108] and more processual relationships between persons and things through the very operation of free flight. But the projects of imperial domination and national security provided a foundation and continuing support for aerostatic operations, offering discursive elements to assist in organising the aerial view into an assemblage that could, most definitely, wage war.[109]

# 2

# LINES OF DESCENT

*Derek Gregory*

It is a queer experience, lying in the dark and listening to the zoom of a hornet, which may at any moment sting you to death. It is a sound that interrupts cool and consecutive thinking about peace. Yet it is a sound—far more than prayers and anthems—that should compel one to think about peace. Unless we can think peace into existence we—not this one body in this one bed but millions of bodies yet to be born—will lie in the same darkness and hear the same death rattle overhead

Virginia Woolf, *Thoughts on Peace in an Air Raid* (1940)

## The Distance of Death

Virginia Woolf composed her brief essay in August 1940 during the Battle of Britain. The bombers in the night sky over London must seem a world away from the drones over Afghanistan and Pakistan, but there are genetic pathways between Woolf's hornets and what the Pashtun call the *machay*, bees that have their own deadly sting. According to Daniel Swift, 'today's Predator drones in Afghanistan and Pakistan are the direct descendants of the Heinkels and Lancaster bombers of the Second World War'.[1] These are very different sorts of war, of course, but there are several senses in which today's drone wars in the global borderlands were anticipated by the advocates of 'progressive' or even 'beneficial bombing' in the 1940s: 'progressive' because air war was supposed to be short, sharp and decisive, avoiding the protracted carnage of trench warfare.[2] In 1942, six months after Pearl Harbor, A.P. de Seversky could already see a future in which bombing would be conducted over such vast distances that intermediate bases—like the United States Army Air Force bomber stations being prepared in Britain—would be unnecessary: 'The entire logic of aerial warfare makes it certain that ultimately war in the skies will be conducted from the home grounds, with everything in between turned into a no-man's land.'[3] By the end of that year, when Germany had successfully tested its first 'Flying Bomb', it was even possible to imagine bombers without pilots. Soon after D-Day in June 1944 Germany launched a barrage of V-1 and V-2 rockets against Britain, and in response the USAAF toyed with directing hundreds of worn-out Flying Fortresses filled with high explosives and a new weapon, napalm, to targets in Germany from accompanying aircraft using mounted television cameras and remote ('robot') control. Only fifteen, unsuccessful missions were flown, but General Arnold asked for further research into remote-controlled, television-assisted aircraft that could 'fly over enemy territory and look through the leaves of trees and see whether they're moving their equipment'. On VJ Day he predicted that 'the next war may be fought by airplanes with no men in them at all'.[4]

There is another, equally powerful sense in which distance threads through the genealogy of bombing. The production and articulation of what is now called the kill-chain typically works to render bombing an abstract, purely technical exercise for those who execute it.[5] Here is Len Deighton describing the target maps used by Bomber Command for its area bombing of German cities under cover of darkness. These were bare-bones affairs, printed in black and magenta so that they could be read in the dim amber light of the navigator's table: 'The only white marks were the thin rivers and blobs of lake, and the roads were purple veins so that the whole thing was like a badly bruised torso.' The cities became 'just shapes, like the ill-defined blurs that passed across the H2S radar tube,' he continues, and 'that, of course, was the whole idea': 'The new grey faceless maps were just one aspect of a new kind of war.'[6] And here is Graham Swift describing a photo-interpreter's bomb damage assessment:

I looked down with a privilege no pilot ever had on target after target … I learned to distinguish the marks of destruction—the massive ruptures of 4,000-pounders from the blisters of 1,000 pounders and the mere pock-marks of 250 pound clusters—and to translate these two-dimensional images, which were the records of three-dimensional facts, into one-dimensional formulae— tonnage dropped as against acreage devastated, acreage destroyed as against acreage attacked (the tallies never included 'people', 'homes') … And as operations progressed, the statistics grew larger, the images more other-worldly, more crater-ridden, more lunar.[7]

These are not novelists' flights of fantasy. A veteran of Bomber Command considered it 'one good thing about being in an aeroplane at war' that 'you never see the whites of their eyes … You drop a four thousand-pound cookie and kill a thousand people but you never see one of them.'[8] 'A German city was always this,' wrote a navigator, 'this hellish picture of flame, gunfire and searchlights, an unreal picture because we could not hear it or feel its breath. Sometimes when the smoke rolled back and we saw streets and buildings I felt startled. Perhaps if we had seen the white, upturned faces of people, as over England we sometimes did, our hearts would have rebelled.'[9] But they rarely did. One Lancaster pilot thought it just as well 'that no picture came into his [the navigator's] mind of shattered limbs, of burning clothing, of living bodies crushed by rubble. He only saw a coloured target-indicator, as he squinted through his bomb-sight and thumbed the release button.'[10] After his first raid another wrote that the fires burning below 'looked like sparkling diamonds on a black satin background … [T]hey weren't people to me, just the target. It's the distance and the blindness which enabled you to do these things.'[11] This was a common sentiment, and Charles Lindbergh saw it as the very diagnostic of modern war, where 'one kills at a distance, and in doing so does not realize that he is killing'. Far from imagining 'writhing, mangled bodies' on the ground below, he wrote in 1944, it was like 'viewing it on a motion-picture screen in a theater on the other side of the world'.[12] Many critics believe that Lindbergh's metaphor has been realised—and radicalised—in today's drone wars, where the pilots of Predator and Reaper aircraft are not thousands of feet above their target in Afghanistan or Pakistan but usually thousands of miles away in the United States. More or less as Arnold had foreseen, they view full motion feeds on their screens, and in that moment, seemingly, war is turned into a videogame.

Critics have identified a host of problems with killing from a distance. The human operator 'is terribly remote from the consequences of his actions; he is likely to be sitting in an air-conditioned trailer, hundreds of miles from the area of battle'. He evaluates 'target signatures' captured by various sensor systems that 'no more represent human beings than the tokens in a board-type war game'. The rise of this new 'American way of bombing', as it has been called, has two particularly serious consequences. First, 'through its isolation of the military actor from his target, automated warfare diminishes the inhibitions that could formerly be expected on the individual level in the exercise of warfare'. In short, killing is made casual. Secondly, once the risk of combat is transferred to the target, it becomes much easier for the state to go to war. Domestic audiences are disengaged from the violence waged in their name: 'Remote-controlled warfare reduces the need for the public to confront the consequences of military action abroad.'

All familiar stuff, you might think, except that these warnings were not prompted by the flight of Predators and Reapers over Afghanistan and Pakistan. They appeared in *Harper's Magazine* in June 1972, the condensed results of a study of the US air war in Indochina by a group of scholar-activists at Cornell University.[13] As they suggest, crucial elements of today's 'drone wars' were assembled during the US bombing of Vietnam, Laos and Cambodia in the 1960s and early 1970s. A key transition from deliberate to dynamic targeting, from fixed to fleeting targets, comes into view during that war, and I will show that this not only reinforced the power of abstraction that animated bombing in the Second World War but also introduced elements that prepared for late modern war in the global borderlands. There were three vital but largely separate innovations: remotely piloted aircraft; real-time visual surveillance; and a networked sensor-shooter system. Exploring their emergence and fusion reveals that the line of descent from Cologne and Coventry to Kandahar has been more tortuous than Swift allowed. And since the roll-call must include Kobe, Nagoya, Osaka, Tokyo and Yokohama—if Hiroshima and Nagasaki are taken to be cases apart—a focus on Indochina also demonstrates that the line of descent was (and remains) profoundly racialised.[14]

*From Germany to Vietnam*

In the 1960s the combined bomber offensive of the Second World War was still the classical model of deliberate targeting, in which (usually fixed) targets are assigned to aircrews before take-off. There were two kinds of strategic bombing. RAF Bomber Command preferred the area bombing of enemy towns and cities by night, and the US Eighth Air Force prided itself on the precision bombing of military and industrial targets by day. At that time the distinction was more rhetorical than real—while the USAAF made much of the superiority of its Norden bombsight, precision bombing was often terribly *im*precise, and the Americans 'judged themselves by their motives rather than their results'[15]—but it assumed much more substantive form in Indochina. The bombing of North Vietnam placed a premium on precision, whereas many of the most devastating attacks on South Vietnam inaugurated a new form of area bombing.

To the US military the first series of air strikes against North Vietnam between 1965 and 1968 (codenamed 'Rolling Thunder') was an interdiction campaign designed to close lines of communication and choke off the supply of men and materials from the North to the Viet Cong insurgency in the South. To President Johnson and his civilian advisers, its purpose was to open up an altogether different line of communication, a way of 'sending a message' to Hanoi through a 'diplomatic orchestration of signals and incentives, of carrots and sticks, of the velvet glove of diplomacy backed by the mailed fist of air power'.[16] These twin imperatives ensured that Rolling Thunder would dance an intricate gavotte between strategic advantage and political calculation, each of which required careful calibration. The Air Force had a global database of potential targets—the 'Bombing Encyclopedia'[17]—and by August 1964 the Joint Chiefs of Staff had pared down an initial list of 451 primary to ninety-four priority targets. They recommended an intensive air campaign to interdict North Vietnam's supply of war materials through strikes on ports and rail links and to degrade its military capability through strikes on command and control centres, airfields and barracks and supply depots and lines of communication.[18] Johnson rejected their advice, and ordered a gradually escalating series of strikes on target lists that had been re-jigged by his administration. The procedure was convoluted. The Air Force and the Navy submitted lists of targets to the Commander-in-Chief Pacific Command (CINCPAC), whose office reviewed and forwarded a revised list to the Joint Chiefs of Staff, who in turn reviewed and forwarded a revised list to the Pentagon. After officials had calculated the probable impact of a strike and the likelihood of civilian casualties, the Secretary of Defence produced a modified list in consultation with the Secretary of State.[19] By this stage the folders for each numbered target had been reduced to a single sheet of paper with just four columns: military advantage; risk to aircraft and crew; estimated civilian casualties; and danger to third-country nationals (Russian and Chinese advisers).

The final target list was decided during the president's Tuesday luncheon at the White House. This followed a meeting of the National Security Council, and those attending were briefed before grading each target. The president reviewed the grades and made his decision, which was delivered to the NSC in the evening and transmitted to CINCPAC through the Joint Chiefs for immediate execution. The instructions included not only the number of sorties to be conducted against each target but also, in the early stages of the campaign, the timing of the attacks and the ordnance to be used.[20] As targets worked their way up the command hierarchy to Washington, their priority order was reversed: from March 1965 the bomb line was slowly advanced northward through an ascending series of 'route packages' as strikes worked their way up from the bottom of the strategic list.[21]

Johnson also stipulated strict Rules of Engagement that prohibited air strikes within 30 miles of the Chinese border, 30 miles from the centre of Hanoi and 10 miles from the centre of Haiphong, and imposed a complex, constantly changing web of regulation whose details had to be incorporated into each day's operational order. The pilots chafed at the restrictions but were adamant that they were scrupulously observed.[22] One report in the *New York Times* concluded that the Air Force was not bombing 'the area targets of World War II and Korea, when civilian losses were largely ignored', and that attacks on barracks and other military targets required—and received—'pinpoint bombing'. Other correspondents told a different story, notably Harrison Salisbury, who reported that 'the bombed areas of Namdinh [in the Red River Delta] possess an appearance familiar to anyone who saw blitzed London, devastated Berlin and Warsaw' and that 'the effects of bombing at ground level seem to have changed little since World War II'.[23] The truth was to be found at both ends of the spectrum and everywhere in-between:

many of the targets were in rural areas, but when residential districts were bombed there was undoubtedly considerable devastation and loss of life. Johnson had already cancelled Rolling Thunder by the time the Air Force started testing its 'smart bombs' in 1968, and even when Nixon resumed strikes against the North in 1972, most of the bombs were still conventional ones. Some of them had been retrofitted with the PAVEWAY laser-guided system, and these were used to attack difficult targets like heavily defended bridges, and to destroy military-industrial installations around Hanoi and Haiphong that were much closer to civilian concentrations.[24]

In sum, Rolling Thunder was carefully controlled and calibrated, an imperfect exercise in what then passed for precision bombing. Although it was plainly a strategic air campaign, however, the missions were flown by tactical aircraft—fighter-bombers—rather than the long-range B-52 bombers of Strategic Air Command. Johnson feared these giant aircraft would send a far more dangerous signal to Hanoi and Beijing. Thirty B-52s had been deployed to Andersen Air Force Base on Guam as a contingency measure; their numbers were later increased and basing switched to Thailand, but until Nixon ordered them into the Linebacker campaigns against the North in 1972, they were restricted to tactical missions in South Vietnam, Laos and Cambodia. They played a devastating role in all three theatres.[25] By April 1965, General Westmoreland, the commander of Military Assistance Command Vietnam (MACV), was so exasperated at the results of strikes by tactical fighter-bombers against the Viet Cong—their lack of concentrated fire power allowed the insurgents to escape—that he urged the JCS to authorise 'pattern bombing' missions by the B-52s. They agreed to deploy them for 'area saturation attacks against target areas known to include VC-occupied installations and facilities for which precise target data to permit pinpoint bombing attacks was not available'.[26] The first air strikes were closely monitored in Washington, which was almost as apprehensive about the symbolic significance of deploying the B-52s in the South as it had been in the North, and in parallel with the procedures developed for Rolling Thunder, target areas had to be proposed by MACV, reviewed by CINCPAC and the JCS, and approved by the Secretary of Defence in consultation with the State Department and the White House. If Westmoreland was irritated by these stipulations, he was delighted at the combination of 'surprise and devastating power' unleashed by the B-52s. Bombing from 25–30,000 feet, they could neither be seen nor heard on the ground, and their strings of high explosive bombs—one of the horribly iconic images of the war—pulverised a wide area.[27]

These Arc Light missions issued in a new kind of area bombing—'bombing forests', one Air Force critic called it[28]—that was no less abstract than the Second World War original. Wright describes a photo-interpreter poring over a roll of aerial photographs 'to find the enemy in the negative', 'totally absorbed into the fascinating realm of carpet bombing, lost among the oddities of the weave'. This demanded extraordinary attention to detail, but these local textures were quickly converted into abstractions: 'Griffin was required to translate pictures into letters and coordinates that were instantly telexed', and 'the data went round and round, and where it came out he preferred not to hear … Wherever he put circles on the film, there the air force would make holes in the ground.'[29] Since SAC was primarily a nuclear strike force delivering devastating destructive power, pinpoint accuracy was irrelevant, and those 'holes'—vast craters—were distributed across an extensive target box. The first attack was against a box measuring 1 mile by 2 miles, and within thirty minutes 1,300 bombs were dropped, 'slightly more than half of them in the target area'.[30] Throughout the targeting process the language of patterns, areas, circles, holes and boxes erased people from the field of view; bombing became a deadly form of applied geometry.

It was no less abstract for those who carried out the attacks. One journalist reported that a B-52 strike was a 'chillingly spectacular event' for those on the ground, but for the aircrew, 'sitting in their air-conditioned compartments more than five miles above the jungle', it was little more than 'a familiar technical exercise'. They 'knew virtually nothing about their targets, and showed no curiosity'. One of them explained that 'we're so far away' that 'it's a highly impersonal war for us'. His crew saw themselves merely as instruments of policy: 'Where they put the bombs is someone else's decision and someone else's responsibility.' 'If we are killing anybody down there with our bombs,' he continued, 'I have to think we were bombing the enemy and not civilians. I feel quite sure about our targeting.'[31]

The sense of abstraction—and misplaced confidence—was heightened by the creation of 'free bomb zones' (or 'specified strike zones') and the introduction of new forms of radar bombing. In August 1965 Westmoreland was authorised to order strikes in five free bomb zones that were 'configured to exclude populated areas except those in accepted VC [Viet Cong] bases'. Within these zones the designation of target boxes dispensed with precise coordinates and detailed intelligence altogether, so that they became black boxes in every sense of the phrase, and approval was given in advance 'for execution when appropriate'.[32] Westmoreland was perfectly clear that 'anybody who remained had to be considered an enemy combatant', and so strikes could proceed 'without fear of civilian casualties'. By then most Arc Light missions relied on radar synchronous bombing, and from April 1966 a new system called Sky Spot was introduced to direct strikes against 'unseen targets' on the command of a distant ground controller tracking the flight on his radar screen. By the end of the year this was the principal bombing method used by the B-52s.[33] According to one pilot, 'bombs were to be released at fifteen to twenty thousand feet in an area where winds were only roughly known, target location was only approximate, and the vectors on the aircraft at the moment of release could not be predicted. It wasn't your father's Norden bombsight by any stretch of the imagination.'[34] Not all air strikes in the South were like this, of course, and tactical aircraft continued to be used to provide close air support. But area bombing clearly remained a common weapon against the insurgency. As it happened, it was also counterproductive: in the most heavily bombed areas popular support for the Viet Cong increased, provoked by despair at the death and destruction and by the perpetual fear of imminent attack.[35]

*Sensing the Enemy*

If there were continuities between the combined bomber offensive and the air wars in Indochina, there were also significant differences. Some of the most important turned on the problem of air intelligence. Standard target maps and even radar were insufficient in a 'war without fronts' where the situation on the ground changed rapidly. In the North, target sets had to be adjusted as surface-to-air missile sites were moved and oil storage depots dispersed; in the South and along the Ho Chi Minh Trail, a dense tissue of roads and paths running from the North through Laos and Cambodia to the South, it was immensely difficult to detect the clandestine movements of the Viet Cong beneath the forest canopy. The fixity of the map was undone not only by the fluidity of the war but also by the land itself: micro-features used for target identification like river channels or sandbars often shifted from one season to the next so that 'the ground never did look exactly like the map or the target photos'.[36] One pilot was warned that 'there are areas of the country where you'd swear that the map and the ground were two different places'.[37] The air war was transformed by three key innovations that sought to provide time-sensitive intelligence: reconnaissance drones; close-in visual surveillance; and the 'electronic battlefield'.

From October 1964 the Air Force launched reconnaissance drones on programmed flight paths over North Vietnam from transport aircraft and recovered them by helicopter off Da Nang. The early 'Lightning Bugs' were plagued by navigation errors but these were reduced when crew on the accompanying aircraft used television cameras to fly the drones. The images—and for a time video—captured on these missions were vital components of target folders, but their effectiveness was compromised by two factors that continued to haunt aerial surveillance long after the war. One was the balance between resolution and coverage: low-level flights (between 200 and 2,000 feet) provided high-resolution but limited coverage, whereas high-level flights (usually at 50,000 feet) opened up the field of view only to have it muddied by cloud and haze. The other issue was the time taken to process and distribute the imagery; recent stills were better than stock photographs, but it could take days for the film to be developed and analysed, and by then potential targets could have been hardened, dispersed or relocated. This was improved by the introduction of a satellite link to transmit the images from Saigon to Washington—for Johnson's inspection—via Hawaii, where the imagery was analysed and the results uplinked back to Saigon for crew briefings the next day. Satellite links are vital for rapid analysis of the imagery from today's Predators and Reapers too, and they have dramatically compressed the kill-chain: but the Lightning Bugs were all unarmed. In 1972 the *New Scientist* predicted that in two years' time 'a fleet of new bombers will attack North Vietnam—or some other country', flown by pilots 'sitting comfortably on the ground in front of TV screens hundreds of miles away', and speculated that 'unmanned drones may be bombing North Vietnam now'. They weren't, but by then Nixon had resumed air strikes against the North, and the drones had become so closely integrated into air operations that the USAF relied on them for assessing the effectiveness of the Linebacker raids.[38]

In the South the Air Force had been systematically destroying the forest canopy to expose the Viet Cong since 1962. The unmarked aircraft of Operation Ranch Hand were described as 'unarmed', but that deliberately ignored the deadly effects of the defoliants they dispersed on ecosystems and populations. Yet opening up the field of view was insufficient: just-in-time intelligence was even more important here because counterinsurgency relies on dynamic targeting, in which cruising aircraft are directed to (usually fleeting) targets of opportunity that emerge in flight, and often involves providing close air support to ground troops suddenly finding themselves in contact with the enemy.

In June 1965 the Air Force initiated its own visual reconnaissance programme using slow, single-engine aircraft. These 'Bird Dogs' were also used by the Army; they could fly as slow as 40 miles per hour, maintaining a tight turn to keep a site in view, and while they were not supposed to fly below 1,500 feet they usually went much lower. 'You can't even see people from one thousand feet,' one pilot noted: 'You can't see anything unless you go down there.'[39] They not only looked for direct signs of Viet Cong presence—campfires, tracks on trails, footprints on shorelines—but also carried out what is now called a 'pattern of life analysis'. This was an informal practice that could not call upon the formidable analytical apparatus embedded in intelligence, surveillance and reconnaissance today, but in much the same way pilots were required to become sufficiently familiar with their local area of operations—'aware of the eating, sleeping, working, traveling and social routine of the people'—so that they would be able to detect 'the slightest abnormality or change in the ground pattern'.[40] 'I was steadily learning my trade,' one pilot recorded.

I knew how many villagers should be in the rice fields surrounding each village. Too many might mean they had visitors. Too few could mean that a VC recruitment campaign was under way, or that trouble was afoot and the villagers had wisely decided to stay home until it was over. New footbridges had to be analyzed to determine what sort of traffic was using them, for the farmers seldom strayed away from their local village.

A comparative surveillance of the bridges and trails leading to the villages would almost always show the amount of foot traffic in the area. It was impossible to hide movement in the wet season, since tracks would show in the mud and elephant grass. I was starting to feel like something out of James Fenimore Cooper.[41]

It was an odd sort of intimacy—at once detached and intrusive—and it is one which continues to characterise US counterinsurgency today. But it provided a far more animated view than conventional mapping or even photoreconnaissance, and when the pilots also served as Forward Air Controllers (FACs) they could feed information directly to ground troops and mark targets for strike aircraft. 'My observational skills had matured,' one FAC wrote, 'and I no longer mistook cemetery plots for bomb craters'—only to add that 'on most days my objective was to see to it that bomb craters did indeed do double duty as gravesites'.[42] Then as now the sensitivities produced through this system were conditional. 'It gets completely impersonal,' another FAC told Schell, and 'after you've done it for a while you forget there are people down there.' Its affinities were with combat troops not ordinary Vietnamese: 'through this [radio] link,' another explained, 'the FAC's war was personalized and he earned the gratitude of the forces he supported'.[43]

In 1966 and 1967 reports from the Institute for Defense Analysis revealed that Rolling Thunder's interdiction campaign had failed: in fact, the flow of men and materials from the North to the South had increased. The analysts proposed the construction of a networked system of ground sensors and strike aircraft to check infiltration along the Ho Chi Minh Trail. The objective of 'Igloo White', as the system was called, was not so much to damage the Trail network—which was readily repaired or altered—but to strike traffic moving along it, and since the targets were fleeting, the interval between sensor and shooter had to be minimised. Visual reconnaissance was limited because the movement was usually at night, and so aircraft seeded thousands of acoustic and seismic electronic sensors along the roads in western Laos (triggered by trucks) and the paths in eastern Laos (triggered by people). When a sensor was activated, designated aircraft—and for a time drones—orbiting over Laos intercepted its identification signal and transmitted it to the Infiltration Surveillance Center at Nakhon Phanom air base in Thailand.[44] There, two large computer screens displayed the sensor field on a grid; activations were filtered by algorithms to eliminate false alarms triggered by animals or heavy rainfall, and when a critical threshold was passed the sensor was illuminated on screen. One Air Force officer said the Trail was wired 'like a pinball machine' that was plugged in each night when the convoys started their engines. 'As the seismic and acoustic sensors pick up the truck movements,' another explained, 'their locations appear as an illuminated line of light, called "the worm", that crawls across [the] screen, following a road that sometimes is several hundred miles away.'[45]

The Assessment Officers and support staff used the speed and direction of activations to predict the movement of the convoy and to designate a target box and time of attack whose coordinates were automatically transmitted to available strike aircraft. James Gibson emphasised that for those calling in the attack the target appeared only as a trace on a screen; when 'technowar' reaches its apex, he argued, 'it turns completely into representation. Indeed, the very name for a target was a "target signature".' Twenty minutes later, when the loudspeakers reverberated with the noise of bombs exploding, the illuminated trace went out. 'The representation disappeared.'[46] An audio track was played to the Electronic Battlefield Subcommittee of the Senate's Armed Services Committee: 'the sounds of a truck park, men talking, gears grinding, men shouting and then the sound of a Forward Air Controller overhead. Next the Senators heard the roar of jets, the crash of bombs and the firing of anti-aircraft guns as the North Vietnamese fought back.' It left at least one listener stunned. 'Technology now permits one to listen in on an attack you are directing from a control center hundreds of miles away. Only a TV picture is lacking.'[47] The absence of such a direct visual image made a substantial difference to the attack. As the officer responsible for the system explained, 'We are not bombing a precise point on the ground with a point target bomb—we can't determine each truck's location that accurately with ground sensors, which are listening—not viewing—devices. Since we never actually "see" the trucks as point targets, we use area-type ordnance [including napalm and cluster bombs] to cover the zone we know the trucks to be in.'[48] Here too the killing fields were reduced to abstract geometries: lines on screens and boxes on maps. But it was not difficult to imagine what would happen once the electronic battlefield incorporated a visual feed. Not only could it trip the switch from area to precision bombing, but it would also solve 'a constant problem of Vietnam and other wars—that some men must go and fight while others watch on television'. The only difference would be 'the placement of the viewing screen'.[49]

*From a View to a Kill*

These three basic elements—remotely piloted aircraft, real-time visual surveillance and a networked sensor-shooter system—prefigure the technical infrastructure for today's drone wars.[50] Each of them has been transformed and brought together in a unified system. The key difference is that the 'viewing screen' now occupies a central place and has become indispensable for those who wage remote war—at present principally the US military (in concert with other NATO militaries) and the CIA.[51] For USAF operations in Afghanistan, multiple screens are dispersed across a transnational network that includes pilots, sensor operators and mission intelligence coordinators at Creech Air Force Base in Nevada, image analysts and intelligence specialists at US Central Command's Distributed Common Ground System at Langley Air Force Base in Virginia, senior commanders, staff officers and advisers at CENT-COM's Combined Air and Space Operations Center (CAOC) at Al Udeid Air Base in Qatar, and Joint Terminal Attack Controllers deployed with ground troops in Afghanistan.[52] For CIA operations in Pakistan most of the network remains in the shadows, but the aircraft are piloted from Creech AFB and directed from the CIA's Counterterrorism Center at Langley in Virginia.

These are dramatic changes, and yet the essentials remain the same. In the 1960s the electronic battlefield was seen as heralding 'the battlefield of the future', and Westmoreland imagined combat zones under a constant surveillance so the US military would be able to 'destroy anything we locate through instant communication and the almost instantaneous application of highly lethal firepower'.[53] Forty years later the use of Predators and Reapers to combine sensor and shooter in a single platform has compressed the kill-chain to such a degree that war seems to have come perilously close to Westmoreland's vision: so much so that one contributor to *Harper's*—this time in 2010—feared that 'we are watching the future of warfare unfold in the skies over the Afghanistan–Pakistan border area'.[54]

The concerns sound familiar too. Critics of the electronic battlefield believed that it posed two great dangers: the expansion of the physical space of war and the contraction of the moral space of war. Military violence would know no global limits and the distinctions between combatants and civilians would be dissolved. Cornell's activist scholars predicted that abstraction would reach a terrifying climax through auto-

Image 2: Judge advocates are now stationed on the combat operations floor of the CAOC, (USAF).

mation: 'Ultimately we can have the machines fighting the "target signatures" with no human beings involved on either side.'[55] Senator Gravel, who was instrumental in releasing the Pentagon Papers to the public, was appalled that American troops were being withdrawn only to leave behind 'an automated war': 'We intend to turn the land of Vietnam into an automated murder machine.'[56] There was also the well-founded fear that the existence of such an advanced technology would produce pressure to use it in 'countering "insurgency"' all over the world, and that the Pentagon planned to extend its 'lethal pinball machine' to the entire planet, which, 'if wired right, could become a great maze of circuitry and weaponry, a jungle from which those who walk off the straight line from home to office to store would be eliminated'.[57] This is not a far cry from the Obama administration's view that its legal authority to use lethal force is not limited to '"hot" battlefields'—that the battle space has indeed become global—so that remotely piloted aircraft, with their extraordinary capacity to conduct targeted killings at a distance, are the new weapon of choice: particularly if they can be fully automated.[58]

Critics were also concerned at the loss of innocent lives. 'Machines have no qualms about killing civilians,' the Cornell group wrote in their first draft, and even the architects of the electronic battlefield acknowledged that its reliance on target signatures, blind bombing and area ordnance raised the spectre of striking what one of them archly called 'misidentified targets'.[59] The reliance on airpower in Indochina had magnified the central dilemma of counterinsurgency—how to distinguish insurgents from civilians—and the electronic battlefield only compounded the problem. For as Senator McGovern warned, 'the sensor which detects body heat, the aircraft thousands of feet in the air, and the computer complex many miles distant, are completely neutral and indiscriminate'.[60] Whether the lack of discrimination was an inherent limitation of airpower or whether, in all too many instances, it was a deliberate decision, sparked fierce debate. Certainly the Rules of Engagement outside North Vietnam were remarkably flexible and riddled with exceptions, and there were many cases where bombing was unambiguously reckless. In any event, as one military historian conceded, from the air 'all soldiers looked alike and guerrillas were indistinguishable from non-combatants'.[61] This remains a serious concern in Afghanistan, Pakistan and elsewhere, but the advocates of today's remote operations claim the issue has been resolved through the introduction of a new political technology of vision.[62]

In South Vietnam the areas that required the closest surveillance were allocated two Bird Dog flights a day to cover around 300 square miles in three to four hours; most had to make do with just one a day. In contrast, the crew flying a Predator or Reaper changes at the end of every eight-hour shift so that the aircraft can remain on station for twenty-four hours or more to conduct persistent surveillance. The aircraft use a multi-spectral targeting system, including an infrared sensor, a daylight TV camera and an image-intensified TV camera to stream full-motion high-resolution video across the network and provide a real-time view of the conflict zone. The introduction of augmented wide-area technologies like the 'Gorgon Stare' and ARGUS-IS promises to resolve the scale/resolution problem of the Lightning Bugs by quilting images from multiple feeds into a tiled mosaic covering an area of 100 square kilometres.[63] This will enhance the ability of analysts to track multiple individuals through different social networks in order to establish a 'pattern of life' consistent with the paradigm of activity-based intelligence that forms the core of contemporary counterinsurgency. The intention is to bring enemies 'out from the shadows' and illuminate the infrastructure in which they are embedded.[64]

But this solves one problem by introducing another. Petrabytes of data are collected each day, and the Air Force is acutely aware of the danger of 'swimming in sensors and drowning in data', and so, unlike the intuitive readings of the Bird Dog pilots and FACs, much of the analysis is highly formalised. Archived images are scanned to filter out uneventful footage and to distinguish 'normal' from 'abnormal' activity in a sort of militarised rhythmanalysis that is increasingly automated. The programmes include modified television software that can tag and retrieve video imagery, and GeoTime, a system that fuses and visualises data from multiple sources ('combining the where, the when and the who') as a three-dimensional array that mimics the time-geography diagrams developed by Swedish geographer Torsten Hägerstrand in the 1960s.[65] In addition to this forensic monitoring, live video feeds are scanned to push time-critical information to flight crews and ground commanders responding to emergent events. The development of 'per-

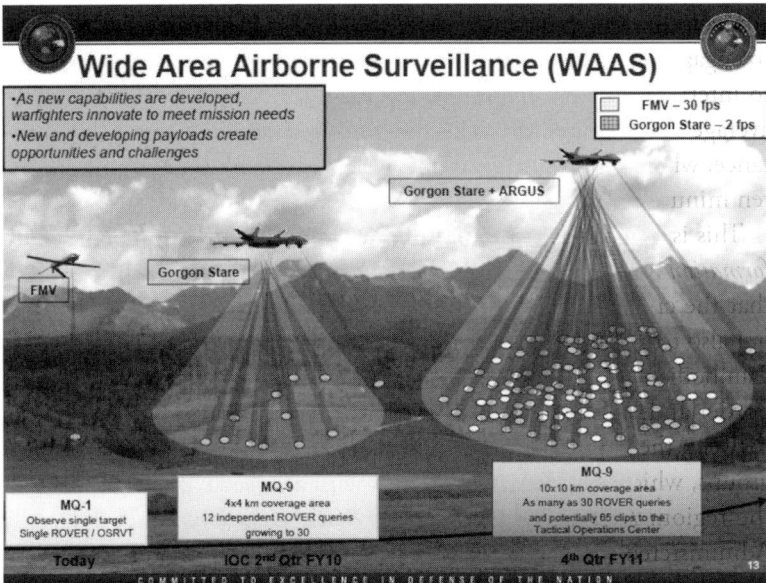

Image 3: Wide Area Airborne Surveillance, (USAF).

sistics' algorithms is seen as particularly promising because these allow surveillance to continue uninterrupted while 'automatically searching [the image stream] with unsurpassed detail for anomalies or preselected targets'.[66] The ultimate objective is the provision of a God's eye view, securing a new condition of what Gordon calls 'hypervisibility' that 'abolishes the distinctions between "permission and prohibition, presence and absence"'.[67]

This may be moonshine, but these new technologies—together with advances in precision weapon systems—have transformed targeting. The US occupation of Afghanistan in October 2001 was spearheaded by a conventional high-altitude bombing campaign, and as the post-invasion insurgency intensified, the air war was ratcheted up by both the Bush and Obama administrations. But Obama has shown a marked predisposition for a less visible, less public war that has involved a dramatic increase in the use of remotely piloted aircraft. They operate in concert with conventional strike aircraft, but their distinctive contribution is to combine hunter–killer roles in a single platform. Predators and Reapers provide 'armed overwatch' for combat troops—streaming live video to a Joint Terminal Attack Controller while maintaining the capacity to drop bombs or fire missiles if required—and close air support to troops in contact with the Taliban and other groups, when the aircraft are cleared to engage emergent targets. This has radically compressed the time taken to intervene in a fire fight: at the start of the Vietnam War it took on average 100 minutes for strike aircraft to respond to a request for assistance, whereas in Afghanistan the average response time is now around ten minutes.[68]

This is a significant compression of the kill-chain, but it is the *transformation* of the kill-chain that is more far-reaching: so much so, in fact, that the criticisms of the electronic battlefield are not only reactivated but also redoubled. Just as the war in Vietnam spilled over into Laos and Cambodia, so the war in Afghanistan has spilled over into Pakistan. The borderlands are porous, and there is a complex web of interaction between the Afghan Taliban, the Pakistan Taliban and other insurgent groups, which have varying relations with al-Qaeda and its affiliates in the region. Afghan fighters regularly seek sanctuary in the Federally Administered Tribal Areas, and the supply lines that provide them with money, weapons and explosives snake across the border. But airpower is no longer aimed exclusively at interdiction. A core role is now *execution*.

The US military uses a Joint Integrated Prioritized Target List to rank target sets in order of importance, but these are no longer limited to areas or boxes, or even physical objects like a training camp, an IED factory or a weapons cache. The targets are now often *individuals*. In Afghanistan the military maintains a subsidiary Joint Prioritized Effects List that identifies 'insurgent leaders' and 'nexus targets' (individuals like drug traffickers with 'proven links to the insurgency') who may be killed or captured. Each week a Joint Targeting Working Group reviews target nomination packets and establishes 'High Value Targets' in concert with judge advocates (military lawyers) and representatives from the CIA and other agencies. Some reports suggest that several hundred, others several thousand names had been put on the list by October 2009.[69] Although Joint Special Operations Command (JSOC) usually takes the lead in these so-called 'find-fix-finish' missions, both conventional and remotely piloted aircraft are often involved too. The capacity for persistent real-time surveillance is key to ground operations, as one Special Operations team explains, since it can be 'tightly synchronized with a finishing force', but the Air Force makes no secret of the fact that its single plat-form enables it to put 'warheads on foreheads' in its own right.[70]

Those who champion these missions insist that they fall within the laws of armed conflict that govern the war in Afghanistan, but extending the kill-capture programme into Pakistan is profoundly problematic because it is outside the war zone. The Obama administration has argued that the United States is in armed conflict with al-Qaeda so that the air strikes are consistent with the right to pre-emptive self-defence, but legal scholars differ sharply on the robustness of the argument. The situation is further complicated by the anomalous relationship between the Federally Administered Tribal Areas and the Pakistani state, and by uncertainty over whether the government in Islamabad has covertly consented to the attacks.[71] Legality is not the only issue, however, since serious questions have been raised about the propriety and effectiveness of targeted killings, and there is widespread public anger at the air strikes in Pakistan. These seem to have been inspired by the Israeli programme of extra-judicial killings in the occupied Palestinian territories.[72] Although the United States originally condemned Israel's assassinations, in November 2002 the CIA used a Predator to kill a suspected al-Qaeda leader in Yemen, and eighteen months later it launched a systematic programme directed at High Value Targets in Pakistan.[73] Here too the target list has been extended, and the list includes a congeries of militant groups as well as al-Qaeda. The nerve centre of the programme is the Pakistan–Afghanistan Department at the CIA's Counterterrorism Center, which prepares target nomination packets for those who are held to present 'a current and ongoing threat to the United States'. The final lists, which are reported to contain two to three dozen names at any one time, are scrutinised by Agency lawyers, endorsed by the CIA's General Counsel, and reviewed every six months.[74] The programme had a slow and uncertain start under the Bush administration, but it has been accelerated by Obama. Soon after taking office he doubled the CIA-controlled Predator fleet to over forty aircraft, and the number of strikes rose from thirty-five to thirty-eight in 2008 through fifty-three to fifty-five in 2009 to 117 to 128 in 2010.[75] This is not a standalone programme. Most clandestine operations in Pakistan still seem to be controlled by the CIA, but JSOC has operated across the border since 2006 and one source claims that many air strikes attributed to the CIA, especially those with high civilian casualties, were in fact carried out by JSOC.[76] The CIA also collaborates closely with the Air Force, which sometimes supplies ('loans'), maintains and arms the aircraft, which are flown from Creech AFB, perhaps by Air Force crews. In addition, the Air Force has recently positioned more of its own aircraft close to the Afghan side of the border so that the CIA can 'hand off' targets to them as they cross over from Pakistan.[77]

We are assured that both military and CIA operations take great pains to minimise civilian casualties, and the numbers certainly do not come close to the totals killed by bombing in the Second World War or the air wars in Indochina. But it is equally clear that civilians continue to be the innocent victims of air strikes—and, it must be said, of ground attacks from militaries and insurgents—on both sides of the border. The numbers are difficult to verify, and all estimates are dogged by controversy: most rely on local press reporting, which is highly uneven, and even the detailed field surveys are inevitably bedevilled by the difficult distinction between combatants and non-combatants. I cannot adjudicate these questions here, but I can show how casualties continue to be caused by what the United States calls 'the most precise weapon in the history of warfare'.[78]

How might such a claim be judged? The accuracy of a weapon is given by the Circular Error Probable (CEP), which is measured under ideal experimental conditions and defines the radius from the aiming point within which a missile or bomb will land 50 per cent of the time. The Predator carries two laser-guided Hellfire missiles, with a CEP of 3–8 metres, and the Reaper can carry fourteen Hellfire missiles or four Hellfire missiles and two 500lb GPS-guided JDAM bombs, with a CEP of 10–13 metres. If all the conditions are satisfied, these are certainly much more accurate than the bombs used in the Second World War, whose CEP was often far in excess of 1,000 metres, and in Indochina, where conventional bombs had a CEP of 130–40 metres and laser-guided bombs 22 metres.[79] Yet this cannot be what American officials mean when they invoke unparalleled 'precision' to endorse the use of remotely piloted aircraft to carry out targeted killings because conventional aircraft use the same weapons. In any case, the most accurate weapon in the world is useless if the target has not been correctly identified and located, and this inevitably becomes much more difficult as targeting contracts from an area or a box to a person. This is the heart of the matter. An editorial in the *Wall Street Journal* trumpeted that 'never before in the history of warfare have we been able to distinguish as well between combatants and civilians as we can with drones', and it is surely this—the principle of *distinction* used to secure a claim to the moral high ground—that the protagonists of the programme seek to emphasise.[80] 'Distinction' is thus more than a military-strategic or even an ethical-juridical concept; it is also a political-cultural construct that is made to do political and cultural work. The advocates of these new platforms insist that they alone make possible the networked fusion of persistent surveillance and high-resolution that at last enables—even compels—those involved in the kill-chain to spare innocent civilians. As I want to show, this is too glib by far. It jibes against the common criticism that drone wars are 'videogame wars' which inculcate a 'Playstation mentality to killing', but I think this is a superficial view too because it fails to engage with the political technology involved in sighting the enemy in this way.

## Sighting the Enemy

These new technologies require many of the skills used in videogames—including rapid hand–eye coordination, multitasking and visual acuity—but this does not automatically reduce war to a videogame. To

61

explain the continued threat to civilians posed by this political technology of vision, I distinguish between near sight, far sight and top sight.

Most of the USAF Predator and Reaper flight crews are based in the United States, from where they control aircraft over Afghanistan via a transatlantic fibre optic cable to Germany and a Ku-band satellite link; the exceptions are the forward-deployed Launch & Recovery crews that use a line-of-sight data link.[81] But the flight crews repeatedly insist that the real-time video feeds bring them right into the combat zone: that they are not 7,000 miles away but just 18 inches, the distance from eye to screen. Insofar as this is a 'videogame war' then it shares in the extraordinary immersive capacity of the most advanced videogames. This is significantly different from the detachment—the 'distance and blindness'—experienced by bomber crews over Germany or Vietnam. And yet the reality-effect this produces may be sufficiently powerful where remotely piloted aircraft are providing armed overwatch or close air support to convert proximity not distance, visibility not blindness, into a serious problem. In such cases remote vision is limited, paradoxically, by *near sight*. Persistent surveillance means that flight crews come to 'know' their areas of operation in a particular way; they interact regularly with troops on the ground through live video feeds and online communications, and the intimacy created by these new forms of military-social networking can predispose them to interpret the actions of others in the vicinity as a threat to their comrades and to precipitate lethal action. 'There's no detachment,' one officer explained. 'Those employing the system are very involved at a personal level in combat. You hear the AK-47 going off, the intensity of the voice on the radio calling for help. You're looking at him, 18 inches away from him, trying everything in your capability to get that person out of trouble.'[82] This is version 2.0 of the 'personalised' war described by FACs in Vietnam, now transformed through new communications technologies and the formalised process of image analysis. From Nevada knowledge of the war zone is indexed by frequency not familiarity—it comes from 'actuarial surveillance'[83]—and Afghanistan remains an alien landscape where common ground is confined to the virtual presence of coalition troops on the screen and in the online chat-rooms. An example will illustrate what I mean. In the early morning of 21 February 2010 a Predator was tasked to track three vehicles travelling down a mountain road in Uruzgan province in central Afghanistan, several miles away from a Special Forces unit moving in to search a village for an IED factory. The Predator crew in Nevada had radio contact with the Special Forces Joint Terminal Attack Controller and they were online with image analysts at the Air Force's Special Operations Command headquarters in Florida. At every turn the flight crew converted their observations into threat indicators: thus the two SUVs and a pick-up truck became a 'convoy', adolescents 'military-aged males' and praying a Taliban signifier ('seriously, that's what they do'). After three hours' surveillance two Kiowa helicopters were called in, and during the attack at least twenty-three people were killed and more than a dozen wounded. Only after the smoke had cleared did the horrified Predator crew recognise the victims as civilians, including women and children. Two military inquiries faulted the flight crew, but the incident was more than a matter of individual responsibility. It was also a structural effect of a culturally divided visual field: a visual interpellation that made it extremely difficult for the crew to see the subjects of the surveillance as anything other than insurgents until it was too late.[84] It is hard to know how common this is, but the risk to civilians in such situations resides not only in the pressures of time-sensitive targeting but also in the video feeds from the Predator immersing its operators in, and to some substantial degree rendering them responsible for, the evolving situation on the ground. High-resolution imagery is not a uniquely technical capacity but part of a techno-*cultural* system that renders 'our' space familiar even in 'their' space—which remains obdurately Other.[85]

Where hunter–killer missions do not involve close contact with ground forces then different considerations apply. In these situations, attacks depend on airborne surveillance and signals intelligence, and in a remark that echoes the controversy that swirled around the electronic battlefield, John Nagl, one of the architects of the current US counter-insurgency doctrine, declared that 'we're getting so good at various electronic means of identifying, tracking, locating members of the insurgency that we're able to employ this … almost industrial-scale counter-terrorism killing machine that has been able to pick out and take off the battlefield not just the top level al-Qaeda-level insurgents, but also increasingly is being used to target mid-level insurgents'.[86] Just as the critics of the electronic battlefield predicted, a reliance on 'target signatures' substitutes for local knowledge of the cultural landscape, but this is now aggravated by the use of formalised methods of analysis that compound the distance effect. Again, an example will clarify the situation. On 2 September 2010 ISAF announced that a 'precision air strike' earlier that morning had killed the Taliban deputy shadow governor of Takhar province in northern Afghanistan and 'nine other militants'. The target had been under persistent surveillance from remote platforms until two strike aircraft repeatedly bombed the convoy in which he was travelling. Two attack helicopters were then 'authorized to re-engage' the survivors. The victim was not the designated target, however, but the election agent for a parliamentary candidate; nine other campaign workers died with him. A painstaking analysis by Kate Clark clearly showed that one man had been mistaken for the other, which she attributed to an over-reliance on 'technical data'. Special Forces had concentrated on tracking cell phone usage and constructing social networks. 'We were not tracking the names,' she was told, 'we were targeting the telephones.'[87] This is a standard operating procedure, part of the militarised rhythmanalysis that I described earlier, and it is not confined to the US military or to Afghanistan. The CIA has been authorised to use lethal force against unnamed individuals in Pakistan on the basis of their suspicious 'pattern of life'. These people—at best ordinary foot soldiers, at worst innocent civilians that 'walk off the straight line'—are known as 'signature targets', and in their anonymity and abstraction they are ghostly traces of the target signatures that animated the electronic battlefield. Since the air strikes intensified one source estimates that twelve times more low-level fighters than mid- to high-level al-Qaeda or Taliban leaders have been killed, and reports claim that the 'vast majority have been individuals whose names were unknown or about whom the [CIA] had only fragmentary information'.[88] Takhar was a Special Forces operation, but the lesson is a general one. Here too the visual field is divided, with abstracted US intelligence separated from what Clark calls knowledge of 'the everyday world of Afghan politics', but the problem was primarily one of *far sight*. Even where there are no ground forces to engage the sympathies of flight crews, the inability to read the intimate textures of the landscape turns out to have catastrophic consequences for the innocent.[89]

Horrors like these are supposed to be prevented, or at least mitigated, by the introduction of a normative armature to targeting: near sight and far sight modulated by *oversight*. In Vietnam, 'collateral damage'—the term was invented during that war—was limited by a political calculus.

Johnson boasted that the Air Force could not bomb an outhouse in the North without his approval, and air strikes in the South required at least nominal clearance from provincial political leaders. But in both cases there were few legal restrictions. International law was conspicuously silent about air war, not least because the process that produced the Geneva Conventions in 1949 was dominated by the states that had the most experience of bombing, while those with the most experience of being bombed—Germany and Japan—were excluded. Operational law remained strikingly undeveloped, and apart from one judge advocate in Thailand who scrutinised target lists in North Vietnam, military lawyers were not involved in the targeting process.[90] The situation changed dramatically after Vietnam, however, and today the risk of civilian casualties is a vital consideration throughout the kill-chain, driven by the requirements of international law, notably the Additional Protocols to the Geneva Conventions (1977), and the prospect of international scrutiny.

USAF targeting is now a quasi-judicial process in which visuality is a central modality. Targets must be positively identified from more than one source, and can only be attacked if a visual 'chain of custody' is maintained. Continuous observation is thus mandatory, so that the persistent presence of a remotely piloted aircraft becomes indispensable. A pattern of life analysis is conducted not only to acquire individual targets but also to determine the presence of civilians; collateral damage estimates are made using a classified 'physics-based' program integrated with the RainStorm precision targeting system; casualties may be mitigated by changing or modifying the ordnance to be used ('weaponeering').[91] This refined process retains the objectivist, calculative logic of earlier kill-chains, but its language shows that it has become a quasi-juridical exercise too. Judge advocates are now stationed on the combat operations floor of the CAOC in order to provide expert counsel to commanders on the 'prosecution' of the target. They are required to consider international law, the Rules of Engagement and any special instructions, but they also have their eyes on the screen, monitoring up to twenty different chat windows at any one time and reviewing onscreen target folders containing imagery and other intelligence.[92] Here at least justice is not blind. According to Schmitt, persistent surveillance ensures 'a significantly reduced risk of misidentifying the target or causing collateral damage to civilians and civilian property', while Beard claims that these virtual-visual technologies introduce 'unprecedented levels of transparency' to the killing space, 'eliminating some of the key excuses that states have long used to escape responsibility for attacks that appear to cause excessive civilian casualties'.[93] In these ways, the high resolution-level of the imagery is translated directly into a refined legal calculus that can supposedly make equally fine-grained distinctions between military and civilian targets and between combatants and non-combatants. Indeed, while he was serving as the staff judge advocate at the CAOC, Colonel Gary Brown claimed that airborne intelligence, surveillance and reconnaissance 'gives us the ability to actually apply [laws of armed conflict] principles (with almost mathematical precision) that were originally just concepts'.[94]

What protocols the CIA follows for its targeting process in Pakistan are unknown—though Etzioni describes what he calls 'a secret matrix' and claims that it offers 'robust oversight'[95]—but if they are similar to the Air Force, then the protection afforded to civilians will be conditional. This is not to say that legal advisers do not intervene to prevent strikes where there is a risk of civilian casualties; they do. But international law allows for civilians to be killed in attacking a military target ('distinction') so long as their deaths are outweighed by military advantage ('proportionality'). This means that sometimes civilian deaths are accidental—the system is far from perfect—but in others they are *incidental* to what is deemed to be concrete and direct military advantage, in which case they have been anticipated and endorsed by judge advocates.[96] When legalisms take centre stage in cases like this, Tom Engelhardt advises us 'to think of magicians', and Brown's invocation of 'mathematical precision' is pure sleight-of-hand; this has to refer to the process of collateral damage modelling rather than legal judgement since elsewhere he concedes that proportionality is 'not a mathematical formula or anything like that' and that the laws of armed conflict contain some 'very wiggly concept[s]'.[97] Indeed they do, but the still sharper point is that the legal armature that secures the process of target validation is not above the fray but is embedded within it. To refer to the 'prosecution' of the target and its 'chain of custody' is to concede that judge advocates are not impartial tribunes, still less defence attorneys, and when Beard asks 'where now is the [military] lawyer's client?' his answer is unequivocal: Creech or the CAOC.[98] The incorporation of judge advocates into the kill-chain evidently does not diminish the privilege accorded to the military in the determination of military advantage;

as Orford emphasises, the relevant body of international law 'immerses its addressees in a world of military calculations' and ensures that proportionality will always be weighed on the military's own scales.[99]

## The Death of Distance

Our understanding of bombing has been dominated by political and military historians who focus on strategy, and social historians who recover the experiences of those who were bombed. These are vital contributions, but the gap between the two—the kill-chain—is too important to be left to buffs and geeks. Too often, focusing on strategy can make air war seem as clinical as its 'progressive' proponents proclaimed, and yet by the time we crouch under the bombs and give voice to the victims, it is too late. We need to understand not only why the thing was done and with what consequences, but also *how*.

In tracing the twisting lines of descent from the Second World War through Indochina to Afghanistan and Pakistan, I hope I have not minimised the differences between them. Swift's descendants of the Lancaster bomber are not being used to destroy cities, and their targets are not the victims of the 'faceless' war described by Deighton—at least not where they seek out known and named individuals. The same cannot be said of those who carry out the attacks, however, and in a powerful critical commentary on the drone strikes in Pakistan two counterinsurgency proponents, David Kilcullen and Andrew Exum, described 'a frightened population' living under a constant threat from 'a faceless enemy that wages war from afar'.[100] Their point is a double one. They allude to the perpetual envelope of terror under which ordinary people are obliged to live (and die), and when we read accounts of air raids during the Second World War—the wail of the sirens, the crowded shelters and the thud of the anti-aircraft guns—we should remember that the people of the borderlands hear no warning, have no place of refuge and have no means of defence. Kilcullen and Exum also invoke a peculiar horror that is seemingly attached to war from a distance. And yet air wars have always been fought from a distance, and war more generally has a complex, developing relation to the spaces through which it is fought. By its very nature, bombing produces an alternation between different spaces. In his classic account of *The Command of the Air* Giulio Douhet noted that 'by virtue of this new weapon, the repercussions of war are no longer limited by the farthest artillery range of guns, but can be felt directly for hundreds and hundreds of miles'. He predicted that in the future 'the battlefield will be limited only by the boundaries of the nations at war, and all of their citizens will become combatants, since all of them will be exposed to the aerial offensives of the enemy. There will be no distinction any longer between soldiers and civilians.'[101] These were prophetic words, and that last sentence has haunted the wasteland of bombing for more than ninety years. And yet, even as it transformed the reach of military violence and rewrote the geography of armed conflict, dissolving the old boundaries between the front line and the home front, when air war was conducted from ground stations rather than aircraft carriers, it also re-established a distance between the relative security of the bases from which aircraft took off and the targets that they attacked. Pilots and their crews were hardly commuters to war, and they faced extraordinary danger as they flew through hostile air space, but if they returned safely it *was* usually to relative safety. Critics make much of today's Predator and Reaper crews driving from home to the (remote) killing zone every day, but this is not the complete break from the past that they imagine it to be.

In short, I think it is a mistake to turn distance into a moral absolute. Pilots and crews in Nevada are 7,000 miles from their targets, but is this experientially any more remote than the B-52 crews flying from Guam to bomb targets 5 miles below them in South Vietnam? If distance is the issue (and I am not sure that it is), at what point does wartime killing become acceptable? In posing these questions I do not mean to say that nothing has changed since Vietnam either: Predators and Reapers do not carpet bomb whole landscapes. Neither do I think Vietnam—or Cologne or Hamburg—should be the moral standard against which we judge today's air wars. To be sure, many writers have drawn parallels between Vietnam and Afghanistan—and, again, there are differences too—but if Vietnam was a quagmire then Afghanistan–Pakistan threatens to become a vortex. If the battle space is now global, and if the United States claims the right to use lethal force against its enemies wherever it finds them, then what happens when other states claim the same right? And when non-state actors possess their own remotely piloted aircraft?

Virginia Woolf's thoughts during an air raid ended with these lines: 'Let us send these fragmentary notes to the huntsmen who are up in

America, to the men and women whose sleep has not yet been broken by machine-gun fire, in the belief that they will rethink them generously and charitably.' I hope that this essay too might trouble the sleep of today's masters of war busily cocking their Predators and Reapers.[102]

3

# AERIAL SURVEYING, GEOPOLITICAL COMPETITION AND THE FALKLAND ISLANDS AND DEPENDENCIES AERIAL SURVEY EXPEDITION (FIDASE 1955–7)

*Klaus Dodds*

In the 1940s and 1950s, the struggles over the ownership of the Antarctic Peninsula and outlying islands were fundamentally three-dimensional.[1] At stake was the control and administration of, quite literally, everything above and below the polar ice, rock and water. Two fundamental issues conspired to make these struggles more challenging—the remoteness of the territory itself (at least from the point of view of British administrators notwithstanding South Atlantic islands such as the Falklands) and uncertainty about the polar continent. In 1950, for example, scientists were still unclear as to whether Antarctica was composed of two parts.[2] Understanding of the ice sheet was modest, and no one knew how deep the ice might extend below the polar plateau.[3] Mapping and surveying, before the Second World War, was sporadic and geographically concentrated in the more accessible areas such as the Antarctic Peninsula and surrounding islands.[4]

# FROM ABOVE: WAR, VIOLENCE AND VERTICALITY

In 1948, the Falkland Islands Dependencies Survey (FIDS),[5] the chief operating agency in Britain's Antarctic territories,[6] were informed that their mission was 'to obtain knowledge of the Dependencies and advance the general development of the FID [Falkland Islands Dependencies] and of the Antarctic region as a whole', and 'to carry out observations and investigations whose results will benefit the whaling industry'.[7] The reference to the 'whaling industry' is significant because it was through the licensing of whaling (in the main by Norwegian commercial companies operating out of South Georgia and the Antarctic Peninsula region) that FIDS was able to fund and implement its surveying programme.

In other words, land-based surveying was being funded by maritime research into whale stocks and prevailing sea/weather conditions.[8] And this research into whaling, predicated on aiding its sustainable harvesting, facilitated claims to 'environmental authority' over the FID. The governor of the Falkland Islands, Sir Miles Clifford, who was also the British polar administrator, noted the following in 1948:

> The only true wealth that this area contained, so far as know today, is still as in the past its marine wealth—its whales and seals; these, as we have noted earlier, could be exterminated by indiscriminate killing and it was this recognition of this danger which decided His Majesty's Government to bring their industries under control and led to the establishment of British sovereignty over the area now known as the Falkland Islands Dependencies. The motive was a purely unselfish one, to conserve the harvest of those seas for the benefit of mankind as a whole.[9]

As geopolitical competition intensified between rival claimants—Argentina, Britain and Chile—this apparently laudable claim to managing the living resources of the Antarctic for the benefit of mankind (*sic*) faced fresh scrutiny. This claim to 'environmental authority', while echoing past episodes of British imperial management,[10] was unable to contain those rival claimant states. For the Argentine and Chilean governments, the Antarctic was an integral part of their national territories due to its geographical proximity and geological connection to the Andes. Argentine and Chilean nationalists were swift to highlight environmental connections such as similar snow conditions in Patagonia and Antarctica alongside similarities in rock specimens. This appeal to environmental stewardship as the basis for imperial authority was rejected and became less significant as the Norwegian whaling industry based in the Southern Ocean began to decline in the post-war period.[11]

The mandate for FIDS was, as a consequence, multifaceted—it was to obtain knowledge about the FID, usually by way of compiling maps and surveys, to secure commercially valuable information for the whaling industry pertaining to weather and oceanography, and most controversially of all, to be on the front line in terms of demonstrating the continuous occupation of Britain's Antarctic territories. So those individuals recruited to FIDS in the post-war years, with specialised training in the geographical and geological sciences, were expected to be not only scientists but also imperial administrators, whether in the form of justices of peace, postmasters or dispatchers of notes of protest should they encounter 'intruders' on British 'Crown Lands'. The intrusive presence of South American parties and their associated polar ephemera were a regular feature of conversation between the FIDS based in their field stations in Antarctica and the Colonial Office civil servants and FIDS scientific staff charged with administration in London. To compound matters still further, the Argentine and Chilean governments were determined to enforce their territorial claims to the Antarctic Peninsula.[12] British administrators turned to air survey with the aim of radically altering the geopolitics of the Antarctic Peninsula.

It was and remains a deeply contested territory, and growing speculation over possible mineral wealth provided further spice, with some commentators claiming that there might be 147 different types of minerals in Antarctica.[13] We need to appreciate some of the geopolitical context that empowered civil servants and government ministers to approve a substantial investment in the aerial mapping of the Antarctic Peninsula in the mid-1950s. What was at stake was the occupational fate of thousands of square miles of ice, rock and sea. With two rival claimants to confront, alongside the uncertain intentions of other major actors including the United States and Russia, the period between 1943 and 1961 was immensely tense as the three counter-claimants invested considerable faith in the map, the survey, the place-name and the photograph to strengthen their particular claim to the Peninsula region. And it was the aeroplane, with its promise of mobility, and thus widespread capacity for enhancing accessibility, that further animated British administrative plans for exercising effective territorial sovereignty.

The Falkland Islands Dependencies Aerial Survey Expedition (FIDASE) stands as a testimony to the faith invested in aircraft and, as Denis Cosgrove reminded us, 'the view of the airman'.[14] Such a view offered exciting possibilities including assembling an epistemological optic that equated the view from the air with a new form of legibility, one that was denied to those mortals tied to the ground. The notion of the synoptic view was an appealing one to those charged with administering a vast imperial territory with precious few facilities such as airports, roads, bases and garages. The infrastructural and institutional 'moorings' that configure and enable mobile subjects and their materials and objects, including planes, were thinly distributed in the Antarctic.[15] As the then director and general manager of Hunting Aerosurveys Limited noted to the Colonial Office in April 1955:

Vertical photography of nearly 50,000 square miles of very mountainous country is a big undertaking anywhere in the world; but when the area to be surveyed is far from the normal air or shipping routes and is entirely without any sort of established air base at all, and moreover enjoys notoriously unreliable weather which may alter in a matter of hours both the nature and appearance of the area to be photographed or the bases used for take off and landing, it is evident that it is likely to be a very difficult and somewhat hazardous undertaking.[16]

As the personnel attached to FIDASE discovered to their cost and frustration, the director and general manager's assessment of the 'scope of the project' was far-sighted. On some days, in the short polar summer season, it provided opportunities to create a series of 'aerial achievements' such as flying for ten hours photographing vast swathes of Antarctic territory. On other days, a combination of low-lying cloud cover, unrelenting polar winds, sea ice (especially problematic for amphibious craft) and the freezing of instruments grounded planes and pilots. A grounded aircraft is not an instrument of power projection, and there were very few 'good' flying days in the two operating summer seasons of FIDASE. The aerial view, in the Antarctic context, was never an assured one.

This chapter addresses a number of themes germane to the 'view from above'. Initially, by drawing upon the work of an interdisciplinary group of scholars such as Peter Adey, Denis Cosgrove and John Law, it considers how British aerial mapping was a complex if fragile assemblage put to work in the name of a colonial project devoted to a form of long-distance control that far exceeded anything else seen in the context of the British Empire. The distances were vast: some 10,000 miles from London and a territory encompassing a total area many times larger than the United Kingdom. Unlike other areas of the imperial portfolio, 'anti-colonial' resistance largely took the occasional form of plaque destruction and base desecration by rival South American polar parties. Thereafter, the chapter turns more specifically to the institutional contexts and geographical dimensions pertaining to the British Antarctic, which help explain how FIDASE was assembled. Finally, the chapter considers how the aerial view in a variety of guises (either flying over with a plane and/or hovering via a helicopter) was frustrated, constrained and even undone by a combination of weather, accidents and other factors such as the behaviour of other parties. Far from offering a clear and omniscient view, the 'view from above' was frequently subject, as noted above, to the vagaries of the weather that frustrated attempts to make the Antarctic legible.[17] But, on reflection, perhaps the 'view from the air' did something else, which was not anticipated at the start of this particular expedition. It established a threshold in which imperial administrators decided to re-calibrate their geopolitical strategies and focus, quite literally, on grounding British sovereignty rather than attempting to project power aerially.[18] Aerial surveying, after 1957, was not revisited as a geopolitical strategy.

*Assembling Long-Distance Control: Britain and the Antarctic*

Previously unknown ice forms have been found, buried coastlines located, maps redrawn, rates of glacial flow determined, and many geographical discoveries have been made. Photo-grammatically speaking, Antarctica has finally reached the status of other continents, having completed the reconnaissance cycle from early captive balloons photography to standard commercial aerial surveys.[19]

In his elegant examination of Portuguese imperial exploration in the fifteenth and sixteenth centuries, John Law makes the case that agents, networks and objects were critical to the implementation of long-distance control, especially in Asia.[20] As part of that project, the Portuguese sought to manage the control and expansion of their imperial trading interests by drawing upon the social, the technological, the physical and the politico-economic. What his study highlighted was the role of specific factors such as the mobility and durability of their vessels which were able not only to navigate return journeys involving thousands of nautical miles but which were also able, where necessary, to project force effectively through the use of cannon-fire. So while having robust vessels was the first prerequisite of long-distance control, it was insufficient if these objects could not be properly controlled. The ships had to be able to function effectively and in a manner which enabled the participants to return safely from their voyages. Effective navigation was essential, and in order to achieve such an objective there was a need to possess a drilled and trained group of people able to understand and use appropriate documentation and instrumentation. As Law noted, 'I believe the theoretical claim—that the undistorted communication necessary for long-distance control depends upon the generation of a structure of heterogeneous elements containing envoys which are mobile, durable, forceful, and able to return—to be well founded.'[21] As he concluded, albeit with a warning about the need for historical and geographical contextualisation, the trio of documents, devices and drilled people is essential to any explanation of long-distance control—and why it endures for any significant period of time.

This study of the Portuguese (and their project of long-distance control) is relevant for other studies, albeit far removed from the immediate Columbian encounter. In a decidedly chillier and more contemporary context, British authorities charged with managing the most southerly territories within the imperial portfolio faced a variety of challenges. The problem facing British civil servants in London, in particular, was how to manage long-distance control in the face of some stark geographical conditions—Antarctica lay over 10,000 miles from London and it was several days sailing from the nearest inhabited colony of the Falkland Islands. Worse still, the size of British Antarctic territories was considerable (involving as it did an area many times larger than the United Kingdom), and sovereignty was challenged by two South American states, Argentina and Chile. So the dilemma, put simply, was this—how to organise some kind of control over a very remote place, which was under threat by nearer neighbours in a polar climate that was punishing on people, infrastructure and machinery such as airplanes. Finally, to make matters worse, funding was always under pressure and liable to be challenged when juxtaposed next to competing priorities either within the UK or other parts of the British Empire. And it only remained a dilemma of sorts, of course, because no one responsible for administering those southerly territories advocated disengaging from the region, not least because of formal claims to territory posited in 1908 and 1917.[22]

Documents, devices and drilled people were an essential element in both creating and subsequently addressing this territorial dilemma. Documents, in this context, include formal declarations of territorial claims such as Letters Patent (1908, 1917) and a vast correspondence generated within government departments led by the Colonial Office and the Foreign Office. Devices, as we shall note below, included a range of objects such as the aerial camera and surveying equipment. Drilled people, either operating in the Antarctic or in London, were essential to the coordination and implementation of British polar power projection. The territories concerned were judged to need constant visitation, assessment and regulation, and that required immense effort on the part of all concerned. So training, and specifically a manner of drill, was important in places like the Antarctic—all the men, and especially the surveyors, had to be able to operate, navigate and survive in a hostile environment. The men appointed to the FIDS in the 1940s and 1950s were often former servicemen and/or trained at Imperial College Field Centre by professional surveyors. Later recruits were inducted within the Directorate of Colonial (and subsequently the Directorate of Overseas) Surveys-approved courses supported by additional teaching from staff attached to University College London.[23] And, as should be apparent from the above, they were all men. Women were not allowed to join the FIDS, unless serving in back-offices in London.[24]

The initial attempt to reactivate British long-distance control in the British Antarctic was a secret naval mission called Operation Tabarin.[25] Named after a Parisian nightclub, it was designed explicitly in 1943 to reassert British sovereignty in the Antarctic, especially the Peninsula region.[26] The strategic rationale was informed not only by a sombre interpretation of South American territorial rivals (and German naval activity) but also a realisation that British activity had tailed off in the aftermath of the 1934–7 British Graham Land Expedition.[27] Prior to that scientific expedition, the Falkland Islands Dependencies (FID) was an income generator for the British Empire. But the source of that income lay offshore in the form of whales rather than mineral resources within the continent itself. By licensing Norwegian whaling companies, British officials generated a valuable revenue stream (paid into a Falkland Islands Research and Development Fund), as whale oil was used in products ranging from margarine to explosives. For much of the inter-war period, a so-called Discovery Committee, based in London, was charged with generating scientific knowledge about whaling stocks in the Southern Ocean. The leitmotif of the Committee was to use scientific knowledge for the rational and orderly development of the FID.

The commitment to such 'rational and orderly development' was a response of sorts to the challenges posed by long-distance control. It was a way of bundling together, in other words, resource exploitation, imperial management and knowledge generation. For British imperial administrators, Antarctica was conceived in fundamentally volumetric terms—involving horizontal and vertical dimensions. The onset of pelagic whaling in the mid-1920s challenged this management strategy. In essence, as British administrators recognised, if whaling occurred in 'international waters' then it was nigh on impossible for the British government to extract licensing monies from Norwegian whalers no longer based at shore stations within the FID. In the late 1920s the commercial and political relationship between Britain and Norway was at stake as arguments ranged over how the whaling industry should be financed and regulated. To complicate matters still further, the governor of the Falkland Islands, Arnold Hodson, was increasingly agitated over the fact that this whaling-based revenue from the FID flowed directly into London-based coffers.[28] There was no diversion of funding to Falkland Islands-based development projects, and he was a committed advocate of further maritime surveying of the FID for the purpose of producing new assessments of resource potential—so here we have someone more interested in developing and enhancing a 'view below the waterline'.

This was not to happen during his tenure as governor because British officials in London sought to concentrate on maintaining their tenure on the control of Southern Ocean whaling rather than divert resources towards additional surveying. This was to prove increasingly problematic as whales and whaling entered into a new era of international control, transnational negotiation and corporate investment in the form of Unilever.[29] In these uncertain times the emphasis shifted, as is reflected in the minutes of the Discovery Committee, towards protecting the sovereign interests of the British state in the FID. By the 1930s, for example, it was recognised that the voyages of British-registered vessels such as the *William Scoresby* were performing two vital tasks: they were collecting intelligence on potential sovereignty threats, while at the same time quite literally demonstrating British authority in the FID, especially its waters. When Sir Hubert Wilkins, the Australian aviator, proposed a series of flights over the Antarctic continent in the early 1930s, the Discovery Committee faced a geopolitical but also an epistemological dilemma.[30] Should funds be diverted towards an adventurer who might showcase the 'view from above', or should they be used to continue to fund the 'view from below the waterline'? Both offered the prospect of extending Antarctica's legibility.

This dilemma, which was multifaceted, only worsened further as the 1930s unfolded. Whaling was entering a period of decline as the price of whale oil fell alongside a diminishing whale stock. In the meantime, Britain and its Commonwealth partners, Australia and New Zealand, made claim to three-fifths of the poorly mapped Antarctic continent. Their legal authority was not widely recognised, and by the end of that decade it was openly challenged by Argentina and Chile. Maintaining authority was, the Colonial Office believed, only possible with a near constant and visible display of sovereignty, and proof of activity, especially during the summer seasons, where mobility, whether in the form of the sledge, the ship or the plane, was more likely. This, in turn, raised the question of whether Britain needed to fund and initiate a new round of activity focusing on the, hitherto poorly explored, continental interior.[31] This need to perform sovereignty led to the state-sponsored British Graham Land Expedition (1934–7), which was the first venture to overwinter on the continent since the Scott era. As part of its purported task to ascertain whether the Antarctic Peninsula was connected to the rest of the continent, the expedition conducted aerial surveys, and produced a rich collection of aerial photographs, later to be archived at the Scott Polar Research Institute at Cambridge. It also served to act as a benchmark for subsequent discussion of the possibilities of aerial survey in the Antarctic.

The dominant mode of power projection was to change by the Second World War. If the interwar period witnessed attention given to whaling and offshore knowledge generation, the 1940s and 1950s marked a shift towards an interest in 'the view from above', not least because imperial rivals such as Argentina and the non-claimant United States were targeting Antarctica with their aircraft and airmen.[32] Increasingly, it was the aeroplane and the aerial view that were understood as essential modes of long-distance control. In the British context, Antarctica was to be 'targeted', with the help of a new generation of ex-servicemen, some of whom served in RAF Bomber and Coastal Command. Men, trained in wartime flying, navigating and surveying, were asked to transfer their skills to a part of the world where bad weather and prevailing sea ice were to be their new adversaries.

Airmen, with wartime bombing and surveillance experience, were not the only element using their transferable skills and experiences. In London and Cambridge, for much of the Second World War, experienced explorers and scientists were charged with creating handbooks for cold weather regions. Pre-war polar veterans such as Sir James Wordie, Neil Mackintosh and Brian Roberts made the dispatch of the secret naval operation, Operation Tabarin (1943–4), possible. Involving the Admiralty, Colonial Office and Foreign Office, a party of British naval officers and scientists landed in Antarctica in February 1944 and established two bases at Deception Island and Port Lockroy. In his report to his superiors, Lt Commander James Marr reported plenty of evidence of prior Argentine occupation including flags, plaques and brass materials certifying the provenance of an Argentine Antarctic territory. Notwithstanding these materials, Marr felt able to conclude that 'Taking the purely political standpoint, we are firmly established in Antarctica at two widely separated points both of which have been surreptitiously visited and claimed by Argentina ... after a break forced upon us by war, the active interest displayed by Great Britain has once again been revived.'[33] When the British media were finally informed of Operation Tabarin in April 1944, a Ministry of Information briefing paper informed readers that 'A major purpose of the expedition now is to act as a countermeasure to encroachments, which the Argentine government has made in the Falkland Islands Dependencies in recent years.'

The idea of the 'counter-measure' is important for two reasons. First, it highlights a tactical response designed to disrupt Argentine encroachments. Operation Tabarin, for example, involved the removal and disposal of all evidence of Argentine activity in places like Deception Island. Second, and more strategically, the idea of counter-*measure* might be seen more in the terms of James Scott's work concerning the calculation, legibility and measuring of territory.[34] In his review of attempts by state authorities to improve human conditions, Scott notes a tendency to do at least two things with potentially disastrous consequences, especially in authoritarian regimes such as Stalin's Soviet Union. The first is the desire to achieve an administrative ordering of nature and society, and the second is a willingness of the modern state to use power to achieve the desired ordering. British Antarctic territories were, in the post-Tabarin era, to be permanently occupied by a string of new bases sweeping along the Antarctic Peninsula towards outlying

AERIAL SURVEYING

islands such as South Sandwich and South Orkneys. The FID was divided into four geographical areas for the purpose of recording survey progress. These were the northern island groups, the Northern Antarctic Peninsula, the Southern Antarctic Peninsula and an area south of 75 degrees South and east of 20 degrees West. Organisationally, the UK field-based activities in the Antarctic were subject to greater review and assessment so that survey material was transferred, with greater speed and reliability, to the Directorate of Colonial Surveys based in the UK.[35]

What Operation Tabarin and its aftermath brought to the fore was a realignment of British strategic endeavour in the Antarctic. In essence, the Colonial Office, as the lead department in this matter, recognised that the mapping and surveying of polar territory was more likely to be effective, and indeed affective, in the prevailing conditions of the 1940s, rather than the earlier effort devoted to regulating the poorly visible whaling industry. This desire to map and survey was informed, in the main, simply by a political and legal desire to respond to the challenges posed by other states and their visiting parties in the Antarctic Peninsula region. The acceleration of mapping and surveying by the British manned airplane was supposed to demonstrate, in a visually arresting way, the impression of precision targeting and the exercise of sovereign control. As Hunting Aerosurveys Ltd report on the first season's work in 1955–6 stated: 'Although the Falkland Islands Dependencies is probably the best mapped area of any part of the Antarctic, the existing charts of the area are not sufficiently accurate that they can be relied upon for purposes of survey navigation.'[36]

Protecting sovereignty rights in the Antarctic was going to require more than simply a one-off naval operation. The South American counter-claimants were not likely to diminish their interest in the region in the light of their investment in expeditions, ceremonies of possession and research base construction. So the question arose as to how to secure British interests in this most remote part of the empire in the context of limited funding and diminishing interest from the Royal Navy. In the aftermath of Tabarin, the Attlee government agreed to the creation of a civilian-based survey operation called the Falkland Islands Dependencies Survey (FIDS). Supported by the Colonial Office, the remit of the FIDS was established in July 1945: 'The promotion of British territorial claims, the administration of the Falkland Islands Dependencies, the deterrence of rival claimants and their surveyors and the pursuit of accurate geographical knowledge.'[37]

Every summer season, from that moment henceforth, a team of FIDS surveyors were dispatched to the FID in order to begin their mapping and surveying projects. The remit was potentially enormous given that the spatial remit encompassed thousands of square miles of rock, ice and water. Using a network of bases established under Operation Tabarin (including Deception Island, Hope Bay, Laurie Island, Port Lockroy and Stonnington Island), the FIDS surveyors were predominantly reliant on dog sledging and support vessels for travel around the FID.[38] Once in the field, the FIDS men were expected to be ever vigilant, especially when they encountered evidence of rival surveying parties. Every FIDS surveyor was expected to be able to issue a so-called 'sovereignty speech' to South American counterparts while going about their surveying business. Materials gathered in the field were transported back to centres of calculation, such as the Directorate of Colonial Surveys, for the explicit purpose of producing new maps of the FID, which it was hoped would showcase the achievements of FIDS.

This need to produce updated and improved maps of the FID mattered because Britain was embroiled in a mapping war with Argentina and Chile. The projection of power, in a very real sense, depended in large measure on producing cartographic projections of British Antarctica. Funded by Colonial Office grants, FIDS was expected to generate 'accurate geographical knowledge' and, ultimately, to update maps of the FID, which could be used domestically and internationally as evidence of sovereign presence. With no indigenous human population, demonstrating the legibility of British territory was all-important. In 1946, a scientific committee was established within FIDS to ensure that any materials generated by the surveyors in the Antarctic were efficiently transformed into maps and surveys. In an era of rationing, and an austerity-hit Britain, the Colonial Office were acutely aware that they had to demonstrate, in the words of then Secretary of State for the Colonies Arthur Creech Jones, 'good value for money'.

The dilemma facing the administrators of the FID was that Argentina in particular demonstrated that the British never enjoyed a monopoly on the production of 'accurate geographical knowledge'. Argentina, especially in the Peron era, initiated a new round of investment in research station construction, mapping projects and general surveying across the Argentine Antarctic territory. Between 1948 and 1952, both countries sought to expand their remit of activities including dispatching warships into Antarctic waters in displays of gunboat diplomacy. The nadir of Anglo-Argentine rivalry emerged in February 1952 when British and Argentine personnel clashed in Hope Bay over the existence of rival bases. For colonial administrators in London, including the Foreign Office's adviser Brian Roberts, the incident was indicative of a very British dilemma. Writing in October 1952, he lamented that 'The Hydrographer told me last week that he thought it was most unsatisfactory that the Admiralty should now have to correct their charts of the Falkland Islands Dependencies from Argentine and Chilean surveys. We must act now if there is to be any hope of stepping up FIDS activity.'[39] Within a month, the Argentine Naval Ministry sent Roberts copies of their new Antarctic maps and announced that the Argentine Air Force would be conducting aerial surveys over the South Shetland Islands. By the end of 1952, it appeared that Argentine 'sovereignty games' were superior, and that if Britain wanted to retain a serious presence in the Antarctic then something extraordinary would be required. As Peter Mott, the leader of FIDASE, recognised in his memoir:

Obviously the situation was getting out of hand and demanded on the British side some further demonstration of their long-established sovereign rights to administer the area. What better way of doing this than to carry out a comprehensive programme of detailed mapping to replace the existing fragmented charts compiled from exploratory surveys, which, in no way provided the detailed information and reliability required for long term administration and resource evaluation.[40]

### The FIDASE and the Limits of Vertical Geopolitical Authority

Aviation was widely understood, in the Polar Regions as well as continental North America (and the deserts of Mesopotamia), to offer exciting possibilities, at least for the administrator and planner, for mapping and surveying remote and previously inaccessible regions. The apparent advantages offered by air travel were not, therefore, difficult to discern, especially when one considers the laborious and, at times, dangerous foot-based expeditions of explorers and scientists. Aviation offered exciting possibilities to both the scientist and the administrator of polar territory. In December 1951, Brian Roberts admitted that 'it was evident that if an air survey was not undertaken soon there would be a strong possibility that either the Argentines or Chileans would get in first and Britain would lose considerable ground in maintaining her position as the sole administrator of the area'.[41] Roberts's endorsement of the efficaciousness of air survey was not unique to the polar continent, as it was widely acknowledged that such an activity was indispensable to imperial planning and administration around the world. The mapping of British imperial assets, including territory and resource potential, was well established in the post-First World War era.[42]

However, initiating an air survey of British Antarctic territories in the aftermath of the BGLE in the late 1930s proved troublesome.[43] While air survey was recognised as a priority by the Colonial Office in 1945, financial and technical restrictions prevented any implementation. Advice was sought from the RAF, and later from commercial firms such as Hunting Aerosurveys Ltd in 1948, the latter of which initially proposed a programme of trimetrogon photography, lasting three Antarctic summer seasons.[44] While a larger investment in air survey was rejected on cost grounds in 1949, there was a small amount of aerial photography collected in January–February 1950 around Deception Island and Hope Bay by an Auster airplane equipped with a hand-held K-20 camera previously owned by the US Air Force.[45]

The decision to fund what was to be called the Falkland Islands and Dependencies Aerial Survey Expedition (FIDASE) was in effect taken in December 1951, after further interdepartmental studies indicated that a combination of oblique[46] and vertical photography[47] would be preferable for some 10,000-plus square kilometres of territory. Roberts's view was widely shared within Whitehall, at least by those who considered Antarctica to be a policy-relevant interest. The key obstacle to be overcome was funding, however. The UK Treasury was, by the mid-1950s, investing over half a million pounds per annum into the FIDS. Endorsed by the interdepartmental Polar Committee, a grant of £250,000 was secured, and a commercial contract awarded to Hunting Aerosurveys Limited. Directed by a veteran of the India Military Survey and the Colonial Survey Directorate (Peter Mott), the commercial contract made explicit three objectives—aerial photography of the Antarctic Peninsula, ground-based surveying of the FID and the completion of airborne magnetic profiles.[48] Before departure, Mott and his team were dispatched to the Colonial Office and the Directorate of Colonial Surveys in order to better understand the prevailing geopolitical context of the Antarctic, and the aerial photography needed for the updating of maps of the FID.

The FIDASE was in a very real sense the last throw of the mapping/ surveying dice, as it was an explicit attempt to accelerate British map and survey production, predicated on the belief that only British airmen and surveyors could deliver the kind of high-resolution detail of landscapes and seascapes demanded by embattled polar administrators. The latter were eager to show that the title deeds to British Antarctic territories were secure, and mindful of the need to demonstrate value for money. As the FIDASE leader, Peter Mott, recalled at the end of a public lecture on the achievements of the expedition:

In these and other ways [the production of maps and air photos], it is hoped that all the money and effort expended on this first British photogrammetric survey in the Antarctic will be justified. Whatever else we achieved, I am certain that our expedition has demonstrated beyond doubt that the long-range photographic aircraft with helicopter transport for the land surveyor are an indispensable combination in overcoming the tremendous obstacles of terrain and climate, if this last remaining unmapped continent is to be surveyed within a foreseeable period of time.[49]

Peter Mott, in his subsequent memoir of the FIDASE, reflected again on the scale, quite literally one might add, of the task facing his team, 'FIDASE was the first attempt ever made to cover any part of the south polar continent with three dimensional (stereoscopic) vertical photography into contoured maps, and at the same time, provide a network of ground surveyed points needed to convert the air photography into contoured maps'.[50]

This construction of a three-dimensional or volumetric understanding of Antarctica was later to prove highly significant not only in producing more comprehensive maps but also in putting to work 'the view from above' in more overtly geopolitical ways. In the 1950s, for example, the overall political priority was to map all British Antarctic territories north of 75 degrees south and mainly at a scale of 1:200,000. Facing competition from Argentina and the United States, maps and aerial photography were as much objects of prestige as they were scientific documents. So the pressure was on FIDASE to deliver—those planes and helicopters had to be able to fly and transport land-based surveyors around the Antarctic Peninsula and surrounding islands. The targeting of Antarctica depended upon the horizontal and the vertical dimensions of FIDASE being coordinated with one another. Without the accompanying ship and the surveyors, the 'view from the air' was not going to materialise.

Image 4: The East Coast Glacier (Peter Mott).

## Flying Canso

The FIDASE was carried out over two summer seasons in 1955–6 and 1956–7. Deception Island was the base station for the expedition not least because of its natural harbour and beaches which were suitable for potential landing strips. Weather permitting, the Canso aircraft and support helicopters flew regular sorties over the Antarctic Peninsula and surrounding islands. The airplane was an essential element in FIDASE not only as a means of transport but also as a figure of authority, both geopolitical and scientific. While the airplane was an integral element in aerial surveying, it also added ballast to the contention that the British were being seen to exercise sovereign authority over the Falkland Islands Dependencies. This mode of travel figures strongly in the published accounts of the FIDASE by Hunting Aerosurveys Limited, the Colonial Office and individual team members, the most notable examples being the diaries of the flying manager and ex-RAF pilot, John Saffrey, kept of his two seasons of flight in Antarctica between 1955–6 and 1956–7. For two summer seasons, Saffrey provided a daily account of geographical position, prevailing weather conditions and FIDASE-related activities.

Reading through the diaries, the reader is struck by the time taken, especially in the first season, simply to prepare the amphibious aircraft for possible rather than actual flight. After constructing moorings and slipways, the aircraft were then dependent on weather and sea conditions in and around their base at Deception Island. Sometimes equipment failure also intervened. As he noted for 6 February 1956:

Both airplanes flew sorties CF-IGJ [the first Canso] from the slipway being airborne 0755 to 1500 and CF-IJJ [the second Canso] from 0810 to having been forced to return by failure of a heater in the navigator's compartment and subsequent extreme cold suffered by Green (–30 degrees). Quite a lot of photography obtained in areas Smith Island, Trinity Island, and Palmer Coast. Both aircraft came up to the buoys after landing because fresh onshore wind made approach to slipway hazardous … today's hard operational flying, taught many lessons, among them: … the heaters in the Canso are insufficient … the operational ceiling of the Canso is 14, 000 feet for several hours … Helicopter landings and pick ups on snow tops are difficult and considerable practice is needed.[51]

Two weeks later, the planes had hardly been able to leaving their moorings, and Saffrey notes on 23 February that further frustration was to follow, 'the high wind had whipped up such a sea in Port Foster [the natural harbour in Deception Island] that Greenshields [a pilot] decided the surface was unsafe for take-off and landing, so flying was abandoned until the wind abated'.[52] This was not untypical—even when cloud cover diminished, the water conditions in Deception Island could derail any plans to fly.

On a good flying day, however rare as Saffrey's flying diaries reveal, the Canso planes were clearly a source of inspiration and pride. In his introductory note to season two, Peter Mott is upbeat about the aerial achievements:

The productive work completed in this second season of operations has been in marked contrast to that of the first season. Whereas in 1955/56 barely 1000 square miles of effective cover was obtained out of a contract area of over 40,000 square miles, the total coverage at the end of the 1956/7 season stands at approximately 33,000 square miles, or nearly 80per cent of the area originally planned to be done. In addition to their work in the Dependencies, the

Image 5: The Canso (Peter Mott).

Cansos were also successful in obtaining complete coverage of the Falkland Islands, an area of approximately 4500 square miles, during October and November 1956, on their way south.[53]

So in this context, the aircraft are literally credited with creating their own sets of mapping achievements.

Peter Mott's photographs of FIDASE include depictions of the pilots and support crew leaving the aircraft carrying large metal boxes containing film magazines collected from ten- to eleven-hour long sorties. Under such benign conditions, Mott reported that a team might return back to their base at Deception Island with nine to ten full magazines of film, having photographed something like 2–3,000 square miles. In total, the FIDASE generated something in the order of 10,000 aerial photographs covering a previously unmapped area of some 35,000 square miles. While that might appear impressive, it is worth bearing in the mind that the Antarctic Peninsula region encompasses some 200,000–300,000 square miles, and that there were many days when flying was not possible due to impenetrable cloud cover and inclement sea conditions in and around Deception Island. But, as Mott and imperial administrators back in London understood, it was vital that the expedition did generate aerial photography, regardless of the prevailing conditions.

When they did get airborne, the Cansos were supposed to fly at around 10,000 feet above the land surface, with the cameras directed downwards through a hole in the floor of the aircraft. By flying forwards in a parallel traverses, it was hoped that each photograph would slightly overlap one another. The skill involved was considerable. The pilot and navigator relied on a series of ground-based beacons, visual reference points and the radar system of the aeroplane in order to track their way along the designated region. Moreover, as the team acknowledged, the ice-covered terrain posed challenges not only to navigation but also to maintaining camera exposure in the face of the dazzling brilliance of the landscape coupled with the depth of the shadows. Flying these missions were men like Peter Davis, a veteran of the Royal Air Force and a former aerial photographer who had served with the Pathfinder Squadron in wartime Germany. Their wartime experiences were deemed invaluable because of the distortions caused by the changing registers of lightness and darkness alongside the ambiguities caused by sea ice, which blurred any obvious distinctions between land and sea. It was as if the Antarctic Peninsula was engaged in some large-scale operation deliberately designed to obfuscate its geographical features.

## Hovering Helicopters and Ground Control

While aerial photography offered exciting possibilities in terms of generating substantial coverage of territory, it did not remove the necessity of identifying a series of necessary points on the ground to enable interpretation, with due account taken of aircraft tilt, the scale and absolute position of the earth's surface in relation to the photo image. Ground control was vital, so aviation, paradoxically perhaps, did something else. It helped to reassert the importance of field science and ground-based surveying in the Antarctic. In that sense aerial photography did not provide a simple technical fix. Generating thousands of photographs of the Antarctic Peninsula was one thing, but it was quite another to generate those photographs into something cartographically and political useful. Obtaining ground control was vital and the helicopters were a vital element of FIDASE because they helped to transport the surveyors to various locations around the Antarctic Peninsula in order to establish survey points and beacons.

Using the air photos, the aerially transported surveyors undertook control survey operations. The aim of the exercise was to establish so-called control points to be used for more detailed surveying operations. These were constructed out of rock some 5–6 feet tall, where rock and stone was available on particular surveying points. If this was not possible then the surveyors constructed a series of artificial Cairns, using a bamboo tripod with red skirting cloth and four empty 44-gallon fuel drums all tied together. The whole structure was held together by wire ropes to ensure that the wind did not blow these structures over. The construction process was a tense affair and the helicopter pilots and expedition leader had to monitor the weather closely. Looking to the sky was an important element of this process because if the wind picked up and/or cloud cover descended then the surveyors might be trapped on islands, and thus immobilised.

While the accounts composed about FIDASE frequently linger on the importance of the aircraft and the Canso in particular, the helicopter, whether it was the Sikorsky S-51 or the Bell 47D, was essential in enabling different kinds of aerial engagements with British Antarctic territories. The helicopter not only moved surveyors rapidly from ground control points but also enabled the surveyors to hover over Antarctic islands such as King George and to carefully choose their drop off and collection points. The 'view from above' was by necessity a rather different one—not organised from 13,000 feet and far more dependent on a nuanced appreciation of the landscape and accompanying weather, especially when the helicopters landed on islands (and their uncertain surfaces) in order to ferry surveyors around the region.

There was plenty of scope for mishaps along the way. As Peter Mott recalled, 'On the second day of the operation, our single Bell when about to land on an ice-capped island, was caught in a violent down draught and crash-landed, becoming a total wreck.'[54] The wrecked helicopter was later retrieved by the Royal Naval guard-ship HMS *Protector*, and brought back to the UK by the expedition ship, the *Oluf Sven*. The view from above, even hovering over the Antarctic, was a precarious one. On other occasions, a shortage of power sometimes meant that the helicopters in inclement weather struggled to gain altitude. As the official report on the second season's operations acknowledged:

In an emergency, or when the clouds are about to descend, it is desirable to have sufficient payload and space in the aircraft to be able to carry surveyors and say 600lbs. of kit in one flight. Even more important is the need for a large

Image 6: A Helicopter Wreck (Peter Mott).

reserve of power to combat the tremendous and sudden downdraughts which frequently are met in a country of high mountains and very strong winds.[55]

Financial planning meant that the surveying party did not carry a spare helicopter in an effort to achieve cost-savings. With no replacement helicopter, the FIDASE had to wait until a replacement could be sent out to the Antarctic. The resulting delay was disastrous: 'This meant that once again we had lost the whole period of optimum weather and would be obliged to carry out the survey programme during the tail end of the season, when the days were growing rapidly shorter and the weather very unsettled.'[56] The control survey work was literally grounded. The dramas involving the helicopters and their passengers (the surveyors attached to the FIDASE) illustrate well how their respective mobility was so often dependent on prevailing weather conditions including air currents, altitude, temperature, cloud cover and sea conditions given the role of the expedition ship, the *Olav Sven*, in both transporting and carrying the helicopters around the Antarctic Peninsula.

As a consequence of the potential for disaster involving down drafts and lack of reserve power, the expedition looked more carefully for potential surveying points. One of the reasons for selecting King George Island for intensive ground control work was its topographic features and relative proximity to the main base at Deception Island. As Peter Mott recalled, albeit from the safety of London, 'In this case we selected the island [King George] as being one of the least rugged. If ever there was proof of the vital role played by the helicopter in controlling this type of terrain, the example of King George Island provided it.' In other places, the 'type of terrain' appeared less forgiving, as he later conceded, 'Had we been able to operate at the start of the season with unlimited daylight and relatively settled conditions, our progress must indeed have been phenomenal.'[57]

*Conclusion*

The FIDASE was an ambitious attempt to mobilise the 'view from above' for the purpose of imperial consolidation. The idea that the Antarctic was well suited to a form of aerial control—large-scale surveying and mapping—proved elusive. While there was much to commend from the point of view of imperial administrators in terms of aerial photography and limited ground control surveying, the fact of the matter is that the organisational and mechanistic abilities and capabilities of the FIDASE leadership and accompanying machinery were pushed to their limits and beyond. The 'view of the airman', whether out of the plane or helicopter window, was by necessity a compromised one. While cloud cover could obscure the view, the landscape itself could confuse pilots, mapmakers and observers alike with its ice sheets, snow-covered islands, grounded icebergs and intermingling of ice, rock and water. Air survey might have appeared to be a relatively cheap mechanism for pursuing imperial control in an era of decolonisation, but it was never straightforward.[58]

Three aspects of FIDASE, however, have proven long-lasting, the first of which was to demonstrate that air survey and mapping regardless of geographical reach and projection respectively was not going to be sufficient to resolve 'the Antarctic Problem'. Argentina, Britain and Chile were still rival claimants and only the eventual signing of the Antarctic Treaty in 1959 brought a modicum of closure to that matter.[59] Second, the FIDASE led to a fundamental review of British place-naming, as air survey revealed that many features noted along the Antarctic Peninsula did not possess a name, in part because they had not been seen before. Between 1958 and 1960, the Antarctic Place Names Committee approved 501 new names for geographical features while also confirming the existence of 274 names already in use. Members of FIDASE were honoured—Mott Snowfield, Bancroft Bay and Saffrey Islands—alongside the implementation of a new typography of place names in the FID based on, inter alia, notable individuals involved with air survey, photography and navigation. The end result was to populate the disputed Antarctic Peninsula and surrounding islands with a vast number of British place names, and this trend intensified markedly after 1960.[60] Finally, and perhaps unexpectedly, the aerial photographs taken by FIDASE have ended up playing an important role in enabling those interested in recent climate change affecting the Antarctic Peninsula to compare and contrast the contemporary landscapes and seascapes with prevailing conditions some sixty years ago. Rather than advance British imperial control, this 'view from the air' offers insights into how these territories might be melting away.

# 4

# NETWORKS, NODES AND DE-TERRITORIALISED BATTLESPACE

## THE SCOPIC REGIME OF RAPID DOMINANCE[1]

*Martin Coward*

On 21 March 2003 the United States and coalition partners launched a much-heralded air assault on Iraq. Referred to as 'Shock and Awe', the assault—targeted at Iraqi political and military infrastructure, as well as morale—was designed to achieve, in the words of those that conceived the strategy, 'rapid dominance'.[2] Although initial strikes aimed at 'decapitating' the Iraqi regime had been launched several days earlier and coalition land forces had already begun the invasion of Iraqi territory, Shock and Awe was framed as the opening salvo of the conflict. According to the US military, in addition to 1,000 air sorties, over 1,300 missiles or guided bombs were launched at targets in Iraq in the initial twenty-four hours of the campaign.[3] Global 24-hour news channels relayed the impact of Shock and Awe on Baghdad in particular: framed against the night sky, the strikes had a particularly strong visual (as well as destructive) impact.[4]

Image 7: Collateral Murder (Wikileaks).

A dual air targeting of Baghdad thus took place on 21 March 2003. On the one hand, Shock and Awe required that the United States and the United Kingdom in particular use their airpower as a means of directly striking the key nodes in the networks that were perceived to sustain Saddam Hussein's political–military complex as well as wider Iraqi morale. This necessitated striking targets remote from any land-based forces. As such it thus required the use of aerial weapons that could travel long distances as opposed to the more territorially limited land-based forces that require proximity to their targets before striking. Whether fired from ships or dropped from planes, Shock and Awe was, therefore, conceived of as a vertical assault in which so-called 'precision munitions' rained down on key locations in Iraq. On the other hand, this assault was conceived as spectacular warfare. The conspicuous display of destructive force—supposedly both precise and overwhelmingly massive—was an intended part of an 'effects-based' operation designed to cripple Iraqi political-military capabilities and decisively influence wider morale. To have an effect on morale, Shock and Awe would have to be consumed through the visual circuits of the rolling news channels. Shock and Awe was thus not only targeting *from* the air but also *of* the air—of the air-waves and airtime of global news.[5]

This representation of violence in and through news media played an important role in the visual economy of the war on Iraq. While the semiotics of such violence is complex, two dimensions are worth highlighting. Firstly, Shock and Awe might be said to represent a simulacra of power: a gesture that constitutes, rather than simply representing, the reality of power.[6] Rather than seeing the destructive force of Shock and Awe as simple projection of force, we might say instead that the idea of American power is an effect of this spectacular display.[7] Rather than understand the video footage of detonating missiles as a projection of extant power (or representation at a geographical distance from that power's point of origin), the billowing clouds of smoke and flame should instead be seen as constitutive of that which they seemingly represent. That is to say what appears to be a second-order copy of power, a token of its exercise at a particular locale, is in fact that which is constitutive of the idea of power. The idea of a power from which violence emanates is thus an effect of the representation of these explosions. The spectacular visual display of Shock and Awe is thus an example, to borrow from Baudrillard, of 'the precedence of models over the Real'.[8]

Beyond this simulation of power, however, it is also possible to see an element of what Kant and Burke referred to as 'the sublime' in Shock and Awe. Far from being a superlative experience (as everyday contemporary English-language usage of 'sublime' might suggest), for enlightenment thinkers a sublime experience was something in which one felt shock, awe, wonder and fear. It was an experience—often of the overwhelming vastness of the natural world—in which there was a disjuncture between the perceived and comprehension: an experience in which what is seen is beyond reason and understanding.[9] Just as mountains and storms might embody the sublime for enlightenment thinkers, so the detonations of cruise missiles as they impact on their targets might for the contemporary viewer. When watching these images it is hard not to experience a disorienting disjuncture between the overwhelming force on display and the ability to comprehend fully what is seen (to comprehend the violence being enacted as well the consequences that would echo from it across time and space). In the relayed images of this assault then one might see a certain encounter with the sublime: a certain fear and awe before the representation of overwhelming force.

Despite the importance of the performative force of the imaging of Shock and Awe in the global news media, a focus on the representational dimension of this violence neglects another question: specifically, whether Shock and Awe represents a distinctive form of warfighting. In other words, a focus on the targeting of airtime neglects the question of the distinctiveness of Shock and Awe's aerial targeting of Iraq's political and military infrastructures. Billed as a transformation in the application of force by the US-led coalition, Shock and Awe was supposed to presage a distinctive form of warfighting which would enable smaller, technologically enhanced forces to exercise greater power than ever before. The success of this strategy is, in the light of the bloody aftermath of Shock and Awe, entirely debatable. However, beyond the question of success lies that of whether this form of air targeting represents a distinctive doctrine for the application of force.

It is this question of the distinctiveness of the doctrinal discourse—or 'military imaginary'—of Shock and Awe that I want to address in what follows. The 'military imaginary' comprises the set of operative concepts and understandings that comprise the grid of intelligibility that is constitutive of a certain way of imagining the battlefield, the enemy and warfighting. The military imaginary is thus both an explicit set of doctrines expounded by military and political sources, as well as a set of assumptions that underpin these documents. Overall the discourse of Shock and Awe is the set of operative assumptions and discursive dynamics that guide—but do not determine—warfighting, such as the bombing of Iraq in March 2003. The relation of military imaginary to warfighting practices is (as I shall show later) complicated, but nonetheless the latter is framed by a set of assumptions contained in the former. As such it is important to see how the battlefield is framed before examining the complexities involved in the translation of that framing into warfighting.

I will suggest that the air targeting of Shock and Awe represents a distinctive re-imagination and re-constitution of the spatial and political-ethical contours of warfighting. I will begin by examining the trajectory of warfighting within which Shock and Awe should be situated in order to understand its distinctive characteristics. Situating Shock and Awe in the context of relevant historical trajectories in warfighting is not intended to suggest that it comprises the logical outcome of these developments. Rather Shock and Awe represents a contingent hybridisation and radicalisation of certain historical trajectories, fusing elements of diverse doctrines and strategies of warfighting in a distinctive manner as a consequence of certain enabling contextual factors. I will suggest that technological developments—and the scopic regime(s) they constitute— comprise the principal contextual factors enabling such hybridisation and radicalisation. I will argue that the military imaginary of Shock and Awe is constituted by, and in turn gives rise to, a particular scopic regime. This scopic regime prioritises en-visioning the battlefield and enemy in a distinctive manner that has both political and ethical entailments.

### Delineating the Trajectory of Spectacular Air Targeting

If Shock and Awe re-imagines the spatial and political-ethical contours of warfighting, it is simultaneously continuous with, and yet distinctive within, the historical trajectories of warfare. Re-imagination entails the reworking of extant imaginaries of warfare in distinctive ways. In the case of Shock and Awe this means the hybridisation and radicalisation of two trajectories in twentieth-century warfighting.[10] In the first place, Shock and Awe's emphasis on the deployment of overwhelming air-power to affect both infrastructure and morale can be seen as continuous with the trajectory of warfighting that reached its apogee in twentieth-century total war exemplified by the Allied strategic bombing campaign of the Second World War. The bombing of German and Japanese cities was conceived in the context of an imaginary of warfighting in which the political, military and social infrastructures of the enemy were understood to have converged in the figure of the enemy nation.[11] The threat emanated, therefore, from the enemy nation per se—its values and social identity as well as its war machine—necessitating that warfighting treat that nation as a whole. In such an imaginary, no proper distinction is made between combatant and non-combatant. Rather, bombing is oriented towards the annihilation of the enemy as a whole: its war machine as well as the society that enables that war machine both materially and discursively.

On the one hand this is to be achieved by attacking political and logistical infrastructures: the institutions, transport infrastructure and industrial capacity that (re)produce the materiel and discourses of war. On the other it is to be achieved by inflicting casualties in order to affect morale, with the aim of sapping the will of a society to continue fight-

Image 8: Photograph made from B-17 Flying Fortress of the 8th AAF Bomber Command on 31 December when they attacked the vital CAM ball-bearing plant and the nearby Hispano Suiza aircraft engine repair depot in Paris. France, 1943. Army Air Forces., 1942–1946 (Wikimedia commons).

ing. This might be referred to as attacking the social infrastructure:[12] the mass of willing bodies without which the imaginary of total war could not exist or persist. The targeting of cities such as Dresden and Tokyo was a logical conclusion of this imaginary of warfighting: the modern city is the nexus in which social, political and logistical infrastructures converge. Moreover, since it is these infrastructures, and not the territorial jurisdiction of the city, which are at stake, airpower is a natural choice. Airpower enabled the Allies to strike at the nexus of infrastructures in the city without having to establish territorial jurisdiction over the city. While that territorial jurisdiction came later, it was not the aim of strategic bombing.

The second trajectory within which Shock and Awe should be situated is the cluster of late twentieth-century warfighting doctrines oriented around the figure of the network. Two aspects of this historical trajectory are worth highlighting. Firstly, the concept of Network Centric Warfare (NCW) was influential in reorganising the warfighting doctrine of America and its allies in the late twentieth century.[13] NCW focused on the manner in which warfighting can leverage the potential of Information Technology (IT). NCW highlighted the importance of the network to late twentieth- and twenty-first-century warfighting doctrine. The network implied in NCW is that comprised of interconnected computing. Implicit in NCW is the idea that interconnection will dominate future warfighting and, if disrupted, can have significant effects. However, NCW doctrines remain essentially concerned with the organisation and deployment of IT in order to dominate the battlefield.

Secondly, NCW's relatively narrow concern with leveraging IT infrastructures to provide battlefield dominance was broadened in recent warfighting doctrines by the deployment of the concept of the network to characterise the structure of enemy forces. Two examples are particularly prominent: John Warden's conception of the 'enemy as a system'

Image 9: US Air Force weapons control officers onboard an E-3 Sentry Airborne Warning and Control System (DoD photo by Gudron Cook, USAF/Released).

and John Arquilla and David Ronfeldt's notion of 'netwar'.[14] Broadly speaking, these writers took the NCW insight that leveraging the interconnection of IT will give battlefield superiority and broadened it to argue that the enemy itself should be conceived of as an interconnected network. Warfighting should thus concentrate on understanding the network structure and disrupting its key interconnective junctions, or nodes. As such, Warden, Arquilla and Ronfeldt reorient ideas about the utility of interconnectivity in creating IT networks that confer advantage on American and allied forces towards conceiving of the enemy as a networked entity whose interconnective nodes represent an exploitable vulnerability. According to these writers, striking vulnerable nodes enables the disruption of entire networks; incapacitating single nodes can thus have network-wide consequences. This notion of achieving greater effects from limited targeting comprises the core of what came to be referred to as 'Effects Based Operations' (EBO).[15] Though now partially abandoned, effects-based thinking is indebted to the trope of the network and the guiding concept that striking vulnerable nodes can have consequences for an entire interconnective system, thus multiplying the effect of single munitions or strikes.

Broadly speaking, effects-based doctrines refocus the target of warfare from control of territory to the key nodes in political, military, social and logistical networks. In doing so, this form of warfighting conceives of the enemy as a dynamic set of circulations articulated around key junctions (or nodes) where multiple flows are gathered and rearticulated: for example, the ministries in which the various flows of the political process of governance converge, or the transport hubs in which the various circuits of 'logistical life' gather and interconnect.[16] Network-centric doctrines aim, therefore, to map the nodes of enemy networks and, by attacking these key targets, deliver an effect that ripples out into wider society. By disabling nodes, network-centric warfare aims to indicate that the various circulations that ensure the continued functioning of society can be crippled from afar. As such, it serves to affect morale and indicate to a population that it should surrender in the face of this ability to 'switch off' societies.[17] The effect munitions achieve is thus multiplied by the effect the destruction of key nodes has on morale.[18]

Of course, neither of these imaginaries of warfare is entirely distinct or novel in themselves. The targeting of strategic infrastructure as well as the multiplication of the material effects of bombing in the morale of the population have been important questions throughout the development of doctrines or imaginaries of air warfare.[19] However, Shock and Awe represents a hybridisation and radicalisation of both trajectories of imagining the strategic deployment of airpower. Shock and Awe takes the infrastructural targeting of strategic bombing and hybridises it with the effects-based, precision targeting of network-centric warfare to create an imaginary of warfare that retains total war's identification of the entirety of social infrastructure as a target set while also aiming to affect morale and capacity decisively through the destruction of the nodes that will have the greatest effect on both. It is thus total in its target set, and selective in its strategic targeting of key nodes.

*The Revolution in Military Affairs as Catalytic Agent*

The hybridisation and radicalisation of total war and network-centric war into Shock and Awe does not occur in isolation as a context-independent mutation. Rather it occurs in the context of a wider rearticulation of the coordinates by which warfighting is imagined: the so-called 'Revolution in Military Affairs' (RMA) that was central to military-political discourse in the last decade of the twentieth century.[20] As many writers have indicated, there have been periodic revolutions in military affairs throughout the history of warfare—something that the singularity inferred in assigning a proper noun (RMA) to the transformations of the late twentieth-century belies. Each of these revolutions decisively marks a radicalisation of the trajectory of warfighting. As such each has a distinctive character, whether it is the introduction of gunpowder or the emergence of the conscript army. The transformation of the military doctrine of advanced industrial states in the last decade of the twentieth century similarly has its own distinctive character.

In broad terms the RMA is shaped by advances in the gathering, processing and representation of data via information technology. Associated with this are advances in weapons technology and logistics that permit the harnessing of these information capabilities to effect greater speed, accuracy, flexibility and destructive firepower. The RMA thus generates a military imaginary distinctively oriented around notions of information, communication, precision, speed and flexibility. This imaginary is the catalytic agent for the emergence of Shock and Awe: it is the discursive condition of possibility of the particular hybridisation of strategic bombing and network-centric war that produces Shock and Awe. We might say that the RMA is the loom on which the warp and weft of strategic bombing and network-oriented warfighting are woven together into the cloth of Shock and Awe. It is worth, therefore, looking in more depth at the RMA to discern the specific inflection it gives to the air targeting of Shock and Awe.

## *The Scopic Regime of the RMA*

The RMA came to public prominence as a consequence of the 1991 Gulf War. Though the idea of the RMA had been in gestation in doctrinal circles for some time before Operation Desert Storm began, this conflict comprised a very public demonstration of a series of new technologies, chief among which were stealth aircraft, guided bombs and cruise missiles. The deployment of these technologies culminated in what was publicly framed as a 100-hour military engagement by land forces to liberate Kuwait. As such, novel technologies were taken to have provided the conditions for a distinctive form of warfighting that achieved rapid dominance on the battlefield.[21]

Central to the US military achieving dominance over their Iraqi counterparts was the deployment of cruise missiles and aerial bombing to create conditions under which land forces would rout (though not comprehensively defeat) the Iraqi army in under five days. The aerial campaign—an air targeting of the military, political and social infrastructure of Iraq—had two distinctive features that are worth noting. First, the prominent deployment of novel technologies such as the cruise missile and stealth fighter coincided with the emergence of 24-hour news channels. In particular, CNN reported from Baghdad as the air bombardment began. Subsequent footage drew attention to cruise missiles as the unfamiliar weapons were filmed en route to their targets. Secondly, this news footage of novel technologies of air assault was augmented by a visual trope introduced, and given prominence, by the US military itself: black and white footage from cameras tracking munitions as they reached their target. Centred on a cross hair, and filmed from an aerial vantage point, the camera footage was introduced to press conferences and released to the media as a constitutive element of an emerging discourse of precision: a narrative of being able to survey terrain and pinpoint targets. While the majority of bombs dropped in Desert Storm were unguided 'dumb' munitions, the so-called 'smart' bombs and missiles received greater attention as a consequence of this prominent visual trope.[22] Given that the footage disseminated by the army showed only successful strikes, the camera footage rapidly built a perception of precision air targeting.

This perception of precision should be set in the context of the wider dynamics of the RMA. By 1991 a number of technological trajectories were converging to enable a doctrinal re-imagination of warfare. Specifically, developments in information technology and communications allowed the capture, processing, representation and circulation of vast amounts of data. The rapid processing, rendering and circulation of data was the condition of possibility for two transformations of military capability considered revolutionary by doctrinal thinkers. On the one hand the processing of data to visualise the battlefield and targets within it in ever finer-grained ways gave rise to a transformation in both precision and lethality.[23] Technologies such as the Global Positioning System (GPS) and laser guidance enabled the guiding of force from distance to precise points in the political military-social infrastructure.[24] On the other, the rapid processing of data permitted greater sensitivity to context and thus a transformation of the flexibility and agility of forces.[25]

The RMA is thus the leveraging of data processing, rendering and communications technologies to transform the lethality, precision, agility and flexibility of the organised force deployed by Western militaries.[26] Underlying these transformations, however, is a more fundamental recasting of the military-strategic imaginary. While the processing and circulation of data is vital to transformations of precision and flexibility, it is the capability to render data that is transformed in the RMA and which thus gives it its revolutionary edge.[27] The ability to represent data, to capture the sense of its circulation and to thus relate terrain, targets and assets to one another transforms the spatial and temporal rhythms of warfighting, in turn leading to faster decision-making cycles as well as having the effect of de-territorialising the battlefield: removing any previous reliance on line-of-sight, for example. This rendering capability is essentially a way of en-visioning the total set of terrain, target and assets within which warfighting occurs.

The RMA is, thus, essentially a scopic regime.[28] It is more, therefore, than a spectacle: more than a very visible display of destructive power. Rather, underneath the spectacular destruction of the RMA is a grid of intelligibility predicated on en-visioning. Firstly the data capabilities that are constitutive of the RMA are a technique for en-visioning a world that holds hidden dangers.[29] Secondly, the data rendering capabilities of the RMA make visible those threats in a form that is of use for deploying forces to the battlefield. Finally, these techniques of envisioning and rendering visible are the condition of possibility of guiding munitions to target on which precision is predicated. The scopic regime of the RMA is thus founded on the distinctive trinity of *seeing*, *showing* and *guiding*.

It is this scopic regime that is the condition of possibility for the emergence of Shock and Awe as the hybrid progeny of strategic bombing and network-centric war. On the face of it, the annihilatory patterns of area bombing seem the antithesis of the precision called for by network-centric warfare. And yet both are oriented towards effects-based impact on the political, logistical and social infrastructures of the enemy. The scopic regime of the RMA allows for the channelling of ever greater force in seemingly more precise ways. In this sense it hybridises the overwhelming destructive power of strategic bombing with the precision logics of network-centric war to generate a military imaginary in which disproportional force is delivered in seemingly accurate ways in order to deliver effects beyond their immediate detonation: effects on infrastructure, on the life supported by infrastructure and on political/social morale. Shock and Awe is thus best seen as the latest iteration of a number of intertwined trajectories in organised violence: strategic bombing, network-centric war and the RMA. More importantly, the RMA should be seen not as a radical rupture, but as a catalytic agent that recasts extant trajectories of warfighting and doctrine in a way that hybridises key features of their military imaginaries.

In tracing these trajectories we see the way in which seemingly disparate projections of aerial force are intertwined and hybridised. Underpinning this intertwining is the catalytic power of the scopic regime of the RMA. It is thus the en-visioning that is characteristic of the RMA which is the condition of possibility of radicalising the twin trajectories of strategic bombing and network-centric war into the single warfighting imaginary of Shock and Awe. It is this scopic regime that I pointed to in arguing that Shock and Awe should not be seen as simply a visual display. On the contrary the public spectacle of bombing that erupted into the night sky over Baghdad in 2003 is underpinned by a more fundamental visual economy, a scopic regime—inherited from the RMA—according to which the battlespace of 2003 Persian Gulf War was en-visioned. Understanding this scopic inheritance of the RMA and the distinctive manner in which battlespace is en-visioned in Shock and Awe will help us understand the distinctive political and ethical challenges such warfare poses.

### Envisioning Battlespace

The scopic regime underpinning Shock and Awe rests on the manner in which the elements of warfighting (targets, assets and so on) can be brought together and rendered in relation to one another. The placing in relation of dispersed entities (actual, electronic and conceptual) is captured in the term 'battlespace'. Though definitions vary, battlespace is defined by the US Marine Corps as:

the environment, factors, and conditions that must be understood to successfully apply combat power, protect the force, and accomplish the mission. This includes the air, land, sea, space, and enemy and friendly forces, infrastructure, weather, and terrain ... Battlespace is conceptual ... not fixed in size or position ... it varies over time.[30]

The US Army added to this definition the 'electromagnetic spectrum' as a key component of battlespace.[31]

Battlespace is an assemblage of objects, forces and concepts which, when taken together, comprise the domain in which organised force is brought to bear. Any such assemblage requires a gathering together of otherwise disparate elements into an articulated whole. While doctrinal manuals attribute the gathering together of the battlespace to the battlefield commander, this begs the question as to the way in which that commander might understand their position vis-à-vis the various elements that compose this assemblage. Moreover, if we extend battlespace from a concept deployed by commanders to a more general term for the designation of the complex assemblage of things and forces that make up the domain within which organised violence is exercised, the question arises as to what might articulate this otherwise disparate set of entities. Scopic regimes such as that which underpins Shock and Awe perform precisely such a gathering function. In the case of Shock and Awe the scopic regime emerging from the RMA is the grid of intelligi-

Image 10: Air interdiction agent for the U.S. Customs and Border Protection (DoD photo by Senior Master Sgt. David H. Lipp, U.S. Air Force/Released).

bility that assembles and articulates disparate elements of the battlespace into a complex assemblage via the twin powers of capturing and rendering data. These twin powers give to the scopic regime of Shock and Awe the capability to identify an element, position it relative to other elements and to render these relations in real time.

As Bennett notes, 'An assemblage is, first, an ad hoc grouping, a collectivity whose origins are historical and circumstantial.'[32] Any singular instance of battlespace is contingent in precisely the way that Bennett alludes to: a historically and circumstantially bound assemblage of elements. This contingency is, however, compounded by the historical and circumstantial aspect of the scopic regime that underpins the contemporary constitution of battlespace. That is to say the battlespace of Shock and Awe has a dual contingency: both the contingency of its elements and the historically specific nature of the scopic regime that constitutes it. We might say that this contingency resonates with what William Connolly has referred to as 'onto-politics'. For Connolly, onto-politics refers to the contingency of any understanding of 'the possible relations humans can establish'.[33] Since any understanding of such relations is historically and spatially specific, it represents a contingent statement of what exists. However, insofar as such statements pretend to generality, they exclude all the other possible relations that any single contingent understanding cannot encompass. Endorsing one understanding over another is thus a political gesture, whatever rationales it is clothed in. The articulation of battlespace is precisely the constitution of an assemblage that renders explicit a series of relations. As such it can be read as an onto-political gesture. Understanding the specificity of the battlespace of Shock and Awe thus requires understanding the interplay of the scopic regime that underpins it and the elements it identifies and articulates as a political gesture. The spatialising dynamics of this interplay—the manner in which elements are articulated relative to one another—are particularly significant for thinking about the nature and politics of contemporary warfighting.

The spatialising dynamics of Shock and Awe resonate with and reinforce the motif of the network that underpins this imaginary of warfighting. The motif of the network is often deployed unreflexively as if it referred to a substantially real entity. However, it is better to see the network as a discursive trope that, through the distinctive manner in which it problematises (in Foucault's sense of the term) the arrangement of entities vis-à-vis each other, is constitutive of a certain spatiality. The trope of the network has been prominent in discourses of globalisation as signal of a shift in spatiality from methodological territorialism towards a de-territorialised geopolitical imaginary.[34] In this sense the trope of the network gathers entities in a relation to one another rather than in relation to political jurisdiction. That is to say, when entities are networked, they are articulated into relations with one another that are significant precisely because of the way they draw these entities together, not because of any reference to bounded units of political jurisdiction. Typically, this means that individuals in geographically distant places can be in a more significant relation with one another than with the communities that share the same territory as them.

The network is thus a trope that envisages ontology as a series of distinct entities related to one another without direct reference to the jurisdiction in which they may be situated. It is thus a trope that de-territorialises the classical perception of the globe as divided into a patchwork of mutually exclusive parcels of political jurisdiction (typically states). However, this is not to say that the trope of the network does not re-territorialise, or spatialise, in its own way. Here, it is worth revisiting Heidegger's discussion of de-severance in *Being and Time*.[35] Heidegger notes that *Dasein* forms relations with certain entities that make them 'closer' in ontological terms than entities which might be judged to be nearer according to measures devised for dividing space. So a picture on a wall might be ontologically 'closer' than the glasses on the nose of the person viewing the picture. Heidegger uses the concept of 'de-severance' to indicate the active manner in which *Dasein* relates entities and thus establishes a certain spatiality (or 'closeness') between them that is of an entirely different order to the abstract space of measurements such as centimetres, kilometres and so on. Similarly, networks establish a sense of closeness that is of a different order to that established by cartography. As such the spatiality of the network is of a different order to that of territorial geopolitics—it is a web of relations without a territorial frame: a series of lines of force drawing entities into a contingent assemblage. Importantly, between the lines of force there is nothing. Where cartography posits space between entities conceived of as occupying a point in a volume, networks posit nothing between the lines of force that gather entities together. As such, the network is de-territorialised.

This contrast between the territorial and the networked resonates through the scopic regime that underpins Shock and Awe, distinctively spatialising this imaginary of warfighting. Two dimensions of this spatiality are particularly significant. First, the scopic regime of Shock and Awe is predicated on an explicit opposition between territorial jurisdiction and networked assemblage. In other words, Shock and Awe is represented as surpassing an imaginary of warfare in which the possession and control of a territorial jurisdiction was deemed to comprise a strategic goal. Instead Shock and Awe conceives of its strategic goals in terms of points that derive their importance from their relations with other points. As such it is a de-territorialised strategic imaginary that lends itself to aerial tactics. While territorial jurisdiction cannot be said to be irrelevant (as the United States discovered when Turkey denied them a foothold from which to launch incursions into northern Iraq), it is not the primary criterion according to which goals are set and targets selected. Indeed, as this imaginary has evolved since the 1991 Gulf War, successive US operations have witnessed elaborate, globe-spanning bombing flights that have defied territorial logic. Thus, for example,

B-52 bombers regularly flew from the UK to bomb Iraq, refuelling as they did so. These flights, along with others from locations as diverse as the continental United States and Diego Garcia, operate according to an imaginary that seeks to strike points in networks regardless of their cartographic positioning.[36]

This opposition of the networked and the territorial is enhanced, however, by a second dynamic which conceives of Shock and Awe as replacing the modern, planar and volumetric conception of organised force with a non-linear, non-volumetric conception of the battlespace. Classical conceptions of modern warfare have conceived of the projection force along a single plane. The trope of a rolling front line captures precisely this imaginary of warfighting: the front line rolls across a planar surface, taking possession and control of all that it encompasses. However, this conception of the projection of force is not simply planar—it is also volumetric. There is an assumption that the plane of military engagement extends upwards defining a volume to be possessed and controlled by force. Air targeting is commonly thus conceived as a vertical support for an essentially horizontal projection of force, bringing firepower to bear on the task of planar movement and territorial possession. While a vertical vector (the plane's ascent, the bomb's drop), air targeting is consequently tied to a horizontal, planar, volumetric grid of intelligibility.

Shock and Awe, however, rewrites this imaginary of warfare, establishing a non-planar, non-volumetric imaginary of warfighting. While terrain may play a key role in battlespace, it is simply another element of an assemblage which is not distributed across a surface, but assembled in data flows. In the imaginary of warfighting that characterises Shock and Awe, the elements of the battlespace assemblage become dynamically related in non-volumetric space. Points are not related across defined distance but in terms of significance. Moreover, they are not spatially fixed, but shift in dynamic ways according to relationships. This gives rise to the sense of the battlespace as being without volume, without edges, without distance.

Shock and Awe thus represents a spatial recasting of the imaginary of warfighting: a succession of the planar by the non-linear, of the volumetric by the non-volumetric. As such the battlespace of recent conflicts is assembled less with reference to territory and more with reference to the political, military, logistical and social infrastructure networks that sustain the existence of the enemy. It is assembled as a set of nodes that are constituted through their relations with one another. Force is projected within this network, rather than across the territory in which it might be notionally located. The question thus arises as to the political entailments of such a projection of force.

## The politics of Network-oriented Warfighting

The politics of the scopic regime of Shock and Awe are located precisely in the opposition between a planar, territorial and a non-linear, non-volumetric battlespace. The battlespace of Shock and Awe is without boundaries in the classical sense of a boundary beyond which reach cannot be extended: articulation of additional elements into the assemblage is always possible. Moreover, no element is proscribed from the assemblage: since precision gives the impression of allowing for the destruction of targets without destroying their surrounding environs, any entities deemed as integral to the war-making capacity of the enemy can be added to the assemblage. This unrestricted character of the warfighting imaginary of Shock and Awe is a direct consequence of the de-territorialisation set in motion by the scopic regime of the RMA. Territory is ultimately defined by its delimited nature, the boundaries that mark out the extent of any juridico-political domain. Without such boundedness, de-territorialised imaginaries cease to observe limits to what might be added to the assemblages with which they are concerned. The criterion for selection is no longer co-belonging within a circumscribed domain, but rather a salient relation with another entity.

Unboundedness is not, however, the only consequence of de-territorialisation. In addition, territory implies the contiguity of locations within a bounded domain. As such a territorialised warfighting imaginary accepts that targeting any given location is likely to have consequences for contiguous locations. However, a de-territorialised warfighting imaginary is blind to such contiguity, perceiving only a constellation of discontinuous entities without any territorial reference. Indeed, this is the underlying sentiment behind the introduction of a term like 'collateral damage': a term used to indicate incidental damage to entities not expected to be encountered during a military operation, rather than to the inevitable damage caused to entities contiguous to a location identified for destruction.

Battlespace is thus extended to permit targeting despite volumetric positioning. However, whereas areas of a plane can be circumscribed, a networked constellation of points defies the logic of such a proscription. As such, the warfighting imaginary of Shock and Awe rolls back the legal and ethical proscriptions that have (nominally, at least) protected certain classes of entities or types of territorial domain during times of war. Indeed, because the trope of the network both gives the impression that battlespace is de-territorialised and that nodes in a network can be targeted with precision, it fosters a warfighting imaginary in which organised force is unbound from the territorial proscriptions that were codified in the various laws of war.

Thus Shock and Awe marks a conspicuous chapter in the targeting of the city. Itself a non-volumetric cluster of related nodes, the city has had an ambivalent relation with warfighting in the last seventy years. Razed by strategic bombers and devastated by house-to-house fighting in the Second World War, the city was increasingly seen in subsequent decades as both a military quagmire and civilian space.[37] The complex, three-dimensional space of the city has challenged planar imaginaries of warfighting. The planar conception tends towards a two-dimensional, flat conception of the battlespace (with aerial support being directed from a vertical vantage point on to a two-dimensional plane). Cities defy such a planar conception, since they are three-dimensional and disjointed. They consequently challenge planar conceptions of the projection of force. This has long been recognised by those discussing urban warfare doctrine as has the manner in which complex urban space has frequently become a lethal quagmire for armies in the twentieth century.

One of the key complexities of urban space is the manner in which it is largely civilian. Only in rare cases are the entire populations of cities displaced by war. Urban warfighting must therefore also negotiate the question of the balance between force and the protection of civilians. While this balance has, in almost all cases, eluded Western militaries, it has been the focus of a legal and political discourse which has led to the strengthening of proscriptions against disproportionate force in the city.

Shock and Awe, however, challenges these perceptions, re-enchanting the city as a target.[38] The perception of precision targeting and the de-territorialisation of battlespace give rise to the sense that the complexity of the urban space can be mastered by destruction of key nodal targets while simultaneously protecting civilians from harm. This re-inflection of the territorial hesitance and ambivalence that has governed the relation between war and the city in the past seventy years is of concern because it can potentially recast the legal proscriptions that have sought to safeguard the urban as a place of civilian life. As the city is extensively securitised, it is clear that the city is rapidly being reincorporated into the battlespace of Western militaries.[39]

## Beyond Shock and Awe: The Excesses of Unboundedness

As was noted earlier in the chapter, the military imaginary of Shock and Awe is constituted by a particular scopic regime: a grid of intelligibility that frames the manner in which battlespace is en-visioned and thus how warfighting should proceed. The military imaginary of Shock and Awe is thus an ideal type that promulgates a normative injunction as to how the battlespace is to be imagined and the enemy attacked. The relation between this scopic regime, its normative assumptions, the manner in which it en-visions the battlespace and empirical instances of warfighting are complex. It is certainly the case that 'nodal' targeting is an accepted practice within US joint force doctrine.[40] As such, the network and the nodes that comprise it can be said to constitute the framework for the empirical application of force. It thus seems fair to assert that the munitions that rained down on Iraq in March 2003 and the military imaginary of Shock and Awe are directly related.

However, this relation is complex, not least because the precision that nodal targeting implies does not translate into the practice of warfighting. As Human Rights Watch have noted, in the first phase of the Iraq War in 2003 '[a]ll of the fifty acknowledged attacks targeting Iraqi leadership failed'.[41] This failure is evidence that, even when delivered with precision, nodal strikes cannot be assured of actually hitting their target and thus of achieving the intended effect upon a network. Moreover, faulty targeting methodology as well as mistaken targeting led to a significant number of civilian deaths.[42] Iraq Body Count have argued that air attacks—of which precision bombs and missiles are a significant element—have been one of the biggest causes of civilian deaths in Iraq since 2003.[43] As such, the disjuncture between the assumed precision of the non-linear, de-territorialised, networked warfighting and the reality on the ground in Iraq are plain to see. Civilian deaths have continued as a consequence of the deployment of airpower—particularly strikes launched from drones—in Iraq, Afghanistan and Pakistan. Indeed, in 2011 the Afghan authorities told NATO to stop using airpower to strike at houses suspected of containing 'militants' due to the high incidence of associated civilian deaths, while Pakistani authorities similarly asked US forces to vacate an airbase from which drones were launched amid controversy about the mounting death toll from strikes.[44] In their documentation of drone strikes in Pakistan, The New America Foundation note that up to 33 per cent of drone strike fatalities are civilians.[45]

Alongside the mounting evidence that nodal targeting rarely achieves surgical precision, there is a refutation of the sense that precise air targeting can achieve the aims that effects-based thinking had promised. In Iraq, the 'surge' that began in 2007 was premised on an increase of land-based force and wider interaction between that force and the communities it sought to secure. Far from relying on a light, agile force, the surge was predicated on providing territorial security, not simply removing key elements of enemy networks.[46] Furthermore, in Afghanistan, General Stanley McChrystal attempted to deprecate the use of airpower in the wake of disquiet about civilian deaths (although this policy was not carried forward by General Petraeus when he arrived in 2010).[47] Concurrently, the US Joint Forces Command has rejected 'effects-based operations' doctrine, arguing that it creates more confusion than benefit.[48]

However, despite the clear evidence that so-called precision targeting leads to significant civilian casualties, and despite the reformulation of US joint force doctrine, the imaginary of Shock and Awe resonates through the various global warfighting operations of America and its allies since 2003. While a comprehensive picture of this resonance cannot be provided here, a single example might illustrate the manner in which the network thinking and scopic regime that guided Shock and Awe continues to provide a grid of intelligibility that guides current practice: drone warfare in the so-called 'AFPAK' region.

Since the invasion of Afghanistan in 2001 the border region between Afghanistan and Pakistan has been characterised as a 'vast, rugged, and often ungoverned' zone harbouring a militant threat that must be eliminated in order to provide both regional and global security.[49] Whether accurate or not, this characterisation has led to the extension of the theatre of operations for NATO forces in the region into this area. However, international norms of sovereignty and domestic political sensitivities in Pakistan make it problematic for NATO forces to cross the Afghan border in pursuit of militants. In response the United States

Image 11: RAF Reaper from 39 Squadron (UK Govt Licence).

escalated the use of drone strikes in the region: in 2010 over 100 drone strikes were launched.

The ratcheting up of the aerial campaign over North West Pakistan is evidence of two important elements of the military imaginary—and hence the scopic regime—of Shock and Awe that reverberate through contemporary American warfighting. On the one hand it is still indebted to the network as an organising schema. Even in rejecting effects-based operations as doctrine, US Joint Forces Commander James N. Mattis retained the concept of 'nodal analysis as it relates to targeting'.[50] Drone strikes are predicated on the idea that the enemy can be reduced to a series of nodal points that are articulated together into a wider whole. Targeting those nodes is thus taken to be the aim of warfighting. Indeed, this nodal analysis has been revitalised by the emergence of social network analysis as a significant methodology in counter-terrorism. Nodal analysis alone, however, is not enough to explain the migration of the air war from Afghanistan to Pakistan. To understand this, we should look to the unboundedness that network tropes give to military imaginaries. This unboundedness echoes through into the drone war over Pakistan. More specifically, these operations are mobilised by the idea that the network of viable targets can be expanded despite territorial limitations. Thus while ground-based forces are unable to intervene in the Pakistani tribal areas, the air war has flowed over the border to target nodes of what is perceived to be a global militant assemblage regardless of their territorial positioning. As such, then, while the doctrinal imaginary of Shock and Awe may have been substantially challenged by both its empirical implementation as well as subsequent development in American and allied warfighting, its scopic regime continues to reverberate through the so-called war on terror.

## Conclusion

The air targeting of Shock and Awe thus provides a key moment in which to understand the manner in which the discursive trope of the network reverberates through the constitution of battlespace in the contemporary era. Beyond the spectacular simulacra of the initial air assault on Baghdad in 2003 lies a scopic regime which suggests that the capture, processing and rendering of data can lead to the precision targeting of the nodal junctions of political, military, logical and social infrastructures. This scopic regime gives rise to a de-territorialised understanding of the infrastructure it seeks to disassemble: an en-visioning of battlespace that lends itself to surgical metaphors implying a bloodless precision. Recast as a de-territorialised constellation of elements shaped by the relations vis-à-vis each other (rather than to territory), battlespace is unbounded. As such, it is able to incorporate elements that might have previously been approached with wary ambivalence. As a consequence, as Graham has so clearly shown, Western militaries are rapidly incorporating the city into their conceptions of battlespace, resulting in an urbanisation of security that will have profound consequences in the era of global urbanisation.[51] The consequences of such a recasting and unbounding are both political and ethical, giving rise to a de-restricted warfare that gradually erodes any hesitancy we may have previously had at striking what might be said to be predominantly civilian locations. Air targeting plays a vital role in this recasting of contemporary battlespace. Air targeting can be perceived as a delivery of destruction that seemingly escapes the constraints of territorial jurisdiction. This perception (whatever the reality) reinforces the de-territorialisation and unbounding at the heart of Shock and Awe's refiguring of battlespace and the consequences this has for warfighting in the contemporary era.

5

# PHOTOMOSAICS

## MAPPING THE FRONT, MAPPING THE CITY

*Paul K. Saint-Amour*

So much information that would otherwise have been concealed from the
enemy was revealed by the all-seeing lens. … No matter how carefully machine-
gun emplacements may have been concealed, their position was often betrayed
by the disclosure of foot tracks made over-night to and from munition dumps.
Dummy trenches and other appurtenances of warfare were detected easily
enough, for the absence of shadow on the completed prints was sufficient to
rouse and to justify suspicion. … [the enemy's] possible future activities (as well
as past and present) [were] laid before [the Allies] like an open book.

Clarence Winchester and F.L. Wills on First World War
aerial reconnaissance (1928)

Mrs. Smith's washing can be seen hanging on the line, so you know Mrs.
Smith's wash day even if you don't know Mrs. Smith.

Sherman M. Fairchild, on peacetime photogrammetry (1922)

# FROM ABOVE: WAR, VIOLENCE AND VERTICALITY

This chapter is about a period of technology transfer—the dozen or so years after the close of the First World War—when wartime reconnaissance techniques and operations were being adapted to a range of civilian uses, including urban planning, land-use analysis, traffic control, tax equalisation and even archaeology.[1] Entailed in this transfer are a series of questions about what happens when technologies developed by a state for use against a wartime adversary are adapted for peacetime use by that same state in relation to its own citizens, territory and municipal infrastructures. In what ways does such a technology residually construct its domestic objects as distanced, derealised and oppositional—that is, as targets in the military sense? What do the civilian uses of the term *target* retain of their military origin? What complexities might we lose by reducing our origin-narratives about particular technologies to strictly military or civilian geneses?

In respect to aerial imaging, our thinking about such questions in recent decades has tended to proceed from certain articles of faith. The plan view, we like to say, totalises by shearing off singularity, complexity, anomaly and dissent in favour of schematic simplicity. Relatedly, distance—especially the growing spatial, optical and technological distance between perceiver and seen—instrumentalises and in some cases dehumanises the seen in profoundly consequential ways. Finally, the growth of this distance has happened alongside an acceleration in perception, transmission and feedback, to the point where what military planners now call the 'kill-chain'—the time elapsed between identifying a target and destroying it—is being massively compressed. This chapter attempts to identify some of the seams between these articles of faith as well as the instabilities within each of them. Without denying that aerial imaging can both totalise and instrumentalise what it sees, it suggests that these operations have recoverable limits and, furthermore, that they sometimes require the very exceptions—the partial, the belated, the disjunctive, the differential, the site-specific, the erroneous—whose absence appears to be their defining characteristic. As a paragon of perfect visibility, I will suggest, aerial imaging depends on certain derangements of vision that it hides in plain view.

I have made a related argument about the use of the stereoscope in aerial reconnaissance interpretation from the First World War on.[2] By pairing vertical images taken hundreds or even thousands of feet apart, this technique vastly expands the virtual interocular base of the viewer, producing exaggerated 3-D effects that are interpretively useful in proportion as they expose the spatial contingency of human depth perception. This aerial *hyperstereoscopy* imparts a fantasy of all-powerful vision by insisting that we always see from somewhere—or, more accurately, from two adjacent but distinct somewheres. And because it asks the viewer to synchronically fuse two images taken in rapid diachronic sequence, the technique insists, albeit more subtly, that seeing is an event rather than a condition: that we always see from some *when*. In turning here to consider the First World War's other major reconnaissance legacy—the aerial photomosaic, in which overlapping images were pieced together into a composite image—I will be discussing a form that works harder to efface its multiplicities of viewpoint even as its diachronicity becomes more difficult to conceal. For to look at a photomosaic is to be pulled in two directions: on the one hand, toward accepting the fiction that its constitutive images synch up in an integrated spatial geometry; and on the other, toward awareness of the fact that they were captured at discrete, often distant, moments. When the illusion of spatial integration fails, it does so principally because the composite image separates into tiles of non-synchronous time. In contrast to the vertiginous spatial depth-effects of hyperstereoscopy, the photomosaic offers the distinct vertigo of *temporal* parallax, one arising from the experience of counterfeiting a spatially self-identical landscape from a constellation of segregated *moments*. A photomosaic is perforce a mosaic of temporalities.

My chapter has a broader theoretical aim as well. In calling attention to the planimetric view's construction from a diachronic series of images and moments, I underscore the extent to which our still-dominant notions about vertical imaging are themselves based on a misrecognition. These notions were endorsed, as we have already seen, by the practitioners of early aerial photography. But they were given theoretical heft by structuralism's fixation on a biaxial model of optical space, a model that poststructuralism attempted to dismantle but more effectively perpetuated. According to this biaxial scheme, the vertical is the axis of order, paradigm, symbolic function, disutility, unimpeded sightlines and disembodied omniscience, whereas to the horizontal belong disorder, syntagm, enunciative function, utility, partial sightlines and exposure to visibility. In Roland Barthes' 'The Eiffel Tower', for instance, the vertical stands outside of the city's history and structure, yet by dint of that exemption it gives spectators a privileged view of both: as the vantage most conducive to 'intellection', the Tower's verticality 'permits us to transcend sensation and to see things *in their structure*', granting us access, in a single glance, to the city's blueprint or x-ray and to something like its deep time.[3] Michel de Certeau's 'Walking in the City' transposed Barthes' parable to New York, with the view from the top of the World Trade Center epitomising the 'pleasure of "seeing the whole," of looking down on, totalizing the most immoderate of human texts'. As against the labyrinth-dwellers at street-level, the Icarus on the 110th floor accedes to 'a scopic and gnostic drive … to this lust to be a view-

Fig. 13

Image 12: Photomosaic made in 1920 of land along the Anacostia River east of Washington, DC. To illustrate the method of photomosaic composition, the edges of the composite image have been left untrimmed and the variation in the constituent images' exposure uncorrected. (Lee 1922: 22a).[4]

point and nothing more'.[5] For de Certeau, the ground-dweller is not only a sign exposed to the city planner's panoptic reading and schematising; she or he may also elude that scopic discipline through the practice of resistant pedestrian speech acts, walking and using the urban grid against the grain of its planners' intentions. Yet despite celebrating the *flâneur*'s jamming of the *planeur*, de Certeau's essay concretised Barthes' biaxial mapping: the vertical remains the viewpoint of power's monopoly on paradigm, even if that power is occasionally stymied, and the horizontal remains the axis of the masses trapped in syntagm, even if they have recourse to resistant enunciative tactics.

Both of these influential essays are of course more complex than my accounts of them. Barthes descants on the 'dialectical nature of all panoramic vision', whose frictionless movement across an exposed landscape nonetheless requires the viewer to struggle to decipher it by locating familiar '*signs* within it' from history, myth and lived experience on the ground.[6] De Certeau makes the claim, although without dilating on it, that the haunting of places by memory 'inverts the schema of the *Panopticon*'.[7] But like so many cognate discussions of power and visuality, both essays' schemata have had more longevity than the exceptions, elaborations and argumentative eddies that complicate them, as the trellis outlasts the bougainvillea it was built to support. One result, ironically, has been that essays interested in developing a model of *differential space* have been annexed to the *abstract space* of highly schematic thinking. Here I use Henri Lefebvre's terms deliberately, as a reminder that the native abstraction of 'theory' makes it even more susceptible than its objects to homogenisation—to the dampening of internal differences, the smoothing of anomalies, the honing off of resistant historicity. For Lefebvre, these differences within the space of both theory and social practice are nothing less than the sites of potential social transformation:

From a less pessimistic standpoint, it can be shown that abstract space harbours specific contradictions. Such spatial contradictions derive in part from the old contradictions thrown up by historical time. These have undergone modifications, however: some are aggravated, others blunted. Amongst them, too, completely fresh contradictions have come into being which are liable eventually to precipitate the downfall of abstract space. The reproduction of the social relations of production within this space inevitably obeys two tendencies: the dissolution of old relations on the one hand and the generation of new relations on the other. Thus, despite—or rather because of—its negativity, abstract space carries within itself the seeds of a new kind of space. I shall call that new space 'differential space', because, inasmuch as abstract space tends towards homogeneity, towards the elimination of existing differences or peculiarities, a new space cannot be born (produced) unless it accentuates differences. It will also restore unity to what abstract space breaks up—to the functions, elements and moments of social practice. It will put an end to those localizations which shatter the integrity of the individual body, the social body, the corpus of human needs, and the corpus of knowledge.[8]

Aerial photography would seem to be the quintessence of abstract space, and indeed it has been cast repeatedly in that role for over a century. But this very role makes the recovery of aerial imaging's differential qualities a matter of consequence both for visual culture studies in particular and for our critical habits of thought more generally. As much as this chapter reads a specific constellation of technologies, institutions and rhetorics within a constrained historical moment, it also offers a parable about our theoretical optics: about the subsumption of theory's differential energies by the homogenising, abstracting processes of capital, and about the stakes of reversing those homogenising processes. Here, to accomplish such a reversal would be to find the differential not in opposition to the vertical (for that would be business as usual) but *within* the vertical, where its presence has long been camouflaged, ignored or misattributed exclusively to the ground. It would be to find within the aerial photomosaic, and within the distance-optics of theory, some of the bird's—or god's-eye view's—most revealing and emphatic self-denunciations.

None of which should be read as an injunction to renounce high-altitude, comprehensive, integrative modes of seeing or thinking altogether. Note that, for Lefebvre, it is abstract space's tendency to *localise* rather than totalise that shatters individual, social and epistemological bodies. Homogenisation, that is, can be a function of the local—both when it abstracts the local from the larger bodies on which it depends and when it extrapolates a false image of the whole from a part. To be immanently critical of the totalising energies within a scopic regime or a system of thought is emphatically not the same thing as giving up on the *question* of totalities, which is to say, on the question of how ideological effects are produced through the effacement of connections between the distant and the proximate. Far from jettisoning total viewing in favour of partial, the kind of immanent critique I have in mind seeks a dynamic, interrogative relation between the two, recognising that the total may be, in one instance, a special case of the partial or, in another, the category through whose occultation the partial produces the myth of its autonomy, normativity or sovereignty.

In what follows, this immanent critique goes by the name of 'applied modernism'. I use the expression for two reasons. First, as I describe here and elsewhere, early aerial photomosaics not only were associated with avant-garde painting during and after the First World War but also shared with much of that painting the premise that distortion was the only route to revelation. And second, the photomosaic partook of a tendency shared by many Western modernisms—literary as well as visual ones—to view the total from the vantage of the radically site-specific, subjective or fragmentary, underscoring in the process how portraits of a given totality can be both generated and apprehended only from viewpoints that are optically and ideologically partial. Hence my earlier point that the total might be understood as a special case of the partial rather than the reverse.[9] In calling the modernism of the photomosaic 'applied', I wish to emphasise the utilitarian contexts of military and urban planning to which the photomosaic was conscripted from the 1910s onward. But I would stop short of implying that other modernisms were, by contrast, 'pure' or 'non-instrumentalised'. In fact, it may be precisely by studying their parallel deployments in military and civilian contexts that we discover the full extent to which supposedly 'pure' modernisms were themselves always applied, always kitted out for oppositional spaces such as the battlefield and the marketplace.

Because the photomosaic foregrounds questions of temporality, my discussion of technology transfer, spatial and ethical distance, and velocity will be focalised through the aperture of the *event*. This will mean asking questions about what events are entrained in the production of photomosaics, from the flight and image-capturing by the camera plane to the rectification and piecing together of the mosaic to the mass-production of the composite image. In the wartime context, we will need to consider, too, the destructive events—the artillery salvoes, bombing runs and infantry advances—that are variously planned and confirmed by way of the photomosaic. In such cases, the eventfulness of the image's production would seem to be eclipsed by the kill-chain it exists to set in motion—that is, by the sequence of events along which a target is identified, its importance evaluated, a decision made and destructive forces unleashed.[10] But what becomes of the status of the photogrammetric event in peacetime, when the image is no longer motivated by the extreme eventfulness of the kill-chain? Does the civilian photomosaic map allow each of its constitutive images to retain its native eventfulness, as both an image taken and a moment captured, without instrumentalising it? Or, to the contrary, is the eventfulness of the photograph effaced, first, by techniques of rectification and, second, by the photo's conscription to the logic of the map, whose utility requires the fiction of its uneventfulness?

## 'Applied Modernism' Over the Western Front

The aerial photomosaic came into widespread use as the photographic correlate of trench warfare. With sites of engagement stretching for many lateral miles, and with those miles of front supported and supplied, in turn, by extensive behind-the-lines networks, trench warfare made massive swathes of territory tactically relevant and therefore subject to reconnaissance overflights by the fledgling air services on both sides. The limitations of the aerial camera and its heavier-than-air platform, however, and the need to use telephoto lenses from an altitude above artillery range, meant that no individual image could capture very much ground at a resolution high enough to be useful to photographic interpreters.[11] In order to provide coverage of significant portions of front line and supporting positions, reconnaissance pilots flew in switchback lanes, making exposures at regular intervals in such a way that each image would overlap about 20 per cent with the adjacent images in its lane, and each lane overlap, in its turn, with the lanes that bordered it. To ensure the verticality of the coverage, the plane had to fly as level as possible; to minimise blur, the pilot had to avoid lurches and vibrations. Once the film had been developed on the ground—often very rapidly, in cramped, lorry-mounted darkrooms—the prints were scaled to one another, fitted together to produce a continuous image and glued to a drawn map of the same scale. The mosaic itself was then photographed, and the resulting image labelled, reproduced at high speed and in massive numbers—again, by workers toiling in factory-like conditions—and disseminated.[12] 'The time and energy saved by this process is enormous', writes Harold E. Porter in 1921, having been a captain in the US Army Air Service during the war. 'A whole county can be photographed in an hour or two, and a mosaic and map made of it in a few days; whereas to do the same work by an ordinary survey might take a couple of months. Besides, there are no surveying gangs in the front-line trenches.'[13]

The interpreters of aerial reconnaissance were not just the end-users of the distributed mosaic images—they also helped to produce them by decoding the landscape and leaving a record of their findings in the reproduced image's key or margin. By pairing adjacent exposures from a given flight line under a stereoscope, they could produce parallactic depth-effects in the areas of overlap. They were trained to observe changes from one coverage to the next, to spot camouflage and decoys by the shape or absence of the shadows they cast, and to decode subtle deformations of the landscape—to read, say, in the bent blades of meadow-grass a sign that several men had passed through, and then to find the new gun emplacement at the end of the track. As Porter puts it, these military knowledge-workers were 'trained to know how things *ought* to look under all sorts of different conditions in a vertical photo', and thereby to detect deviations from the expected norm. Against the uneventful baseline of the condition ('how things *ought* to look'), they awaited signs of an event—of some new initiative by the adversary that would require the equal-and-opposite event, the counter-event or event-in-kind, of targeting. From the interpreter's point of view, the target had less to do with tactical value or meaning than with eventfulness: it was a break in the placid surface of 'conditions' that irritated the kill chain into an equally eventful restoration of placidity. This condition/event binarism tended to produce an indexical and starkly decisionist heuristic according to which the eventful, event-provoking thing either was or was not present. Allan Sekula's account of this heuristic in military intelligence is worth quoting at length:

Simply put, the problem was to decide what was there and to act on that decision before 'whatever it was' moved. If the entity in question fell into the category of 'enemy', its destruction by artillery fire, or by other means, was ordered. The value of aerial photographs, as cues for military action, depended on their ability to testify to a present state of affairs. The photographic sense of 'having been there', identified by Roland Barthes, must submit to the demands of 'being there.' … The meaning of a photograph consisted of whatever it yielded to a rationalized act of 'interpretation.' As sources of military intelligence, these pictures carried an almost wholly denotative significance … Within the context of intelligence operations, the only 'rational' questions were those that addressed the photograph at an indexical level, such as 'Is that a machine gun or a stump?' In other words, interpreting the photograph demanded that it be treated as an ensemble of 'univalent', or indexical, signs—signs that could only carry one meaning, that could point to only one object. Efficiency demanded this illusory certainty.[14]

These were images without ambiguity, intrinsic interest or a future beyond the short-lived horizon of intelligence-gathering—images wholly subordinated to their denotative function and its date-stamped temporality. Once such an image had either been superseded by more recent coverage or assisted in the eradication of any unwelcome events it discloses, its sole reason for being was also eradicated; it is as if the reconnaissance image were immolated alongside the target, indivisibly from the target, in the name of the 'conditions' it helped restore. (If such were the case, any eventfulness inhering in the image's production was indeed consumed by the eventfulness of what it denoted.)

Image 13: '"CUBIST" country', from *Characteristics of the Ground and Land-marks in the Enemy Lines opposite the British Front from the Sea to St. Quentin*. (Branch Intelligence Section of the G. H. Q. Wing, RAF 1918).

For Sekula, this rationalist, instrumentalising conception of the image is 'illusory' because the same image may be resignified in another, peacetime context—imbued, for instance, with a rhetorical structure, a claim on the beautiful or the glamour of an authorship-relation. I suggest, however, that instead of being celebrated during the war exclusively as the 'triumph of applied realism' (the phrase is Sekula's), aerial reconnaissance was understood by First World War practitioners not only as accurately denotative but also as crucially defamiliarising, revealing objects only in proportion as it deformed both the geometry and the temporality of human vision. We have already seen how the open book of the enemy's intentions became legible to photo interpreters only after they had attempted to rectify the errors inherent in the image-taking. Interpreters who used hyperstereoscopy did not just correct for errors but actively exploited the delirious depth-effects inherent in their technique. And if a revelation that comes of abandoning the spatio-temporal norms of visuality sounds remarkably like the pre-war avant-garde, at least a few souls in the RAF seem to have made a similar connection in labelling varieties of landscape '"CUBIST" country' and '"FUTURIST" country' in a 1918 atlas designed to familiarise new pilots with the vertical views of the territory behind German lines. Scholars referring to this text in the past have done so in order to confirm something about the reception of avant-garde painting—that its terms and visual modalities had become household words, or that its contemporaries saw the Cubist and Futurist painters as having anticipated the estranging rectilinearity of the land seen from an airplane.[15] However, face to face with the '"FUTURIST" country' page (as against its 'CUBIST' counterpart), we can now see that it contains not a lone aerial photograph but two overlapping ones—a rudimentary photomosaic, in other words—and that the composite photographic image is accompanied by a swatch of cadastral map depicting the same area. Whether '"FUTURIST" country' resembles paintings by Boccioni, Balla or Carrà any more than '"CUBIST" country' does those of Picasso, Braque or Gris—and whether the difference between these disorienting modes could be of any practical help in *orienting* young pilots—to call a *de minimis* photomosaic 'Futurist' in a photo atlas is to engage in a kind of photographic metacommentary. It reads, by the light of Futurism, not just the denatured view of the earth from above but also the particular technique of the photomosaic, with its projection of discrete moments of seeing onto a unified picture-space, its dependence on a technologised circuit of production.

It apprehends, too, the high speeds that attend that circuit: velocities of airplane, shutter, photo development, interpretation, reproduction, dissemination, decision and assault; velocities, too, of innovation and industrialisation in all of the foregoing. Such a reading may not deliver the photomosaic from its instrumental place in a military command arc. But it does decouple that instrumentality from a straightforwardly denotative optics, making indexicality something that can be hallucinated precisely and exclusively from the site of its liquidation. Far from being the triumph of applied realism, photogrammetry was understood by its wartime proponents as the triumph of applied modernism.

*1920s Civilian Air Surveys*

I have argued that aerial photography's departures from the conventions of a perfectly scaled orthographic map were a chief source of its use

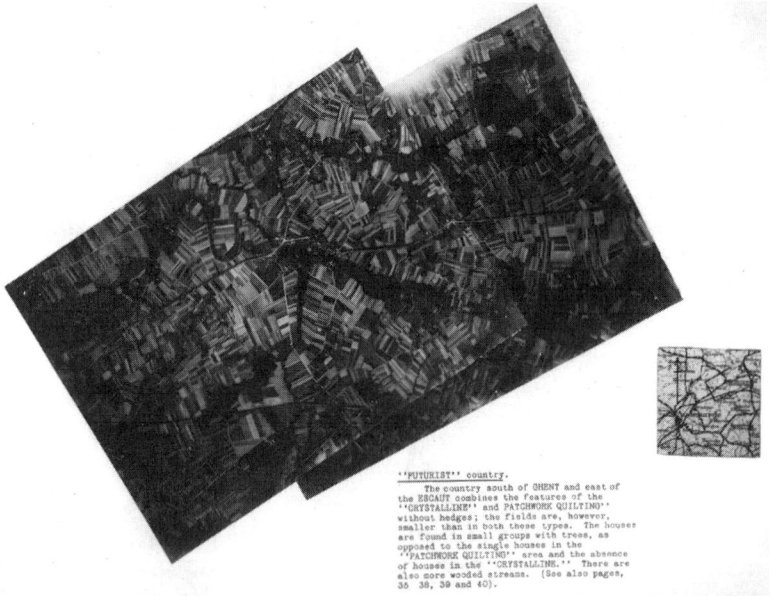

Image 14: '"FUTURIST" country', from *Characteristics of the Ground and Landmarks in the Enemy Lines opposite the British Front from the Sea to St. Quentin* (1918).

during the First World War. When that technology was transferred to civilian practices after the war, however, the distortiveness that made it tactically useful in war appeared unseemly in peacetime, as if avant-garde optics could only be tolerated in the nakedly oppositional context of battle. The result was a widespread reapplication of the language of realism to aerial mapping. But this resurgence of realism-claims was not just a way of making a wartime technology safe for civilian use. It also registered the fact that the precision mapping demands of civilian photogrammetry required, if anything, more severe manipulations and counter-distortions to make aerial photomosaics conform to the conventions of cadastral maps. Elaborated to the point of spawning a discrete profession, these photographic and trigonometric rectification techniques needed a compensatory super-realism—a set of exaggerated claims about the accuracy, transparency and self-decryption of the medium—as a rhetorical counter-distortion. But while the disfigurations, discontinuities and site-specificities I have called 'applied modernism' could be banished to both the optical and rhetorical peripheries, they could not be eliminated. It will be our task in what follows, then, to read the modernism at the margins of peacetime photogrammetry.

Most wartime photomosaics were assembled at speed, with minimal attempts to hide the seams or the exposure, scale and parallactic differentials between neighbouring photos. But these gaps and differentials could be significant. To begin with, the mosaic's constituent photographs were seldom perfectly vertical—that is, seldom taken with the vertical axis of the camera's lens exactly perpendicular to the ground or projection plane—because camera-bearing aircraft were easily thrown out of level flight by turbulence. Fluctuations in the plane's altitude caused changes in photographic scale from one exposure to the next. And even if the plane flew at a perfectly consistent altitude, scale remained a problem. In any photograph, there are as many scales as there are depth planes. This continuum of scales is advantageous in most photographic contexts, as the greater scale of foreground objects contributes to the impression of their proximity over background objects, giving the viewer non-parallactic depth cues. But in aerial photos that would be used to rectify existing maps, a consistent scale was essential. In vertical photos of topographically extreme terrain, the scale in which valleys appeared could differ noticeably from that of mountaintops; likewise, in an urban environment, with streets versus the tops of skyscrapers. While the relief displacement—the radial 'leaning' of objects away from the vertical photo's nadir—may have been essential to aerial stereoscopy, it played havoc with the planimetric and orthographic conventions of most maps, in which projection lines are perpendicular to the projection plane rather than converging toward the perspective centre. And the shadows of tall objects could produce areas of *pseudoscopy*—the reversal of relief effects—on a photomosaic if they were oriented improperly. As Herbert E. Ives put it, 'Even when "corrected" and retouched, aerial photographic mosaics could present weird effects: in maps of cities, buildings sometimes look concave instead of convex because the shadows are falling away from rather than toward the viewer, particularly if the map is conventionally oriented with North corresponding to the viewer's "up".'[16]

An unrectified or 'uncontrolled' photomosaic, then, could look like an old street with its cobblestones akimbo, the choppiness and radial warp of its constituent units drawing attention to the composite nature of the surface. But when greater standards of photogrammetric accuracy were called for, as was the case with the civilian aerial survey projects from 1918 on, new procedures arose for correcting (or, in the weirdly redundant term, *orthorectifying*) photomosaics. Fluctuations in the tilt and altitude of individual exposures were calculated and then counteracted with the use of more and more elaborate projection printers. The same device could also alter the scale of a particular depth plane, so that areas of the photo depicting an extreme depth or height (e.g. valleys or high plateaus) could be reprinted in a rectified scale and then glued over the corresponding unrectified area. Relief displacement could be minimised by using only the central part of a given exposure rather than its more radially displaced perimeters. But even a scrupulously 'controlled' mosaic was, at best, a geometric entente between the perspective projection of its constituent photographs and the conventional planimetric projection to which the mosaic as a whole aspired.[17]

One method of controlling photomosaics was to plot the photos to an identically scaled line-map of the terrain, whereby discrepancies between mosaic and map would guide rectification of the photos. Here we arrive at the photomosaic's mimetic *mise-en-abîme*: a medium supposedly capable of correcting and even supplanting less accurate, less informative planimetric line-maps relied on those same maps to point up its optical waywardness. If lone aerial photographs distorted the terrain, they at least did so in much the same way the eye did, reproducing the relief displacement, the scale-differential at varying altitudes, even the flatness of the unaided aerial view. In the controlled, orthorectified context of the photomosaic, hundreds or thousands of fitted prints were made to represent not a humanly perceptible view of the terrain, but an imaginary view exempted from the situated optical traits of parallax and scale differentials. Finally, having been rectified, controlled photomosaics were retouched in order to conceal the composite mode of their production. Tonal variations among adjacent prints were regularised by the application of a red dye to the negative; lines from shadows cast by the edges of overlapping prints were painted out.[18] By eliminating its seams, the mosaic's makers gave it an artificial unity of perspective, uniformity of scale and temporal simultaneity. A photographic form that revealed camouflage had itself been camouflaged, its contingencies disguised or levelled so that it resembled an idealised schematic as seen by a disembodied observer. Canonised by intelligence personnel, surveyors, police, entrepreneurs and city planners as the utmost in photographic realism, the mosaic was considered 'accurate' only in proportion as it bent the rules of optics away from perception toward orthogony. Aerial photomosaics ended up reconsecrating as ideal the very cartographic abstractions they were meant to correct or supersede.

After the Armistice, some of the reconnaissance workers trained during the war remained in national air corps, mapping colonies, mandates and protectorates (as in the British case) or arguing that photogrammetry research and development should continue despite post-war military budget cuts and disarmament (as in the United States).[19] Others applied their military training in civilian mapping projects, both privately and publicly funded. In France, a law passed shortly after the signing of the Treaty of Versailles required all cities above a certain size—several hundred in all—to be surveyed aerially by civilian companies. Territory laid waste during the war was also surveyed by the French Army Aerial Photographic Service, in many cases because property lines had been effaced and local archives destroyed in the fighting.[20] Whereas French aerial photographers were put to work in a reparative relation to the destructive work they had abetted during the war, US reconnaissance veterans were sometimes described as bringing the war's adversarial energies to American cities. Nelson P. Lewis, a consulting city planning engineer with the Russell Sage Foundation, extolled the virtues of a repurposed photogrammetry that retained aspects of its wartime strategic uses:

We know that aerial surveys and mapping were of the greatest possible use during the World War in locating points within the enemies' lines which were vulnerable to attack, but we have found that this same method of aerial photographic surveying will disclose the logical points of attack for those who are bent, not upon destructive but upon important constructive work, upon the better utilization of natural conditions for commerce, for homes or for wholesome recreation.[21]

During the 1920s, dozens of similar articles appeared in *The American City Magazine* and other urban planning periodicals; many of them partook of Lewis's difficulty in swerving from a logic of targeted attack to a claim of 'constructive work', as if uncertain how to honour the returning soldiers and the war technologies in which they had been trained while reassuring non-combatants that both could be absorbed safely into civilian life. A less ambivalent point of view lamented photogrammetry's adoption by the military: '[the] mapping use of airplane photography has been to a certain extent set back by the war, for the reason that certain scientific views [e.g. the chemical development of colour-sensitive emulsions that could help reveal camouflaged objects], which are not relevant to commercial photography rather held the foreground'.[22] Two incompatible portraits of aerial photogrammetry emerged: the triumphant yet domesticable war technology versus the civil technology hampered by military misappropriation. That there were, by 1920, warring camps in respect to aerial photography's originary conditions reveals the extent to which that origin-narrative mattered—and the power of stigma and glamour that a technology could carry once it had been marked by war.

The catalyst for this debate was the post-war emergence of companies such as Abrams Aerial Survey Corporation, Aero Service, Brock and Weymouth and Fairchild Aerial Surveys, which began to make increasingly detailed photomosaic maps of US cities. Fairchild's 1921 and 1924 maps of New York, in particular, were widely discussed and publicised in newspapers, aviation weeklies, scientific monthlies and city planning journals, eliciting rapturous celebrations of the new form's beauty, accuracy and myriad uses; the latter were said to include recording city growth, informing current zoning and future planning, correcting errors in extant line maps, revealing traffic problems and untaxed buildings, even allowing police inspectors to 'note the location and details of every roof exit, scuttle hole and skylight'.[23] Some revealing moments of ekphrastic rapture came from the photographers themselves, who wrote puff pieces for their own projects. Here is Sherman M. Fairchild, who would become the most celebrated civilian figure in interwar aerial surveying, describing the enormous photomosaic map of New York City that his company completed in 1924:

The map pictures the city with the minutest detail. It shows every structure from contractor's temporary tool shed to skyscraper; back-yards, gardens and parks with every tree and bush visible; avenues and alleys, streets and unrecorded footpaths; big league ball parks; water-front clubs, with their yachts and motor boats; the boardwalk of Coney Island, and the crowds of people appearing like small black dots. Even the congestion of traffic on busy thoroughfares is clearly shown.[24]

There is not, although there should be, a name for this kind of bird's-eye reverie that would master the landscape by comprehending its constitutive objects. It is related to the exhaustive census one associates with dollhouses and model train layouts, in which *every* feature—'*every* roof exit, scuttle hole, and skylight', '*every* structure', '*every* tree and bush'— of the thing itself has been faithfully miniaturised at the proper scale and without loss of resolution. It captures the intimacy between the vertical or planimetric view and urban planning, an intimacy the journalist and travel writer Lowell Thomas recognised in explaining why he preferred to see a city from the air for the first time: 'Instead of coming in through a lot of dirty railway yards and uninteresting factory and poorer residential sections, you get a perfect panoramic view, a view that once and for all puts a plan of the city in your mind's eye.'[25] Yet even in these euphoric passages, two distinct views of the city have been awkwardly fused together: one that faithfully replicates and one that schematises and aestheticises; a realist view that apprehends things in their material tangibility and a socially hygienic view that sees them, as Barthes said, '*in their structure*'.[26] Suspended uneasily between these two views are the city's inhabitants, for as any dollhouse builder or model train enthusiast knows, human figures are a problem for the miniaturist of built environments: they are at once required by a realism that operates in their name and yet unwelcome, a noise that interferes with the signal of structure. The 'small black dots' in Fairchild's reverie are oddly cognate with the dirt, industry and poverty Thomas prefers to avoid by approaching cities

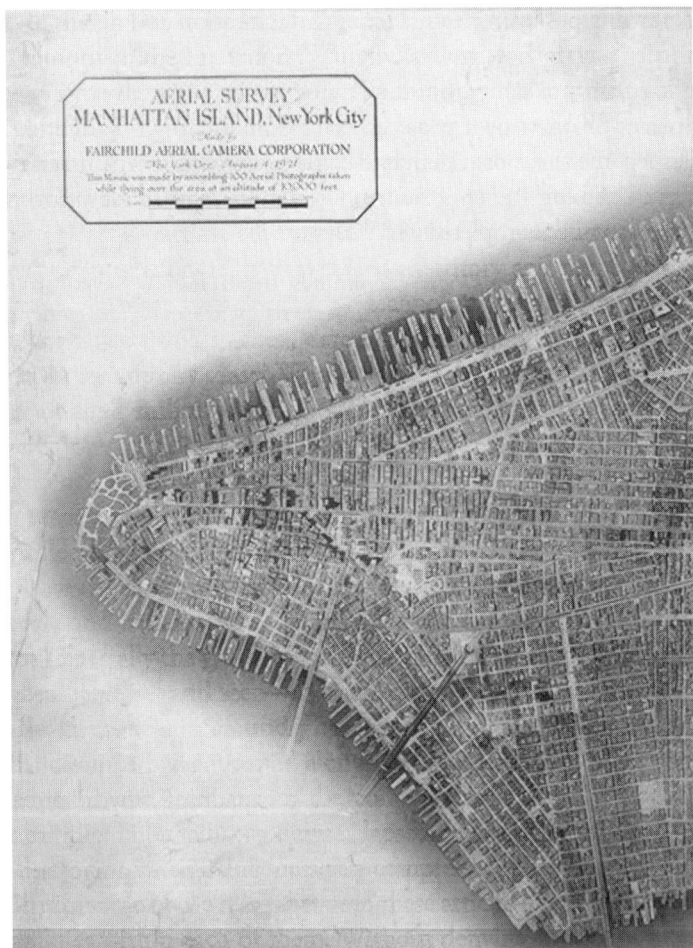

Image 15: Unsigned, 'Practical Aerial Photography'.

from an altitude that effectively depopulates them. Better to infer the presence and behaviour of human organisms from their 'unrecorded footpaths'—those deviations from structure that constitute a new structure—than to encounter humanity in plain view. Better to deduce Mrs Smith's washday from a glimpse of her linen than to know Mrs Smith.

But a photomosaic map expansive and detailed enough to reveal a rooftop laundry line could not of course reliably inform one of the day of the week on which that particular rooftop had been captured. Fair-child's boast about aerial surveys—'Mrs. Smith's washing can be seen hanging on the line, so you know Mrs. Smith's wash day even if you don't know Mrs. Smith'[27]—not only extends the wartime figuration of the photographic interpreter as Sherlock Holmes, it also participates in a fantasy of simultaneity and temporal precision that was increasingly belied by the production of aerial photomosaics as they grew in scale and resolution during the 1920s. Like reconnaissance coverage of the Western Front, early post-war surveys such as Fairchild's 1921 Manhat-tan mosaic could be assembled from photographs shot during a single flight ('New York Mapped by Sky Camera in 69 Minutes!' declared one headline). But the larger scales subsequently demanded by urban plan-ners required many more photographs per square mile: whereas the 1921 map consisted of 100 photographs taken by a single plane, the 1924 map combined 2,000 photos taken by three planes over the course of several months.[28] The protracted period of image-capturing owed not just to the sheer number of individual photos involved but also to the fact that, as one of Fairchild's accounts put it, 'Few days are suitable for photographic mapping work' at such a high-level of detail.[29] Shots could be ruined by low-flying clouds, haze, smoke from the soft coal used in nearby factories or high-flying clouds whose shadows darkened patches of the city. The shoreline had to be photographed at low tide for consis-tency's sake and to capture as much dry land as possible. But protocol favoured midday photography to minimise shadows and the pseudo-scopic effects they could produce, so shoreline runs were confined to clear days when low tide fell no later than 2 p.m. Days with snow on the ground were ruled out, so the camera planes were grounded for much of the winter, during which the rectification and assembly of mosaics were the agency's principal focus. In the completed mosaic, the still containing Mrs Smith's washing lines might be adjacent to a photo taken only seconds later on the same day, or it might be blended with a neighbouring photo taken months later at a different time of day, its slight shadows leaning in a different direction, its foliage in another seasonal phase. Far from presenting an aerial view of New York at a discrete moment—on a given Tuesday, a given washday—the Fairchild map stitched and smeared together photos taken from a variety of places and instants under a variety of conditions. As much as it was annexed to the cartographic fiction of an eternal present, it was in fact a miscellany of moments, a highly disciplined crazy-quilt of the city in time.

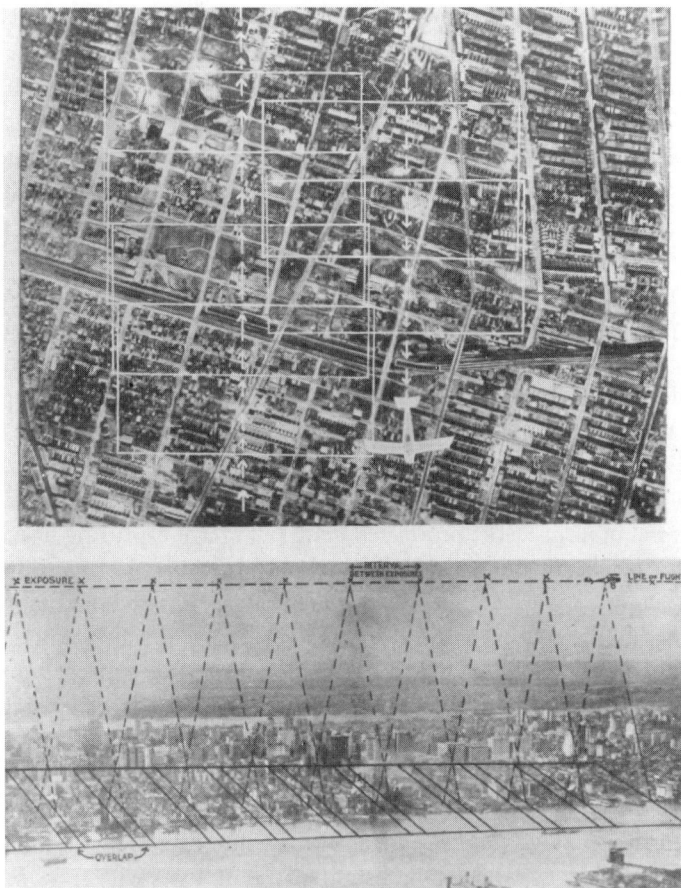

Image 16: Overlapping exposures and flight lanes in urban photogrammetry, (Unsigned 1922: 116). Original caption reads: 'A diagram, showing how thirty-two square miles of New York was [sic] plotted and photographed in sixty-nine minutes with a Modern Aerial Mapping Camera. Below is a cross-section diagram, showing the method of procedure.

Fairchild's boast actually leads us to a question: what happens to clock- and calendar-time in a city whose photogrammetric self-portrait is a concatenation of thousands of moments, many of them flung far from one another? Can any discrete thing be said to have *happened*, or to have left a legible trace of its happening, when it has been almost indetectably merged with both near-simultaneous and radically non-simultaneous images? What—in the silent ligations of such an image, or in the city that takes such an image as its self-portrait—is the status of an event? To be sure, the celebrants of aerial city-mapping invoked certain events—break-ins, riots, traffic jams, natural disasters and human-made emergencies—that photomosaics could help abate or prevent. In this, the proponents of 1920s aerial mapping as an urban planning tool simply replicated the adversarial stance of wartime reconnaissance, which established a baseline of 'condition' against which to measure unwelcome events it would then eliminate or forestall. These nods to eventfulness constructed it as the exception that validated the rule, the rule in this case being the uneventful time of the unblemished grid. Yet despite this seeming lock-step between planimetric seeing and the timeless state of the condition, there remains in both the discourse and the practices of Fairchild et al. a captivation with events that cannot simply be adduced as proof of the aerial survey's utility, events in excess of exception-that-proves-the-rule rhetoric. These are exceptions that prove no rule: workaday ones, such as the hanging out of a stranger's laundry, or more momentous ones, such as the Carpentier-Dempsey championship fight, which a Fairchild plane photographed from the minimum legal altitude of 2,000 feet over Boyle's Thirty Acres in Jersey City on a rainy night in July 1921.[30] We might think of such glimpses, such stunts, as eventful flickers on the periphery of the mosaic gaze, and as testimony that that gaze, for all its affinities with the general and the structural, was made entirely out of flickerings of the particular.

Earlier I described as 'applied modernism' this flickering quality of the aerial photomosaic, with its patchwork temporality, its indelible discontinuities and its subtle exposure of orthography as a conjuring trick that summons an impossible image from resistant pixels. The photomosaic testified, in short, that seeing was a situated act: not a condition but an eventful apprehension of events, along a sightline in relation to which other visible objects are perspectively displaced. However we think of this particular modernism—as a cunning assemblage of technologies, as an anti-Cartesian correlate of avant-garde painting, as a scopic regime of late modernity—we can at least insist on its opposition to the schematising modernism of, say, Le Corbusier, for whom, as for Lowell Thomas, the best way to see and to plan a city was from the air. Note, however, that the applied modernism of the photomosaic does not oppose the schematic and—let's say it—totalising modernism of Le Corbusier and others from the ground, via some emphasis on local performances and pedestrian enunciations that evade the all-seeing aerial eye.[31] Irreducible to either the street-level practices of the *flâneur* or the totalising gaze of the *planeur*, the mosaic disrupts critical and spatial reflexes stabilised by the likes of Barthes and de Certeau—reflexes by which we project all authority, distance and spatiality on to the universal vertical axis and all resistance, locality and temporality on to the horizontal.[32] Instead, the photomosaic shows us how the always-situated optics of vertical seeing can reverse verticality's misrecognition as the necessary axis of the planner, the bomber, the sovereign. It asks us to consider what forms of derealisation and coercion depend and thrive on horizontality.[33] And it suggests, more broadly, that an ethics of situated perception might oppose the totalising view from nowhere without ceding to that view, or to those viewers, the project of comprehensive portraiture.

As of December 2009, the largest spherical-panoramic image in the world was Jeffrey Martin's 18.4 gigapixel photograph of Prague.[34] Shot with the help of a robot platform at the top of the Žižkov TV Tower, the image is oblique in varying degrees, approaching vertical only when one navigates so as to look at the ground at the base of the tower. Nonetheless, such images may be the closest we can come to experiencing the delirium of Fairchild-era aerial surveyors as they first shot and then assembled the first detailed photomosaics of US cities and got lost in the plenitude of detail. In fact, Martin's 2009 image is itself a photomosaic: 'Hundreds of shots were shot over a few hours', says the accompanying description; 'these shots were then stitched together on a computer over the following few weeks.' That stitching was done by sophisticated software called PTGui, specially designed not only to fit digital stills into navigable 360x180-degree panoramas but also to correct *automatically* for tilted and rotated images and to blend seams, match tone, mask anomalies: to do digitally, in seconds, what interwar aerial surveyors did over many hours using backlit ground glass, special printers, stretchable photographic paper, dodging and burning, scissors, pins and glue. Yet because its hundreds of constituent stills were shot over the course of several hours, the Prague panorama—like Google Earth, at high enough resolutions—is vexed by weird effects of temporal parallax that descend directly from interwar aerial surveying. To take a single, haunting example: not far from the TV Tower, west by south-west, a roundabout called

Škroupovo náměstí encloses a circular park, on whose perimeter walk-way a grey-haired man sits on a bench reading a book. Just in front of him are two pedestrians clad in baseball caps, blue backpacks and jeans. Both carry multi-coloured plastic shopping bags. Two pedestrians, five strides apart—who are clearly the same person photographed twice. (Maybe it is you.) Martin's digital SLR camera, turning incrementally on its automated tripod, has caught one person in two consecutive images a second or two apart, and PTGui has stitched them together so the seam between the two images is invisible. But the software does not automatically paint out optical twins, and Martin has either missed this instance of inadvertent doubling or mischievously let it stand. Yet it is not the pedestrian's street-level viewpoint that gives us this anomaly or Easter egg, but the pedestrian as captured from a high vantage whose supposed immunity to time founders on those two proximate bodies and falls to pieces. Verticality undone by verticality. So there you are—

Image 17: Detail from Jeffrey Martin's 18.4-gigapixel panoramic digital photo-mosaic of Prague, (Martin 2009).

having walked unwittingly in a few strides through the wall between two moments, two photographs—in two places at once. Against the static background of Prague seen in its structure, an *event*.

# SECTION TWO

# AERIAL AESTHETICS, DISTORTION AND THE VIEW FROM BELOW

6

'CONCEALING THE CRUDE'

AIRMINDEDNESS AND THE CAMOUFLAGING OF
BRITAIN'S OIL INSTALLATIONS, 1936–9

*James Robinson*

*Introduction*

When thinking of 'flight' and of the 'sky', one finds oneself instantly
catapulted into a world of dreaminess and fantasy, of imagining 'the
world up above' from 'the world down below'; one dreams about the
freedoms of being in the air, of becoming ungrounded from the terrestrial domain, with the imagined sights and sounds of flight fuelling a
desire to experience the boundlessness of the air. Aerial geographies,
spaces and practices seemingly elicit a wide range of feelings and affective responses. Within the contemporary world, the act of flying has
become very much conceived as an 'everyday' experience, with several
commentators drawing attention to the commonality of the flying experience. For example, Adey et al. argue that 'like other invented necessi-
ties—cars, television, internet—it is impossible to imagine life without

flight. Air routes provide the backbone of the global economy; the airport is the key entry point to any world city; flight enables goods, knowledge and people to *flow* across national boundaries.'[1] Coinciding with this, the experience of flight has been conceived as often dull and banal, even an experience which on occasion frustrates, angers and causes unease. Corn, for instance, writes that today, 'flight has become something we take for granted. Leaving on a jet plane, even for an exotic destination, kindles little romance or excitement. A flight is something to be endured, a forced confinement in a stressful and often uncomfortable environment'.[2]

Geographers, social scientists and historians have long expressed an interest in the 'geographies of the air', from work which has explored air travel and its effects upon global geographies to studies which have examined the spaces of 'aeromobility',[3] wherein 'the airport terminal has become the focus for much of this thinking, drawing analysis of what it means to inhabit these spaces as sites of alienation, strange encounters and inequality ... [as well] as a place of home, relative stasis and dwelling'.[4] Emerging alongside these interests has been a body of work which focuses on the concept of 'airmindedness'. Generally speaking, it has been contended by several commentators, such as Wohl and Fritzsche, that airmindedness was a state of mind which emerged in the 1920s and was fostered not only through witnessing the spectacle of flight itself, but also through multiple forms of material culture and aero-memorabilia.[5] Within the academic literature, discussions of airmindedness have tended to focus on its more positive dimensions, and the ways in which aviation technology was perceived as being able to transform the world for the greater good; as Corn notes, 'to be airminded, as contemporaries used the word, meant having enthusiasm for airplanes, believing in their potential to better human life and supporting aviation development'.[6] But as Edmunds observes, by the 1930s, it 'became something else. It became a way of thinking about the world which included the use of air transport and the threat of air power as every-day realities.'[7] It was not a simple enthusiasm for or an appreciation of aviation which the founders had sought, it was the unthinking use of aviation as a tool in the same way that other technologies are tools for shaping or relating to the physical world.'[8] In these terms, Adey argues, airmindedness can be 'conceived as a moral geography that worked to define particular ideas, beliefs and behaviours acceptable for one to attain citizenship of the nation, locality or city'.[9]

In this chapter I explore the ways in which airmindedness affects (and has affected) how we comprehend, connect with and transform our grounded spaces, practices and presences. Drawing upon a case study of British civil camouflage efforts during the late 1930s, I illustrate how imaginary and cultural conceptions of aeronautical technologies and the ways in which aviation was seen to enable new ways of interacting with terrestrial surfaces facilitated the modification of the British landscape during this period. In the first section I critically explore the relationship between the practices of civil camouflage during this period, and the ways in which its performance became entangled up in knowledge about the air and the aerial view. Following on from this, I historicise the specific British preoccupation with aeronautical technology, drawing upon a wide range of cultural constructions of the air to expose some of the specific characteristics and traits of the aeroplane itself which instilled fear into both the British populace and civil defence planners during this time period. Consequently, I examine how these anxieties about the aerial threat culminated in the emergence of the civil camouflage project, with concealment strategies being devised primarily to defeat the enemy eye in the sky. In the final section, I explore the attempts of civil camouflage practitioners to conceal an oil installation at Hamble, demonstrating how vertical visualities and 'thinking aeronautically' saturated all stages of the practice of concealment.

## Civil Camouflage: An Airminded Technology

During the 1930s and 1940s, a nationwide programme of civil camouflage was instigated and practised by the British government in an attempt to conceal and hide strategically important features in the 'Home Front' landscape. The term 'civil' or 'static camouflage' embraced a wide variety of artificial and natural features in the landscape, from water surfaces and white horses, through to 'factories, power stations, public utility undertakings and centres of communication, [with] the word factory being used to include any site on which war materials or essential commodities are produced'.[10] During this period, static camouflage was very much influenced by the rise of aviation, and, in many ways, can be conceived as an airminded technology. It was a technology and a practice which embraced the aeroplane and which encouraged its practitioners—the camoufleur—to 'think', 'act' and 'see' aeronautically.

147

Moreover, it acknowledged contemporary assertions that the aeroplane had enabled 'the emergence of a new form of spatial consciousness', facilitating new ways of interacting with and seeing the world.[11] For instance, Hauser has drawn attention to how Britons were increasingly utilising aviation and aerial photography to understand the British landscape in new and novel ways, positioning 'hidden' archaeological remains alongside modern structures.[12] By the 1930s, Britons had come to realise that the proliferation of such vertical visual experiences, conduced by the expansion of the uses of the aeroplane, were exposing the modern landscape in different ways from those of grounded perspectives. Mitchell Schwarzer, in his work *Zoomscapes*, has highlighted how the aeroplane has transformed the ways in which architecture and the built environment are engaged with, experienced and critiqued from the air. Indeed, he has remarked how some features which are 'prominent' when viewed from upward-looking and horizontal viewing positions become redundant and meaningless from the aerial perspective, while other characteristics and features become greatly emphasised; for instance, 'spikes and domes, pyramidal and stepped roofs do not dominate our impression, as they would in the view from the ground. Flat roofs come into their own in the aerial view, their smooth expanses packed with heating, air conditioning and ventilation equipment.'[13]

Civil camouflage, however, became much more than engaging in these new vertical visualities; it was a technology which encouraged the proliferation of such ways of seeing, coupling the imagining of the ground as viewed from the air with the emotive and affective responses to aviation. With respect to civil camouflage, these emotions were namely those of fear and anxiety, corresponding with the rising concerns of civil defence planners. Emanating from this, camoufleurs were encouraged to engage not only with vertical visualities but also with forms of knowledge of the aerial body and aerial practices. Through the adoption of an airminded mentality, static camouflage techniques produced their own moral geographies, shaping the behaviours of people on the ground and the ways in which they began to interact with places, spaces and with each other. Hence camouflage not only culminated in new ways of critiquing landscape—it also began to transform it. Within the academic literature on airmindedness, the airport is often taken as the dominant terrestrial expression, with airports being constructed as the embodiments of avian metaphors, as a 'meeting place of sky and earth', where a 'coterminous' relationship between the sky and the ground is forged.[14] Through an examination of civil camouflage, however, it becomes clear that the privileged position of airport spaces should be unsettled and destabilised, with other everyday, mundane spaces, places and landscapes being affected and transformed by airmindedness. In many ways, camouflaged spaces provide one such example through which we can attempt to look at how airmindedness can be translated in unique and distinct ways on the ground.

*'The Bomber Will Always Get Through': Airmindedness and Interwar Britain*

Death, so long a ground-floor tenant, has, as it were, moved to the top storey and made himself comfortable at home amongst the thunder and lightning of the clouds.

L.E.O. Charlton, 1935.

Throughout the interwar period, Britons marvelled at the wonders of aeronautical technology. During this time, writers such as David Edgerton have accentuated 'English enthusiasm, indeed over-enthusiasm, for the aeroplane.'[15] Indeed, like other airminded countries, interest in such technology was being promoted through engagements with aeroplanes at aerial spectacles and events, celebrations of the aerial exploits of Britain's airmen and women and the development of avant-garde, all-metal monoplanes. In 1929, for example, Alan Cobham, the renowned British aviator, visited 110 cities across the UK in his plane *The Youth of Britain* as part of his campaign 'Make the Skyways the Airways', providing an estimated 10,000 school children with their first experience of flight. Furthermore, an emerging programme for the development of civil airports also nurtured this passion in the aeroplane; Fearon notes how, in 1929, the Royal Institute of British Architects instigated a national campaign for aerodrome building, which 'by means of competitions, exhibitions, lectures, radio talks and many other devices, succeeded in conveying to the general public…the continental achievement in the development of airports and air services'.[16]

However, despite these efforts at promoting and fostering a wave of optimism in aviation, airmindedness within Britain was much more complex. Rather than being envisaged as some sort of 'winged gospel' as Corn has demonstrated with the case of the United States, enthusiasm

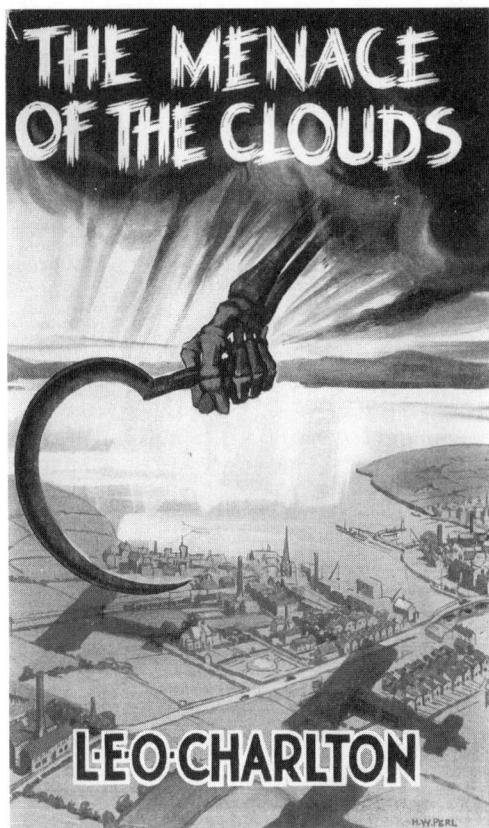

Image 18: The cover artwork for L.E.O. Charlton's *The Menace of the Clouds* (1937).

for the aeroplane within Britain was tempered by the realisation of its true potentialities. More specifically, there were some who conceived of the aeroplane as an 'invention of the devil', as an unstoppable 'death-dealing technology' capable of obliterating human civilisation. The imagining of the aeroplane in such a way is quite uniquely conveyed in the cover of *The Menace of the Clouds*, a book by the critic Air Commodore Lionel Charlton on air defence policy;[17] here H.W. Perl's illustration is presented in a red tone, with the hand of the reaper reaching out from the clouds, taking a swipe with his scythe at the exposed population below. Even more menacingly, the shadow which the reaper casts on the ground is the silhouette of a formation of bombers. In order to account for the emergence of such macabre and morbid representations of the aeroplane, it is important to draw attention to three particular dimensions of aeronautical technology which contemporaries emphasised in the production of such constructions.

Firstly, paranoia over the capabilities of aviation to cause destruction stemmed from imaginings of the technology of the aeroplane itself. In many ways, it could be argued that apprehensions about aeronautical technology were part of a long-standing, deep-seated uneasiness about the benefits of modern technology. Charlton's work, for example, expressed the view that:

it is gravely doubtful whether any great mechanical invention, considering the crazy civilisation which enfolds us, has really conduced to the ultimate happiness of mankind ... Steam, electricity and the internal combustion engine, ... have much to be said of in their favour; and yet the power of progressing with great or moderate rapidity from one place to another, by land, by sea, or by air, has not noticeably resulted in a kindlier commingling of the races.[18]

Certainly, what is represented here is the feeling that technological progress has and will always possess a dark side; rather than bringing people together, technology was considered to exacerbate and intensify tensions between nation states. Aside from the general concern about technological development, however, there were also specific elements of the aeroplane which contemporaries believed made them extremely dangerous. For Charlton, the potency of the aeroplane lay in its ability to be modified:

fresh from the maker's hands, for whatever purpose fashioned, be it for polar navigation, for mail transport, for private pleasure, or for a record-breaking purpose, they are instantly adaptable to deal death and destruction over a wide area, and to do so have only to fulfil the function for which they were contrived and leave the ground.[19]

This adaptability served to further reinforce the potency of this technology.

Following on from this were the ways in which the aeroplane made use of space and, subsequently, reconfigured spatial relations. The ability of the aeroplane to use a space which afforded it great freedom was conceived as posing an immensely complex problem for defence planners; Charlton writes of how 'the freedom of space above in all its immensity, and the power of evasion which that space bestows … must inevitably secure them the measure of their success'.[20] Such imaginings of aerial spaces as open, enabling unhindered movement, led to the contemporary assertions that 'the bomber will always get through', a line famously delivered to the House of Commons by Stanley Baldwin on 10 November 1932. For defence planners, this utilisation of a boundless space was perceived to have profound geostrategic implications, refashioning the ways in which space would shape the future conduct of conflict. The greater freedoms of the air extended conflict vertically, shifting emphasis away from 'traditional' strategies which were dependent upon horizontal trajectories and topographical features. The aeroplane, therefore, was constructed as a technology which produced new 'geographies of vulnerability',[21] facilitating 'the bodily removal of the shield of protection provided by the forces in the field, behind which, even though the people cowered, they were at least comparatively safe'. For defence planners, the aeroplane challenged and undermined Britain's geostrategic advantage of being an island, and effectively signalled the end of a dependency on the Royal Navy for protection.

Finally, the affective realities and cultural constructions of an aerial attack itself also fuelled anxieties over the destructive capabilities of the aeroplane. For many Britons during the interwar period, the enduring memories of aerial bombardment during the First World War were still pertinent. Aerial attacks on Britain had commenced with the Zeppelin and Schutte-Lanz airships, and had been followed by attacks in May 1917 by Gotha aircraft bombers, which had exposed great difficulty in dealing with this threat. In the first attack of this nature, 'defending machines from Hendon ascended and attempted to get on terms with the invaders, … the enemy escaped scatheless and with their formation in tact'.[22] Further events on the continent throughout the interwar period, such as the devastating attack on Guernica during the Spanish Civil War, reconfirmed to many the annihilation that the modern aeroplane could deliver. These realities of aerial destruction were additionally reaffirmed through popular mediums: the new 'talkies' testified not only to the visual but also the aural experience of an aerial attack. For example, the feature film *Things to Come* written by H.G. Wells, and produced by Alexander Korda, is saturated with sights of obliteration from the air; the scene where an attack is conducted on the city of 'Everytown' is described by Robert Wohl as:

one of darkness, screeching sirens, ineffectual anti-aircraft fire, and ambulances careening through the streets. Gas bombs are dropped; pedestrians run about in all directions, their eyes filled with panic; buildings crumble; masses of people force their way into a tube station, like frightened animals taking refuge in their burrows in the ground.[23]

Scenes such as this reinforced established imaginings of an aerial attack and continued to underpin contemporary assertions of the devastation which an aeroplane could inflict.

In terms of the response to such a threat, counter-attack had been proposed as the only solution: 'the only defence is in offence', voiced Baldwin, 'which means that you have got to kill more women and children more quickly than the enemy if you want to save yourself'.[24] Although this had been advocated as the only option by the Air Staff, other departments questioned the effectiveness of such a decision. Instead, government departments such as the Home Office proposed the use of alternative strategies, combining active and passive forms of defence. The adoption of civil camouflage was one such proposition.

*Countering the Aerial Threat: The Emergence of the*
*Civil Camouflage Project*

On 9 July 1935 the Home Office distributed the first government circular on Air Raid Precautions (ARP), signalling 'an invitation to local authorities and to private employers to cooperate with the Government in creating A.R.P. machinery; and to the public to learn the rudiments of protection and to volunteer for A.R.P. duties in their districts'.[25] This represented the first attempt by the British Government to devise defensive measures which would work to lessen the effects of an aerial attack. As part of these measures, it was decided that civil camouflage could be deployed as an ARP strategy. Civil camouflage was conceived as 'the only possible means of giving protection to the majority of industrial installations, since physical protection was out of the question on the score of the immense cost involved'.[26] Work began on the development of civil camouflage with the setting up of the Camouflage Sub-Committee of the Committee of Imperial Defence in October 1936, which was 'to undertake the direction and control of research into and experiments in connection with new methods, or the improvement of existing methods, of camouflage in its application to the needs of ... passive defence'.[27] This was followed by the creation of the Camouflage Branch at the Royal Aircraft Establishment, Farnborough, on 14 December 1937. Under the supervision of Colonel Francis Wyatt, a veteran camoufleur who had been responsible for the British Camouflage Division in France during the First World War, this Camouflage Branch was assigned the task of investigating 'the principles of camouflage and [to] decide how far they could be met by the use of paint applied in appropriate patterns'.[28]

From the outset it was acknowledged by those working for the Camouflage Branch that static camouflage was to be a practice and a technology which would be directly influenced by aerial spaces and practices. It therefore became necessary to integrate aerial knowledge and understandings of vertical visualities into the heart of civil camouflage practice. As a first step, it was highlighted that static camouflage should be devised to operate against the aerial observer only. This marked a significant change from First World War conceptions of camouflage, in which the purpose was to deceive the aerial photographer as to the position of field guns, army movements and so forth. Although the Sub-Committee recognised 'the impossibility of camouflaging large installations ... with anything like the same effectiveness as can be done with minor field defences, owing both to their size and to their fixed positions in relation to known and often conspicuous landmarks such as rivers, railways and canals', the Sub-Committee, nevertheless, felt that 'the object of static camouflage should ... be, not an attempt to conceal the target completely, but so to disguise it as to mystify and mislead the pilots of attacking aircraft'.[29] The emphasis on the bomber body meant that it was essential for knowledge of the bombing run and of the specific visualities of the bomber body to be collected. Through the gathering of intelligence from RAF sources, and from discussions with Air Staff representatives at Sub-Committee meetings, it was argued that camouflage had to render a target inconspicuous from an aeroplane 'approaching at 180mph from a distance of 4 miles and flying at an altitude of 5,000ft'.[30]

More importantly, however, was the recognition that 'the bomber's view was quite different—fleeting and more oblique' than that of the expert analysing aerial photographs. It was noted that, from his elevated position in the air, the airman would see and critique the landscape in a different way to the grounded observer examining the world around them. Mirroring the assertions of Schwarzer, it was contended that the vertical and oblique visualities of the bomber body were likely to accentuate some natural and artificial features over others. Civil camofleurs were, therefore, encouraged to appreciate and acknowledge the significant ways in which the aeroplane transformed the means of seeing and interpreting the landscape. For example, in a summary report on camouflage research, it is emphasised that:

in comparison with view from the ground, the view from the air is the bird's eye view not only in the sense of range and comprehensiveness but also in respect of the angle of observation. The airman sees more but he also sees less than the man in the street. Objects which from the ground appear well concealed may be conspicuous from the air, for reasons which are difficult to appreciate by an observer unpractised in air observation. The converse is also true; a well known landmark on the ground may be inconspicuous from the air.[31]

Differences in colour, tone and texture, form and the presence of shadows, were also identified as characteristics which led to the giving away of features in the landscape when viewed from the air, often enhancing the prominence of artificial features when compared to their natural surroundings. Other clues in the landscape were also identified. Roads, for instance, were highlighted as constituting a guiding mark for enemy aircrews: 'in England, for example, the so-called Roman roads which run straight over the countryside, disregarding contours, are valuable guides to navigation and identification; they may appear as long, continuous, uniform and regular lines, marked rather by the hedges and trees which border them than by the surface of the actual track'.[32] In addition, railway tracks, ground scars, water surfaces and the presence of steam and smoke were also highlighted as prominent features in the landscape which could aid the bomber crew in navigating towards their target, as well as facilitating the identification of the target itself.

In order to deal with these features, a variety of solutions were proposed, many of which go beyond the scope of this chapter. For the period under consideration, it was proposed that buildings would be generally treated with paint-based solutions, with two different camouflage strategies being put forward: imitation and disruption. Imitation entailed producing an effect whereby the structure would be painted to appear in harmony with the surroundings which necessitated drawing upon neighbouring features, either natural or artificial, in order to achieve this. Disruption, on the other hand, was a simpler version of 'Dazzle' which had been used to conceal ships and artillery guns during the First World War and entailed distorting the shape of the building through sharply contrasting colours and tones of paint. In 1938, however, when the Camouflage Department commenced its work, the effectiveness of these two design schemes remained hypothetical as they had only been trialled on models. In order to judge the effectiveness of these two design principles when viewed from the air, the Camouflage Branch recognised the need to carry out full-scale experiments.

*Camouflaging Oil Installations and the Hamble Experiment*

From the outset, the Camouflage Sub-Committee was consciously aware of the importance of testing out new and existing techniques of camouflage on actual industrial buildings. It was contended that such experimentation was 'essential if the efficacy of camouflage when viewed from the air was to be determined'.[33] At the end of their first meeting on 30 October 1936, the Sub-Committee requested that the Air Raid Precautions Department (ARPD) 'submit a list of possible targets for air attack, the camouflage of which they consider to be of the most importance and which would be suitable for initiating experiments in the subject'.[34] On 19 November 1936, this request was fulfilled by the ARPD, who submitted a memorandum listing typical factories which they believed would benefit from camouflage work being carried out on them. This list consisted of a large variety of industrial undertakings, from motorcar, aircraft and munitions factories, through to blast furnaces, gasholders, power stations and railway works. At the following meeting, this list was discussed with a view to selecting a feature which could be used to gain valuable research experience and build up confidence in the use of camouflage as a means of defence against the aeroplane. At first, interest was expressed in the camouflaging of gasholders, with these being described as being 'in the majority of cases, outstanding marks, and even if not of vital importance in themselves, act as leading marks or pointers to other vital spots'.[35] Furthermore, it was expressed that 'the public was very concerned about the danger of widespread damage in the event of gasholders being bombed. Moreover, where industry was dependent on gas, gasholders would be a very vital target.'[36] Despite this initial interest in attempting to camouflage a gasholder, the Sub-Committee concluded that 'it would be desirable to tackle the easier problems first, leaving such things as gasometers, which were obviously more difficult, until more experience had been obtained'.[37] Eventually, the decision was taken for the new experimental establishment to undertake 'full scale trials on an oil farm and some factory of national importance'.[38] Subsequent indecisiveness to select a particular factory that suited the needs of experimental work would result in the Camouflage Branch only trialling camouflage principles on a single oil installation by early 1939.

Of all the modern features within the British landscape during the latter half of the 1930s, oil installations posed several unique design challenges for the camoufleurs and there was a great deal of scepticism as to the extent to which effective concealment could be applied to such features. It was argued that oil tanks posed 'two great difficulties': first, their strong cast shadows, and second, their geometric lines.[39] In addition, oil tanks were rather large features, being approximately 40 feet (approximately 12 metres) wide and 20 feet (approximately 6 metres) in height. In terms of colour, the majority of UK-based tanks were painted in a standard Navy Grey, which was quite light in tone and had a tendency to increase their conspicuousness when set against the often darker tone of the surrounding landscape. An additional critique of these structures was their geographical situation. This was an issue which was raised by Colonel Turner, in a letter to the Sub-Committee, who wrote that:

I would like to state very definitely that I am of the opinion that it is not possible to camouflage tank farms of the usual type at any reasonable cost. I consider that camouflage and defence must go together and that in many cases it is far cheaper to defend or protect a building than camouflage it. This is especially the case where natural features give away a site completely even if camouflage is adopted and is found to be otherwise completely successful.[40]

In the case of many oil installations, a large number were located within estuarine areas, in which reflections from water and wet, shiny surfaces coupled with the distinctive shape of these physical environments would aid in identifying and revealing their geographical location. In some cases, such as the oil installations at Thameshaven and Shellhaven in London, any attempts at concealment were abandoned on account of this.

Despite these challenges, the Sub-Committee proceeded to pursue the agenda for the camouflage experimentation on a relatively small oil farm. The 'desiderata' were that the potential site 'should consist of three or four tanks of the normal commercial size, and to facilitate observation the installation should be isolated and in fairly open country'.[41] Several sites were suggested at locations such as Gosport, the Medway, Falmouth, and Killingholme, Lincolnshire. Eventually, the decision was made by the Sub-Committee to commence experiments on seven oil tanks at Hamble, outside Southampton, and they proceeded to obtain permission from their owner, Shell-Mex, to begin camouflage work.

Before the work was carried out, Wyatt had carried out an examination of a booklet *Methods of Concealment of Oil Fuel Tanks by Using Colour Only* (1923) with the aim of discovering a camouflage design 'which when applied, will ... merge the unmistakeable foreign appearance of the unconcealed depot in harmonious continuation with the natural relief of the background, and so present an unbroken ground picture to the eye of the airmen'.[42] Throughout the booklet, several methods of imitation were proposed. For example, the idea was advocated that tanks be 'dressed up' as residential buildings to make them merge into their urban surroundings. For more isolated or rural locations, other imitative strategies were suggested, from the carrying over of nearby service roads over the tanks in an attempt to divert attention from them, through to the application of an allotment design. In this case, it was argued that 'viewed from the air, the location would have a chess board appearance—small rectangles of different tone'.[43] Colours used as part of this design had to be in keeping with the natural surroundings, and therefore they had to consider seasonal variations. Of all the imitative techniques suggested, this allotment method was the only one which had been employed experimentally. In the early 1920s, an allotment pattern applied to Admiralty tanks at Gosport had proven to be ineffective, with the design making 'the tanks if anything more conspicuous'.[44]

In the light of Wyatt's investigations, preparation work began on the Hamble site. At the Sub-Committee meetings, various solutions were suggested, from dusting the surfaces of the oil tank with powdered slate to reduce glare, through to the erection of vertical screens to break up their cylindrical form. In the end, it was decided that experimental work would be confined to the use of matt paint, with the intention of 'test[ing] theory as to the kind of pattern to be adopted for large objects of this nature and also to acquire information as to the suitability of various paints'.[45] Although confined to paint-based solutions, an attempt was to be made to explore the addition of textural material to the paint mix-

Image 19: An illustration of the concealment of oil tanks by imitation of allotments. (Reproduced from *Methods of Concealment of Oil Fuel Tanks by Using Colour Only*, 1923, The National Archives).

ture. At Hamble, four paint types were trialled: ordinary matt paint, two types of sanded paints and a sanded distemper, which was 'used to see whether such a material would last sufficiently well as a paint to be used in an emergency only'.[46] A final method, which was also proposed, was the application of coloured sand to a drying varnish finish. In a move away from the imitative methods which Wyatt had been researching, a disruptive pattern of grass green and earth brown was to be applied.

Before the application of the scheme to the site, however, it was argued that aerial observations of the site were essential to obtain an understanding of how the oil farm appeared from the sky, as well as to identify prominent landmarks in close proximity to the site. Following the reconnoitring of the site by Wyatt, it was noted that 'treatment of the buildings alone will not provide adequate camouflage'.[47] In the area surrounding the oil tanks at Hamble, it was remarked that there was the Armstrong aircraft factory and the Hamble aerodrome, both of which, it was argued, would require 'landscape gardening' to ensure the complete success of the scheme. It should be highlighted here that despite these observations no attempt was made to conceal these additional features. Assessment from the air also followed the application of the scheme, where a judgement of the efficacy of the design was made through the adoption of the same critical vertical perspectives as the bomber body. For the evaluation of the Hamble camouflage scheme, Wyatt was able to enlist the services of a Heyford Bomber crew from Coastal Command to carry out a 'normal bombing run' over the site and make assessments of the location. These assessments were to demonstrate the negligibility of the camouflage efforts which had been undertaken.

In the review of the work which had been done, it was highlighted that these flights had provided useful insights into the visual appearance of the different techniques deployed when viewed from the air. The simple matt paint treatment was exposed as particularly dangerous: in the Second Progress Report of the Camouflage Department it was commented that 'this was proved as expected, a failure as it shone in all lights'.[48] Following on from this, aerial observations showed the two sanded paints and the sanded distemper to be 'good', although Wyatt questioned the durability of such methods: he wrote that 'it remains to be seen how they will weather'.[49] It was, however, the coloured sand on varnish treatment which was revealed to be by 'far the best', but concerns about its expensive cost raised issues about the economy of its more widespread use. More importantly, however, was the overall impression that these aerial assessments gave of the Hamble site. Due to the fact that the aerodrome and aircraft factory had not been considered in the camouflage design, as well as some of the tanks on the farm itself, the effectiveness of the camouflage remained questionable; the crew of the reconnaissance plane spotted the location from 3 to 4 miles from the site and described in their account how they had been 'guided onto the target by the aluminium tanks but doubted that they would have picked them up otherwise if they had been off course'.[50] In further critical reflections of the work which had been done, the challenges of concealing an operational industrial undertaking had also been uncovered. Wyatt, in particular, was not entirely satisfied with the 'fastness' of the different paint types used. By this, he meant the time taken for drying; subsequent discussions with Shell-Mex 'attributed this to a combination of salt air and heat—the tanks which contain heavy oils are kept at 100°F (approx. 38°C)'. In short, the Hamble experiment, while confirming the importance of assessing features from the air to identify conspicuous characteristics and verify camouflage effectiveness, also revealed that a significant amount of work needed to be done if the issue of camouflage for oil farms was to be resolved effectively.

*Conclusion*

Throughout this chapter I have demonstrated how a study of the everyday practices and spaces of civil camouflage during the late 1930s can not only contribute to our understandings of the historical and cultural constructions of our airspaces, but also the effect that aerial spaces and practices can have upon our everyday horizontal topographies. Indeed, it has been argued that terrestrial expressions of airmindedness can be found beyond the confines of the airport, with our everyday grounded geographies being affected by the emergence and sustained utilisation of aeronautical technology. Moreover, the ways in which we imagine, or are affectively and emotionally moved by, these spaces can have an effect on the ways in which we choose to understand, critique and transform our surroundings. Through a case study of civil camouflage, I have illustrated how the fear and anxiety caused by the imagined presence and destructive capabilities of the aerial bomber provoked the emergence of the camouflage project in 1936, with airmindedness becoming deeply engrained in the camouflage process. Indeed, civil camouflage becomes a response to a perceived threat not just from the aerial destruction, but of also being viewed from the air. Furthermore, I have explored how the proliferation of vertical visualites and new ways of sensing and interpreting the landscape from the air provoked civil camouflage practitioners into transforming and modifying industrial structures to merge them into their natural surroundings. In doing so, I have argued that terrestrial forms themselves become reflections of airmindedness, as expressions of the feelings and associations we make with our aerial spaces. It is clear that we need to continue to develop our understandings of our aerial geographies and expand our insights into how they move us emotionally and affectively on the ground. Moreover, it is important to recognise that aerial spaces and practices have far-reaching consequences upon the ways in which we come to understand and transform our everyday experiences on the ground, from shaping the ways in which we interact with others to the ways in which we choose to constitute our built environments.[51]

# 7

## FLYING INTO THE UNKNOWN

### CINEMATIC CULTURES OF WAR AND THE AESTHETICS OF DISAPPEARANCE

*John Armitage*

The pilot Jean-Marie Saget declared recently in an interview with *France-Soir*: 'At that time being a test pilot meant really *flying into the unknown* … but now we've got to deal with another frustration; it's a pity, but we can't fly the planes of competing companies because of the commercial competition. *I've never taken off in an F15*, I'm sorry to say. The company pilots, on the other hand, fly everything, because they've got be able to make comparisons … they're privileged characters'. At the moment of leaving and climbing into his Mirage 4000, Saget adds as a kind of goodbye: *I'm going to the other side*!

Paul Virilio, *The Aesthetics of Disappearance*, (2009) [1980][1]

### *Introduction*

Film and cultural theorists of war do not usually associate either with the name Paul Virilio, the contemporary French 'critic of the art of technology', or with his conception of the 'aesthetics of disappearance'.[2]

Image 20: Predator 2, from Joy Garnett's Predator Series (Garnett, 2013).[3]

This chapter examines Virilio's theoretical contribution to the debates over contemporary cinema and war by considering several of his most powerful texts, including *War and Cinema: The Logistics of Perception* (1989 [1984]), *The Aesthetics of Disappearance* (2009 [1980]), *Desert Screen: War at the Speed of Light* (2002) and *Strategy of Deception* (2000).[4] The theme of Virilio's texts is centred on the aesthetics of disappearance, and cinema as the 'escalation' of 'the modern technical beyond' in the world's advanced cultures. Virilio's texts are therefore about how diverse ways of perceiving and coping with the realms of cinema, aesthetics and disappearance are appreciated and incorporated into postmodern culture, a concern Virilio shares with other cineastes, aestheticians of disappearance, as well as theorists of war such as Jean Baudrillard.[5]

*War and Cinema, The Aesthetics of Disappearance, Desert Screen*, and *Strategy of Deception* are gradually becoming some of Virilio's most extensively read, culturally important, influential and contentious books. Originally published in French in the 1980s and 1990s, they are at last attracting comments from theorists in various subjects. Indeed, they are not only providing the setting for many contemporary explanations of the aesthetics of disappearance but also provoking debates that are influencing how subjects like logistics and the philosophy of perception carry out their research.

Perhaps the principal claim of these texts is their description of the aesthetics of disappearance, cinema and the escalation of the modern technical beyond as 'flying into the unknown', just as an equivalency was emerging between cinematic visual sensory perception and the evolution of machine guns, battlefield observation balloons (see Kaplan in this volume), balloons outfitted with an aerial-mapping telegraph, camera-kites, camera-pigeons, camera-balloons; in short, views from above.[6] Before presenting an explanation of what Virilio means when he employs concepts such as the aesthetics of disappearance or the escalation of the modern technical beyond, it is vital to grasp how this assertion stems from *The Aesthetics of Disappearance* in particular. Consequently, the purpose of this chapter is to offer a foundation for an appreciation of what Virilio means by defining cinematic cultures of war and the aesthetics of disappearance as *flying* into the unknown.

*Cinema, War and the Aesthetics of Disappearance*

Cinema, both moving films and the entire network of institutions and people engaged with film production, indicates, for Virilio, an instant of technological and psychological shock comparable to that of weaponry. Cinema is, therefore, a doorway to his thoughts on modern and postmodern war in *War and Cinema, Strategy of Deception* and *Desert Screen*. Virilio argues that the techno-psychology of shock that is cinema and war have cultivated a deadly mutual dependence. But how can we consider such lethal interdependence? What does cinema do to film and cultural theory and war? And what kinds of film and cultural theory or war are feasible after the invention of cinema?

Virilio's examination of these questions in *War and Cinema* commences not with the birth of cinema itself, though, but with the case proposed by that German Nazi cineaste and Reich minister of propaganda, Paul Joseph Goebbels, who, during the Second World War (1939–45), initially tried to ban the showing of films in colour to the German populace. In an instance of ministerial propaganda, this 'patron' 'of the German cinema, banned the showing of the first film in Agfacolor, *Women Are Better Diplomats*, on the grounds that the color was depressing and of wretched quality'.[7] By contrast with 'American Technicolor', Virilio notes, 'the German process struck Goebbels as nothing short of *shameful*'.[8] Virilio indicates how, later, Agfacolor's new colour stock was enhanced and how 'it enjoyed enormous success in occupied Europe':

In 1943, to mark ten years of Nazi cinema and the twenty-fifth anniversary of UFA (Universum Film AG), J. von Baky solemnly presented *The Adventures of Baron Münchhausen*, a high-budget Agfacolor film with a large number of very accomplished special effects.[9]

For Goebbels' wartime double anniversary of Nazi cinema, then, the only UFA film acceptable for showing was one concerning the escapades of a well-known fictional fantasist which was given the expensive Agfacolor film treatment complete with numerous consummate special effects. Nevertheless, Virilio reminds us that UFA 'had been founded during the First World War, in 1917, and that in the following year it became the main [state and arms industry subsidised] complex of cinematographic production, distribution, and development in wartime Germany'.[10] 'At the height of total war', therefore, 'it seemed to Goebbels and to Hitler himself that the rescuing of German cinema from black-and-white would provide it with a competitive edge against the tonic power of American productions.'[11]

Concentrating on the banning of screening films in Agfacolor to the German public by a Nazi cineaste is perhaps a strange opening for a book on war or cinema. Why does Virilio reflect on and react to Goebbels' military and political propaganda? After all, there is considerable historical evidence concerning this 'patron' of the German cinema, and the contentions of those who sought to ban the showing of the first film in Agfacolor have correctly been disclosed as politically motivated and technologically determined.[12] So what is Virilio attempting to accomplish by writing about Goebbels' quarrel with American Technicolor?

Virilio's argument regarding Goebbels is primarily a comparative one. It invites readers to consider the problem of how to react to the war over national cinematic processes by evoking Goebbels' contention, which obviously aimed to refute American technological superiority. Additionally, it shows the capability of national shame to 'improve' Agfacolor stock, and introduces the question of whether military success may well be tied to cinematic success. Equally significantly, however, it permits Virilio to broach numerous issues about the nature of war, politics and writings on cinema generally. Certainly, Virilio gives no credibility to Goebbels' standpoint. Nonetheless, his position concerning Goebbels' thinking exposes several of the possible inadequacies of theoretical perspectives on film, culture and war from the point of view of a conventional explanation of cinema financing or filmic special effects. For Virilio claims that film and war are not only linked to the state and to the arms industry but also to cinematographic colour, and that the 'powerful mimetic faculty of wartime American cinema' in particular 'was a kind of perceptual luxury' that was 'quite distinct from other forms of spectacle and entertainment'.[13] Indeed, for Virilio, cinema is an intangible yet recurrent indulgence that we find difficult to forgo.

Virilio argues that the Nazi leaders appreciated cinema extremely well: 'they placed actors and directors under military discipline right from the outbreak of hostilities, any absence from the studios being regarded as an act of desertion and punished accordingly'.[14] Yet the shift from monochrome to colour in order to compete against the tonic power of American productions was not Goebbels and Hitler's only problem with German cinema. For Goebbels and Hitler also 'had a contemptuous relationship with the cinema people',[15] many of whom were not Nazis but communists and/or Jews who finally met a tragic end by way of suicide or deportation to Nazi concentration camps.

We might object at this point that Virilio rehashes or exaggerates well-worn connections between war and cinema. Film historians, for example, have long recorded the coercive ways in which costly German movies continued to be made until the end of the Second World War. Moreover, it is well known that these final pictures were still being projected in the only remaining Nazi citadels even when many cinemas in the cities of the Third Reich were being bombed by the Allies.[16] But this is not all that is at issue. In fact, Virilio is making a more multifaceted argument that extends far beyond any concern with expensive Nazi

films being shown at the end of the Second World War in 1945. Goebbels and Hitler, he suggests, even when 'staring military collapse in the
face', still wanted their movies to be the 'greatest of all time, spectacular
epics outrivaling the most sumptuous American super-productions'.
Hence, in Virilio's estimation, during the Second World War *something
new occurred in cinema*, which was the Germans' 'obsession with the
American perceptual arsenal'.[17] Not only was there an effort to read
America's magazines and newspapers, correspondence and the international press, but there was also an attempt to *surpass its filmic universe.*

Simultaneously, within the United States, the production and circulation of propaganda movies was increasingly commandeered by the military. Furthermore, and somewhat surprisingly, Spanish surrealist film
directors such as 'Luis Buñuel could be found in 1942 shooting docu-
mentaries for the US Army, while Frank Capra moved from his interwar satires (most notably with Harry Langdon) to the ponderous
didacticism of *Why We Are Fighting* (1942–45).'[18] Even more blatant
were 'the songs and dances of Fred Astaire' that were merely 'disguised
calls for a new mobilization'.[19]

Yet, beyond any proof or data that can be cited concerning the
'aggressive colors' of these American films, frequently regarded by Europeans as signs of American vulgarity, there is an 'energy of the visible'
radiating from them, an energy of the visible given out from the United
States itself that 'made them into veritable "war paintings"'.[20] If we focus
on the cinema of the Second World War, Virilio claims, this energy of
the visible can be characterised as a form of 'logistics', as a kind of perceptual, militarised and immediate 'charge', in the sense of 'excitement'
or 'arousal'. The mission of this energy of the visible was to supply and
instil cinema audiences with new military vigour, to transform instantly
and to 'wrench them out of apathy in the face of danger or distress, to
overcome that wide-scale demoralization which was so feared by gener-
als and statesmen alike'.[21] However, American film, like war, is not
simply one more magical 'capacity for movement', as Virilio calls both
film and war, in the ongoing history of human perception.[22] It is also
directly linked to total war and to the economy. Put differently, following the Second World War and the dropping of atomic bombs by the
United States and its Allies on Japan at Hiroshima and Nagasaki in
1945, Virilio investigates the politico-economic and cultural repercussions for modern and postmodern war and cinema of the full flowering
of the aesthetics of disappearance.

The aesthetics of disappearance? To be clear, for Virilio, the concept of the aesthetics of disappearance refers to the mediated technological effects typical of the contemporary arts. Whereas the ancient aesthetics of *appearance* was based on lasting material supports (wood/canvas in the case of paintings; marble in the case of statues etc.) the present-day aesthetics of *disappearance* is founded on temporary immaterial supports (plastic/digital storage in the case of films etc.). Contemporary images therefore do not so much appear (except as a function of human cognition) as continually disappear. Modern-day images thus apparently move across but actually and repeatedly vanish from the fundamentally immaterial support of the screen as part of a cinematic sequence. In the case of film, such images disappear at twenty-four frames per second or, in the case of special effects, at sixty frames per second and above.

In Virilio's interpretation, therefore, cinema, war and the aesthetics of disappearance become the basis for a new sort of film and cultural theory, for a barely describable wartime sadness directed at a future that might yet require an examination of cinematic destruction, a deliberation on the endurance of the aesthetics of disappearance, and lessons in grief amid the no-longer mediated debris of nuclear war. Beyond the Second World War and the Cold War, the Korean War, the Vietnam, Gulf, Kosovo and Iraq wars, all of which, of course, remain militarily vital, the now ageing German and US wartime propaganda films nevertheless continue as a form of logistics. As such, these films are a kind of perceptual, militarised and instantaneous 'charge' to be experienced by today's masses, from the United States and Europe to Latin America, Africa and Asia, many of whom find themselves seized by and responding to the terror and legacies of nuclear war, as well as its subsequent and seemingly unending 'deterrence' which, according to Michael Madsen's feature documentary film on nuclear waste, *Into Eternity* (2010), will last for the next 100,000 years.

## *The Aesthetics of Disappearance, Cinema and the Escalation of the Modern Technical Beyond of Flying into the Unknown*

The interpretations of cinema in Virilio's writings investigate the impact it has had on war and present-day cultural existence. For Virilio, cinema and war are not just capacities for movement that have long passed by their respective nineteenth-century and ancient development, following which military psychology and physiology or perhaps the inventor Etienne-Jules Marey's chronophotographic rifle, 'which allowed its user to aim at and photograph an object moving through space', can carry on as before.[23] Rather, if we are to think cinematically, particularly following Marey's military research into movement, then that thinking has to change. In a significant chapter from *War and Cinema*, 'Cinema Isn't I See, It's I Fly', Virilio explains the influence of cinema on the concept of *aerialised war*.[24]

Our feeling for cinema, the methods by which we systematise and elucidate our perception, are all essential to us, to our cultural preconceptions, to our understanding of ourselves and others, and also to our aerialised wars. Cinematic technologies thus relate the narrative of how, as individuals, we became *weaponised*; they situate our current weaponisation as a component of a photographic continuity and indicate our likely chronophotographic or weaponised futures. Such cinematic technologies can thus be understood as resembling the logics of targeting and, similar to the technology of photography, can take numerous physical, moving and spatial forms. In 'Cinema Isn't I See, It's I Fly', Virilio explains some of these forms. In fact, he associates them with the most important wars, inventions and weapons of the past century and a half, inclusive of battlefield observation balloons, balloons outfitted with an aerial-mapping telegraph, camera-kites, camera-pigeons, camera-balloons, chronophotography and cinematography aboard small reconnaissance aircraft.[25] Indeed:

By 1967 the US Air Force had the whole of South-East Asia covered, and pilotless aircraft would fly over Laos and send their data back to IBM centres in Thailand or South Vietnam. *Direct vision was now a thing of the past*: in the space of a hundred and fifty years, the target area had become a cinema 'location', the battlefield a film set out of bounds to civilians.[26]

Contemporary attitudes to cinema are therefore bound to systems of war and military values, each labouring towards their own totalising, aerialised and data-led idea of *indirect vision*. The examples Virilio provides in the chapter include the First World War, wherein 'D. W. Griffith was the only American [civilian] film-maker authorized to go to the front to shoot propaganda footage for the Allies.'[27] Additional illustrations involve the targeted late nineteenth-century 'motion demonstrations' of Billy Bitzer, in which he bound his cine-camera to the 'location' of 'the buffers of a locomotive traveling at full speed'; the German film director Carl Dreyer, who strove to 'create an *artificial* unity of time by means of a *real* unity of place, and thereby illustrating Walter Benjamin's observation about a kind of cinema which was able to "present an object for simultaneous collective experience, as it was possible for architecture at all times'"; and, finally, at the close of the nineteenth century, 'Oskar Messter, not having a camera', using 'the room in which he lived as a camera obscura by blacking it out and leaving only a tiny hole at the street side'.[28]

In *The Aesthetics of Disappearance*, Virilio explains these processes of systematising cinema in relation to the conceptualisation of a movement or *escalation* to indirect vision, to, effectively, what he calls the modern 'technical beyond' of flying into the unknown.[29] For Virilio, as broadly illustrated by the quotation at the beginning of this chapter, the modern technical beyond of flying into the unknown refers to the West's distinctive compulsive scheme and forecast toward a specific aesthetic technique at or to the further side of the physical threshold or scope of contemporary culture. It is characterised by a sort of mysterious special knowledge or quasi-religious understanding and anti-natural extreme derived from special effects or fantasies produced for cinema and television by props, camerawork and computer graphics and so forth. In short, according to Virilio, contemporary advanced cultures utilise cinematic technologies to 'transport' themselves (in the sense of joy, delight or ecstasy) elsewhere, to, in fact, anywhere except where they are or what might be called 'the known'. Furthermore, for Virilio, this modern technical beyond has been the coordinating principle for major wars, innovations and weapons systems since the nineteenth century. Yet, as Virilio highlights, the modern technical beyond of flying into the unknown is not, in truth, an *age* of technical ideas, but, instead, a *form* of technical thinking or projection to the 'far side' of the known, a mode of thought that can be portrayed through the indirect vision of the modern technical beyond of flying into the unknown that he defines in *The Aesthetics of Disappearance*.[30] Besides, as he maintains in the latter work, in the era of the aesthetics of disappearance the modern technical beyond of flying into the unknown has *increased* its effectiveness, and can now project its sensibilities globally. This denotes that the meanings of cinema enclosed within it must be reconsidered in the age of the aesthetics of disappearance, and this is what the chapter, 'Cinema Isn't I See, It's I Fly', at least in part, endeavours to accomplish. Consequently, in this chapter, Virilio reflects at one point on the issue of how soldiers during the First World

War (1914–18) were organised to kill enemy soldiers without seeing 'whom they were killing, since others had now taken responsibility for seeing in their stead'. 'What', asks Virilio 'was this abstract zone that Apollinaire accurately described as the site of a blind, non-directional desire?'[31] This is a very important question for any effort to debate the political and cultural relationships between, for instance, the soldiers themselves, who could identify this abstract zone 'only by the flight-path of their bullets and shells …' For here was 'a kind of telescopic tensing towards an imagined encounter, a "shaping" of the partner-cum adversary before his probable fragmentation'.[32] Virilio's response to the question, though, is that we should persist in paying attention to how sight is organised and to how it 'lost its direct quality', how it 'reeled out of phase', and to how 'the soldier had the feeling of being not so much destroyed as derealized or dematerialized', with 'any sensory point of reference suddenly vanishing in a surfeit of optical targets'.[33]

'Cinema Isn't I See, It's I Fly' provides numerous explanations as to why humanity is increasingly structured around such an awareness of cinematic seeing. It catalogues the various developments of the modern technical beyond of flying into the unknown, and alludes to their capacities for movement in the twentieth century that have hurled them into overdrive. Virilio thus offers many capacities for movement that have become a veritable 'logistics of perception' (his term for the military provision of imagery which, from the First World War onwards, developed into the counterpart of the military provision of ammunition, thus establishing a new weapons system derived from the amalgamation of means of war transportation [airplanes etc.] and cameras), thanks to the way that they escalate the logics upon which the modern technical beyond of flying into the unknown is set up. For example, Virilio argues that the artilleryman's attitude and logic of thinking became equivalent to the camera operator's regarding the understanding of reality: the use of 'lighting reveals everything'.[34] Hence, by 1914, and by the utilisation of lighting for military-cinematic movement and escalation, the reality and rationality of anti-aircraft artillery was such that guns were being combined with searchlights to form 'camera-machineguns'.[35] Here, then, the logistics of perception becomes unreal or irrational as it escalates into an indirect or cinematic vision of warfare. Likewise, in 'refining and often misusing' the then revolutionary concept of the *carello* or 'traveling shot', early filmmakers such as Giovanni Pastrone 'showed that the camera's function was less to produce images … than to manipulate and falsify dimensions'.[36] The travelling shot thus confirmed that 'the first difference between cinema and photography is that the viewpoint can be mobile … and share the speed of moving objects'.[37] After Pastrone, therefore, what was confirmed '"false" in cinema was no longer the effect of accelerated perspective but the very depth itself, the temporal distance of the projected space'.[38] Lastly, years afterward, Virilio argues, 'the electronic light of laser holography and integrated-circuit computer graphics would confirm this relativity in which speed appears as the primal magnitude of the image and thus the source of its depth'.[39] All of these instances corroborate the important coordinating principle of the modern technical beyond of flying into the unknown, verifying the system by which it shapes its construction of a cinematic movement or escalation towards a model of indirect vision.

These capacities for movement, for 'cinematic self-propulsion', thus become so many 'logistics' or symbols of the power and accomplishments of the connections between cinematic vision and every important aerialised war, technical advance and development in arms since the mid-nineteenth century. The basic values of the modern technical beyond of flying into the unknown are confirmed by its capacity for movement, by its 'fleeting aerial perspective' which Virilio calls 'dromoscopic', a speeding escalation that indicates the potentialities of the age of the aesthetics of disappearance, of the era in which vision, cinema and aviation *become one*.[40] The question Virilio implicitly broaches in the chapter is how to challenge the authority and activities of the modern technical beyond of flying into the unknown. He contends that, by 1914, there were numerous possibilities, and, in choosing between them, humanity decided on aviation as 'the ultimate way of *seeing*'.[41] Indeed, 'contrary to what is generally thought, the air arm grew out of the reconnaissance services, its military value having initially been questioned by the general staffs'.[42] This is why the concept of the age of the aesthetics of disappearance can apply simultaneously to, for example, the different imaging techniques of reconnaissance aircraft or to the supplying of mobile ground troops with visual information. In other words, the escalation of the modern technical beyond of flying into the unknown through its capacity to, for instance, manage and mobilise artillery barrages, or to take photographs, compels a reassessment of the purpose and organisation of cinematic investigation. Yet, the aesthetics of disappearance, cinema and war studies assume numerous diverse forms according to the philosophical and military ideas and aims of film and cultural theorists. How, then, should we approach and, more importantly, challenge, the modern technical beyond of flying into the unknown and the logistics of perception? Prior to relating Virilio's explanation of the aesthetics of disappearance and the logistics of perception, I shall delineate another important postmodern examination of cinema and the aesthetics of disappearance that will act as a useful comparator to Virilio.

### Cinema and the Aesthetics of Disappearance: Jean Baudrillard

One postmodern philosopher of the aesthetics of disappearance, the late Jean Baudrillard (1929–2007), contemplated the transformations of cinema, technology and the investigation of reality and history in the present period.[43] Baudrillard suggests that the significance of considering cinema lies in its relation to the primitive enjoyment that is the aesthetics of disappearance. Nevertheless, he does so from a different, almost anthropological, viewpoint and with divergent goals to those of Virilio.[44] Baudrillard argues for a turn to a postmodern examination of cinema, and he asserts that there is 'a kind of brute fascination' with cinema that is 'unencumbered by aesthetic, moral, social, or political judgements'.[45]

As a theorist of postmodern film and culture, Baudrillard contends that familiarity with cinema is vital for any deliberation on the cultural politics of aerialised war. Actually, he insists on the need to consider the concept of cinema as immoral, as a postmodern technical beyond whose analytical or primary political power resides in this immorality. Hence, cinematic moving images:

above and beyond all moral or social determination ... are sites of the *disappearance* of meaning and representation, sites in which we are caught quite apart from any judgment of reality, thus sites of a fatal strategy of denegation of the real and of the reality principle.[46]

Put differently, for Baudrillard, a consideration of the postmodern technical beyond is imperative if we are to reflect on the cultural politics of moving images and cinema with a view to thinking about the importance of how they simultaneously disappear yet somehow meaningfully occupy and represent our real, if increasingly indirect, everyday visual lives.

In his important lecture, 'The Evil Demon of Images', Baudrillard explains the aesthetics of disappearance and its culture of mushrooming images as a turn towards a possibly never-ending cinematic thinking that relishes a conception of cinema images as meaningless, as entities that can 'ultimately have no finality and proceed by total contiguity, infinitely multiplying themselves according to an irresistible epidemic process which no one today can control'. For the aesthete of disappearance, Baudrillard claims, cinema has become accelerated through images. Both the foundation for our culture and for our aerialised wars, cinema is mired in a 'mad pursuit of images, in an ever-greater fascination, which is only accentuated by video and digital images'.[47] So, for

Image 21: Predator 3, from Joy Garnett's Predator Series (Garnett, 2013).

instance, the movie *The Last Picture Show* (1971) portrays a 1950s vision of American manners and the atmosphere of small-town USA. But what makes European audiences feel somewhat doubtful is that the film is 'a little too good, better adjusted, better than the others, without the sentimental, moral, and psychological tics of the films of that period'.[48] Baudrillard, then, is surprised at his discovery that *The Last Picture Show* 'is a 1970s film, a perfectly nostalgic, brand new, retouched, a hyperrealist restitution of a "50s film"'.[49]

For Baudrillard, these are some of the essential problems of the aesthetics of disappearance. In a culture of postmodern remakes, cinema is nothing other than films that are 'better than those of the period' they imitate.[50] The point of Baudrillard's lecture is therefore to evaluate the aesthetics of disappearance as the *defeat of the cinematic imaginary* and to advocate a shift to a postmodern analysis of the simulated objects of the modern technical beyond.

Baudrillard's other relevant writings pursue a similar although not identical course.[51] He perceives the aesthetics of disappearance during the Gulf War of 1991 between the United States and its allies and Iraq, for example, as practically effacing the very possibility of contemplating cinema or the mass media as any kind of modern technical beyond. Baudrillard's work during the 1990s thus implicitly represents critiques of the aesthetics of disappearance, such as Virilio's, as rather ineffectual when faced with the delirium-inducing simulated or virtual cinematic imagery of the postmodern Gulf War. In *The Gulf War Did Not Take Place*, Baudrillard proposes that, during the war, he was 'reminded of *Capricorn One*', the film in which 'the flight of a manned rocket to Mars, which only took place in a desert studio, was relayed live to all the television stations in the world'. In other words, cinema and the mass media, considered by Virilio as depictions of movement towards indirect vision (what Virilio identifies as the modern technical beyond of flying into the unknown), have for Baudrillard turned actual aerialised war into its virtual double. Indeed, by means of the accelerated and increasingly sophisticated capacities for movement in operation during the Gulf War, the Iraqis, according to Baudrillard, were simply depicted 'as a computerized target' while the concealment of satellite information by the United States gave the *impression* of an instantaneous 'clean war' whose overriding purpose was in fact aimed at satisfying American duplicity.[52] (Baudrillard's own orientalist 'hollowing out' of the Iraqis and Islam during the Gulf War has been constructively critiqued by Almond).[53] In the era of the aesthetics of disappearance and cinematics, virtual war and television screens, Baudrillard argues:

> everything tends to go underground, including information in its informational bunkers. Even war has gone underground in order to survive. In this forum of war which is the Gulf, everything is hidden: the planes are hidden, the tanks are buried, Israel plays dead, the images are censored and all information is blockaded in the desert: only TV functions as a medium without a message, giving at last the image of pure television.[54]

Contemporary cinematics and the mass media thus render all capacities for movement immediately subterranean and instantaneously informationalised as they shift inevitably underground towards new military formations or further interments. Through such capacities for movement and with the accelerated aerial war in the Gulf apparently screening everything ('the planes are hidden'), neither war's survival nor any sense of its reality is guaranteed today. War somehow 'disappears' into the endless increase of interconnected tanks and entombments as cinematics and the mass media gorge uncontrollably on the military's perspective and propaganda disguised as explanation. In this paradoxical upsurge of military hardware and mass mediated 'discussion', where speeding US jets and Iraqi tanks appear to be only mirages or the equivalent of buried treasure, the reality of whole countries' capacity for movement vanishes, leaving mass audiences, as fascinated media 'consumers', with endlessly repeated, 'US military approved', and simulated yet instantly unmemorable cinematic images. For, as Baudrillardian scholars Toffoletti and Grace argue, when cinema and the terror of war are drained of their symbolic qualities, the outcome is audience indifference.[55] Here is the root of Baudrillard's infamous 'hypothesis' that the Gulf War 'would not' and 'did not take place' as, for Baudrillard, the simultaneously and ostensibly unrestrained yet actually suppressed media reporting, alongside the stunning deployment of imagery and information that appeared to show no injuries, blacked out an entire desert. For Baudrillard, then, the incessantly multiplying speculations, rival explanations and TV images put forward by media 'experts' with airtime and space to fill heralded the age of the 'medium without a message', the age and image of 'pure' TV, in which hyper-efficient arms industry weaponry and simulations, US military misinformation and 'reality' combined to produce an ultimately 'hidden' war.

*The Aesthetics of Disappearance and the Logistics of Perception*

Counter to Baudrillard's hypotheses, Virilio argues for the importance of persevering with reflecting on and writing cinema despite the escalation of the modern technical beyond. Distinct from Baudrillard, Virilio is not suggesting a turn to a postmodern analysis of the simulated objects of the modern technical beyond. Unlike Baudrillard, therefore, Virilio does not argue a case for effacing the very possibility of contemplating cinema as any kind of modern technical beyond. As the case concerning cinema, war and the aesthetics of disappearance introduced in the opening section of this chapter sought to demonstrate, the work of the critic of the art of technology and the aesthetics of disappearance is not to conjure up some new technical beyond subsequent to the modern technical beyond of flying into the unknown. Rather, it is to revisit repeatedly those capacities for movement that have produced contemporary conflicts, postmodern mass culture and national sentiments to discover in them our obsession with perceptual arsenals and our continual attempts to surpass those that already exist.

To continue to discuss cinema following the escalation of the modern technical beyond, Virilio again turns to his own thinking. At this point, Virilio develops an explanation of cinema and aerialised war that persists in imagining them as *interdependent* rather than merely as a haphazard sequence of unconnected capacities for movement. However, he repudiates the creation of a 'dromoscopic' or accelerated account of cinematic movement or military escalation that lies at the heart of the modern technical beyond. This, then, situates Virilio's version of cinema and aerialised war somewhere *other* than Baudrillard's cinematic and media-centred hyper-reality and his turn to an understanding of the postmodern Gulf War as one that did not take place. What permits this idea of cinema and aerialised war as interdependent without making it into a modern technical beyond is Virilio's distinction between notions of *actual* space and time, which concern explicitly 'old' modes of direct representation, and technologically revolutionary concepts that perform an aerial reconnaissance role. For the latter are modes of indirect vision or disappearance involving instantaneous or 'real-time' information. In other words, for Virilio, cinema and aerialised war, movement and escalation, are all ideas that create the system wherein particular physiological objects and our own actual bodies are often forgotten as their traces become accessible to a host of new capacities for movement such as vibration sensors, cameras and thermo-graphic pictures that identify objects by their temperature and so on. In short, as time-lags are lost in real time, real time itself breaks the constraints of chronology and becomes cinematic. The aim of Virilio's consideration of cinema and aerialised war is to broach the issue of how military information allows the past or the future to be interpreted and how human activity, its heat and light, can be extrapolated and escalated in time and space as specific capacities for movement.

Virilio maintains that without a conception of the escalation of cinema and aerialised war, our exploits in this realm would merely seem to be a directionless path involving, for instance, the fitting of spotlights under bomber wing-tips or landing gear. He argues that the link between cinema, aerialised war and capacities for movement must be underpinned by an awareness of:

the problem of knowing which, in the technical mix of chrono/camera/aircraft/weapon, would gain the upper hand in the making of the war film, and whether the topological freedom due to the speed of the engine, and later to its fire-power, did not create new cinematic facts incomparably more powerful than those of the camera motor.[56]

Put differently, if we cannot debate the relations between diverse capacities for movement in cinema and aerialised war, all we are left with are the seemingly subversive effects their actions produce on people and weapons, human perception, aesthetic form and creativity. When Virilio endeavours to explain the problem of 'the technical mix' in 'Cinema Isn't I See, It's I Fly', he contends that the significant point is not to enter into a discussion about cinema *or* aerial war. Rather, as he also suggests in *Desert Screen*,[57] war has always been 'linked to perceptual phenomena', or, more accurately, in *War and Cinema*: 'the war machine appears to the military commander as an instrument of representation, comparable to the painter's palette and brush'.[58] Cinema and aerialised war for Virilio then are involved with the importance of pictorial yet militarised representation, or *the militarisation of aesthetics*. Unlike the modern technical beyond, Virilio's stance on cinema and aerialised war is concerned with the escalation of the problem of the technical mix and indirect vision since, for him, they give rise to an *interdependent* drive towards an increasingly militarised technical beyond. What is necessary, therefore, is an understanding of the implications of the fact that 'the pilot's hand automatically trips the camera shutter with the same gesture that releases his weapon'.[59] As indicated by Virilio, this understanding must take the form of assessing the implications of capacities for movement that function in relation to cinema, logistics and escalation.

Virilio situates such capacities for movement that function as a logistics of perception in the relation between war, weaponry and the human eye, which took place at the beginning of the twentieth century when air forces began to cultivate their armed philosophy. What Virilio declares is vital regarding these capacities for movement is not their individual relation to war, the lightning advances of military weaponry and technology, and the human eye, but their *combined* relation to the 'violent cinematic disruption of the space continuum'.[60] For the growing interdependence between war, weaponry and the human eye 'literally exploded the old homogeneity of vision and replaced it with the heterogeneity of perceptual fields'. As Virilio writes in *Strategy of Deception*: the 'logistics of perception *on all fronts* has won out over the logistics of weapons targeted along a particular front'.[61] What acts as a form of logistics, of the necessity of escalation, in such capacities for movement are not the individual militarised, weaponised or human visual actions themselves but the 'explosion' metaphors that they collectively engender in those who are immediately involved in art and politics. The fact that filmmakers 'who survived the [First World] war moved without any break in continuity from the battlefield to the production of newsreels or propaganda features and then "art films"' shows the import of these filmmakers' 'armed eye'.[62] For Virilio, what this movement points to is nothing less than that such filmmakers 'were themselves merely being hijacked by war'.[63] That is why, for these battle-hardened warriors, the fusion of war, weaponry and the human eye amounted to a sort of logistics wherein 'they thought that, like airmen, they formed part of a kind of technical elite'.[64] This 'final privilege of their art' not only showed filmmakers 'military technology in action', but also demonstrated to them how to serve up such capacities for movement to the public as innovative technological special effects and explosive spectacles, or what Virilio calls the 'continuation of the war's destruction of form'.[65] These ground-breaking technical special effects, exploding displays and the militarised obliteration of form, of course, persist to this day in films like *Avatar*, the 2009 American epic science fiction film written and directed by James Cameron, which introduced numerous pioneering and military-derived visual effects techniques including photorealistic computer-generated characters that were created using new motion-capture animation technologies.

What is critical for Virilio's explanation is that cinema and aerialised war are thought of as interdependent, but not portrayed as a seductive modern technical beyond. All that can be deduced is that *there is continual escalation*. Virilio elucidates this by way of his discussion of 'air reconnaissance operations for the US expeditionary corps during the First World War', which produced so much war information in the form of photographic prints that the 'photograph ceased to be an episodic item'.[66] From the First World War onwards, then, the photograph turned into 'a veritable *flow of pictures* which fitted perfectly with the statistical tendencies of this first great military-industrial conflict'.[67] Yet Virilio's description of photo-cum-cinematic escalation repudiates the setting up of a model for cinematic or military development from which an 'end of cinema and/or war' argument can be inferred and which questions any single viewpoint or method for contending with cinematic or military events. All we are left with is the understanding that there is photo-cum-cinematic escalation, that war is central to it and that the ongoing militarisation of aesthetics must be countered. This is not Baudrillard's cinema and the aesthetics of disappearance, and it is considerably more substantial than his idea of a postmodern philosophy of cinema and technology, reality and history.

Thus Virilio's description of the relation between war, weaponry and the human eye, of the logistics of perception, necessitates ripostes from critics of the military without pre-establishing what form those ripostes must take. One line of riposte or critique concerning the logistics of perception might be a focus upon the 'general interpretation mania' depicted in 'Cinema Isn't I See, It's I Fly', a mania that is imposed by our sense of *apartness* from cinema as an indirect, non-logical, form of perception.[68] Virilio's concern with the logistics of perception is thus primarily about escalation and about filmmakers being hijacked by war. The logistics of perception therefore involves the escalation of militarised organisations and frameworks that represent themselves as global phenomena through the medium of cinema. In truth, for Virilio, such military perceptual logistics have become the points at which the developmental and intensive ways of organising cinema and aerialised war are *no longer questioned*. Rather, they have continued to evolve into new kinds of logistics of perception. One such innovation is the 'future vision' of a '*logistic of electro-optic perception*' that allowed US soldiers in the Gulf War, 'under shelter at their consoles, to take enemy prisoners without having to move, solely by the panic produced by the overflight of ... an airplane without a pilot, an aerial reconnaissance *drone*, with a three-meter wingspan, equipped with a simple video camera'.[69]

Similarly, like the cinema, wars and the aesthetics of disappearance discussed in the first section, cinema and aerialisation during the Second World War, as a mélange of pictorialism and logistics, were deeply *ambivalent* cine-military systems. Even so, these systems' coded information and propaganda still managed to find their way into the pressurised cockpits of Allied bombers. Here, as later in the Gulf and Kosovo wars, the logistics of perception need the bombers' cockpits to 'become artificial synthesizers that shut out the world of the senses'.[70] Yet the results of military-technological separation were so acute and long-term that Allied bomber command responded by lightening such feelings associated with 'the dangerous passage of its armadas over Europe by painting brightly-colored cartoon heroes or giant pin-ups with evocative names over the camouflage'. 'In a kind of CB system', Virilio writes, 'honey-tongued female announcers not only assumed radio guidance of the crews but also helped them through their mission by blurring the image of destruction with jokes, personal confidences, and even songs of love.'[71] What Virilio is emphasising here in *War and Cinema*, as well as in *Desert Screen* and *Strategy of Deception*, is that the bombers' artificial synthesisers, required by what he has labelled as the logistics of perception, identify *new technical vectors of the beyond* or new fusions of cinema with an inexhaustible variety of communications technologies and seductive human voices, every one of which allows different military technologies and aerial possibilities to appear. The logistics of perception are then seemingly humanity's current 'mission', which involves the blurring and, ultimately, the explosion of the image that the female announcers' funny stories, intimate secrets and love songs sought to prevent. The logistics of perception are thus also a kind of historical and contemporary sensation that demands the escalated yet accurate reproduction of audio-visual effects. The capacities for movement themselves that are the foundation of the logistics of perception are used to 'soften' our exposure to nuclear and other explosions, but the logistics of perception represent the truth of their existence by way of, for example, Stanley

Kubrick's precise duplication of such audio-visual effects 'when he used Vera Lynn singing of "We'll Meet Again" to soften the long series of nuclear explosions that conclude *Dr Strangelove*'.[72] The mission of the critic of the art of technology is therefore to try to interpret these logistics of perception much in the way that we might try to understand a widely viewed film like *Dr Strangelove*. There are no predetermined laws for these readings that spell out their implications beforehand, and no explanation is ever conclusive. Alternatively, explications should constantly be inspired by and receptive to a sharp awareness of reality and go directly to the increasingly uniform core of the ever-proliferating war image, recognising that, after the logistics of perception, 'nothing is left but the recording of successive states of discharged matter'.[73] The general idea is always Virilio's military and cinematic conceptions of, for instance, unlocking that 'record of a faraway voice which sings of the desire for reunion', of discovering new ways to overcome the fact that now, after Hiroshima, such 'desires for reunion' are all but 'physically impossible, only this time for everyone and for evermore'.[74]

*Conclusion*

In conclusion, Virilio argues that the age of the aesthetics of disappearance indicates the era in which global cinema in effect became weaponised. He maintains that there are particular capacities for movement whose influences on specific ideas of global escalation make them into logistics of perception or the escalation of the modern technical beyond. In *War and Cinema*, *Strategy of Deception* and *Desert Screen*, the key example of this is cinema. Here, Virilio claims, the modern technical beyond of flying into the unknown escalates because of the fatal interdependence of cinema and war. However, this and other logistics of perception are open to more than mere narratives of the aggressive colours of American films. Rather, they necessitate responses from across a variety of disciplines to, for example, issues of nationalism, and, crucially, to the question of whether military and cinematic victories during wartime have now become one.

One such contemporary response, from the discipline of visual culture, imagines the sky as the 'ultimate public media display'.[75] Making a contribution to 'screenology', a 'hypothetical branch of media studies that would deal with the history of screens as both material realities and discursive entities', Erkki Huhtamo's 'archeology of the screen' uncovers various obscure real and/or discursive attempts to 'turn the sky into a kind of super-screen'.[76] Discussing numerous cultural fields, Huhtamo significantly centres not only upon cinema and warfare, but also upon Albert Speer's light spectacles in Nazi Germany and their consequences for contemporary film and cultural discourses. Drawing on Virilio's reflections on Speer's experience of the night-time mass bombing of Berlin in 1943,[77] Huhtamo, like Virilio, repeats the need to differentiate between reality and spectacle in film and cultural discourses and to recognise that the reality of death and devastation can also be distorted into sinister visions or twisted expressions of the modern technical beyond.

Contrary to other film and cultural theorists, such as Jean Baudrillard, who advocates a turn to a postmodern account of cinematic images and war, and whose writings present the aesthetics of disappearance as primeval pleasure, Virilio insists on the importance of examining the logistics of perception. He argues that cinema and aerialised war especially should be thought through the idea of escalation since the contemporary logistics of perception point to the increasing *uniformity* of cinematic and military systems, and that it is this homogeneity that offers the focal point for the critic of the art of cinematic technologies, cultures of war and the aesthetics of disappearance.

8

# PROJECT TRANSPARENT EARTH AND THE AUTOSCOPY OF AERIAL TARGETING

## THE VISUAL GEOPOLITICS OF THE UNDERGROUND

*Ryan Bishop*

The image always comes from the sky—not from the heavens, which are religious, but from the skies, a term proper to painting: not heaven in its religious sense, but sky as the Latin *firmamentum*, the firm vault from which the stars are hung, dispensing their brightness.

Jean-Luc Nancy, The Ground of the Image (2005)[1]

Why is there nothing rather than something.

Jean Baudrillard, Impossible Exchange (2001) [1999][2]

In Book III of his hilariously savage satire *Gulliver's Travels*, Jonathan Swift turns his rapier rhetoric on a host of speculative theories and theorists receiving large financial support from the wealthy and powerful in the emergent eighteenth century.[3] One bit of speculative research rendered as applied concentrates on magnetism and finds form as military technology. The magnetic principles expounded by William Gilbert and exemplified by his 'terella' or 'little earth' showed the ways in which magnetic principles could overcome gravity by using its power and pulls against itself and thus achieve for magnetic bodies set in opposition against the earth the capacity to levitate. Clearly, for Swift, and paraphrasing Salman Rushdie, his appropriation of these principles intended to have levity triumph over gravity. Swift takes Gilbert's 'terella', a ground magnet contained in a glass that floated above an oppositely charged loadstone, and creates his own little earth: the flying island of Laputa.[4]

The island could fly and hover and menace its ground-dwelling enemies from the air through numerous means, some more benign than others, from blocking their access to sunlight or rain and thus causing environmental disaster (rather like contemporary 'weather war' strategies) to crushing the enemy below, 'which makes a universal Destruction both of Houses and Men'.[5] Putting aside the derogatory Spanish name of the island, a few prescient elements can be noted in the satire: the support of scientific inquiry for the mastery of nature turned to potentially exploitative ends, the shift from scientific speculation to application (from science to technology), the application for military use and, more importantly, the removal of the ground from the earth to become a platform in the sky, thus rendering sky as earth to control the ground from the air. The threat on the ground, as it historically has been, often comes from the sky, from its exposure to whoever can visually survey and control it from above.

Tempting though it is to dwell on Swift's flying island and its multiple manifestations in the trajectories and genealogies that can be traced from the seventeenth century to the present, we shall simply use it as an allegory of sorts for initially considering the ways in which science and technology reconfigure the imaginaries such that they can decontextualise the ground from the earth, make the sky into a staging ground for surveillance and attack, and repeat Heidegger's famous dictum that the essence of technology is nothing technological—it instead resides in the immaterial, the noetic influences and shapes that render the world possible and malleable. The physical constraints of nature become those areas that certain forms of techno-scientific inquiry wish to erase or turn to their advantage as made manifest in the aerial technologies and aerial aesthetics deployed by various militaries, especially through various opto-electronic devices operating from above.[6] The purpose is to gain the ground and control the ground for one's own ends, and to do so the sky proves essential. When Peter Adey eloquently asserts that 'both the ground and the air reside in vertical reciprocity', he is surely correct, but this does not mean that reciprocity is by any measure equal nor does it diminish the human desire to use the latter to subdue and dominate the former.[7]

For all of their appeal to the hubris of mastery, however, what aerial surveillance and targeting provide is both revelation and ignorance: a view we did not possess previously and a reminder that there is more there that cannot be seen, either due to technological limitations or physical limitations apparently insurmountable by technology. The triumph of the surface is unavoidable in the visual domain, and aerial surveillance and aesthetics are almost completely dominated by that which constitutes the visible. But the visible, like the tactile, can only engage surfaces. The most emergent and immediate dimensions of the object, those which are graspable by hand and eye, predominate. Yet the surface also always presumes a depth underpinning it: the aerial view always implies and depends on the subterranean invisible. And depth is accessible only by sound and sound waves.

Thus we find a new major Department of Defense initiative called 'Transparent Earth', one that would not be out of place in Book III of Swift's *Gulliver*, a nascent effort by the Defense Advanced Research Projects Agency (DARPA) to read beneath the earth's surface. More than half a century of advanced satellite episcopy has rendered the surface of the earth consistently and constantly accessible to vision and resulted in the concomitant defensive move of underground weapons systems, battlements and sites. Deep underground military bases (DUMB) provide the military's highly developed systems and technics of aerial observation with their biggest challenge yet, for they cannot be seen or interpreted. To read beneath the earth's surface, the US military is investigating numerous strategies, including harnessing lightning (natural and artificial), radio signals and complex algorithms to 'see' through other sensorial means.

The underground plays an increasingly large role in cutting-edge US military research and strategy, though it takes its symbolic, discursive and material leads from the foundational moment of the current nexus of military, technological, academic and aesthetic inquiry: the Cold War. 'The underground is the paranoid aspect of the Cold War, the dark space beneath the symbolic order reigning above,' Tom Vanderbilt writes. 'It

Image 22: Generation of artificial lightning (DARPA Defence Sciences Office).

is a paradoxical netherworld of both security and insecurity, the place in which we seek shelter, store our possessions, hide our weapons.'[8] The underground marks the new frontier of surveillance and military intelligence challenges, and it does so due to the overwhelming success of the drive to map and always watch the surface of the earth, in 'real time', anywhere. The visual capabilities and capacities of the unblinking eyes in the skies have driven those being watched underground, away from sighted surveillance and in protection from the sighted weapons that almost inevitably follow from aerial sighting: the divide between seeing and targeting being a slim one.

Current military plans to obtain vision below the earth's surface deploy synaesthesia, the connection of two or more senses and an aesthetic mode much favoured by avant-garde artists at the turn the century. The commonly held division between technology and aesthetics becomes difficult to maintain when considering how each side mobilises the senses, thus revealing aesthetic experimentation as a way to render more precisely the vast surveillance and war-making machines of the early twenty-first century.[9] Synaesthesia held a special place in the work of many early twentieth-century avant-garde movements as an attempt to undermine the increasingly rationalised and separated domains of the senses as well as their extension, modification or control by various prosthetic technological enhancements. An important irony emerges because the very military technological trajectories that led the way in hardening the divisions of the senses and reifying the sensorium also led the way in optoelectronic and teletechnological synaesthesia, especially through the conversion of sound patterns into visual data and images of terrain at long distances (e.g. underwater), or of moving objects such as planes, missiles or tanks. The modes of resistance within the artistic movements and aesthetics, then, merely served to give these forces their next logical step, even if it was illogical (illogical insofar as it is the domain of avant-garde artistic experimentation).

The use of combinatory senses to render a visible image of that which could not be seen (the underground) provides yet another attempt to remove the ground of error for military observation and control, reinscribing the desire of mastery operative in the view from above. Telecontrol and action-at-a-distance are hallmarks of the use of the air to control the ground, but the ground and the underground provide obstacles for mastery while revealing the underlying assumptions driving cutting-edge military technology as well as the ways in which these technological innovations reinforce and intensify these assumptions, desires and goals: in this instance, to see through the earth's crust while also standing firmly on it by being a viewing subject on the ground, in the air and underground simultaneously.

*Surface Readings: The Ground of the Image*

Sky is to the 20th Century what landscape was to the 19th Century.[10]

At the peak of our technological performance, the irresistible impression remains that something eludes us ... that, in effect, it is not we who are winning out over the world, but the world which is winning out over us.[11]

Till God, or kindlier Nature,
Settled all argument, and separated
Heaven from earth, water from land, our air

# FROM ABOVE: WAR, VIOLENCE AND VERTICALITY

From the high stratosphere, a liberation
So things evolved, and out of blind confusion
Found each its place, bound in eternal order.
The force of fire, that weightless element,
Leaped up and claimed the highest place in heaven;
Below it, air; and under them the earth
Sank with its grosser portions.[12]

To exemplify one standard way of understanding and interpreting the surface of the earth, John Beck discusses a retaliatory strategy to be used by the US against Japan after the Pearl Harbor attacks. The plan, proposed to President Franklin Delano Roosevelt by a dental surgeon and inventor named Dr Lytle S. Adams, required using millions of bats that would be dropped surreptitiously behind enemy lines—not any old bats, to be sure, but ones laden with incendiary bombs. These would then house themselves until night came before fluttering from their diurnal slumbers to wreak fiery havoc on the wood and paper homes of the enemy. The traditional architecture of Japan had long been noted by the US military, especially the aerial bombing evangelist General Curtis LeMay, as being particularly susceptible to fire bombs. Bats had already produced radar, thus deploying synaesthesic vision from sound in military technology, so why not have them produce waves of apocalyptic flame to engulf the civilian population of Japan? Adams believed that ten planes could carry two million of these blind and winged 'fire starters' and thus was born Project X-Ray, with simulations of the raid conducted in 1942 and 1943 in the deserts of Texas, New Mexico and California.

Beck's excellent study of the wholesale militarisation of the desert south-west of the United States, entitled *Dirty Wars*, reports that that the dentist's proposed deployment of these most evocative of nocturnal mammals had an additional psychological impact because of the shocking effects resultant from a coordinated attack apparently carried out by nature (though actually by nature's creatures made man's minions). The stirring of the very Plutonic bowels of the earth, Beck argues, not only anticipates another important unleashing of the earth's inner fury—that is, the uranium necessary to build the atomic bomb a few years later that was, unlike the bats, unleashed on Japanese civilians—but also gestures toward a deeper interpretation of what lies beneath the surface of the ground we walk on and view from the air. The strategy is one that taps into, literally, the mystery and awe of concealment, the strategy of what is hidden from view (i.e. our necessarily always redundantly surface

190

view). In these instances (i.e. the bats and the bomb), the elements nestled underground cause anxiety and fear about the unknown lurking beneath our feet and homes while also reminding us that we are astride powers latent, invisible, vast.

The hermeneutic strand of reading the earth's surface that is mobilised in this admittedly strange example of a warped foreshadowing of the bomb renders it far less odd than it might initially seem. It is the one that interprets the vertical–horizontal axis of space in the following manner.[13] The horizontal plan offers the promise of movement, openness, light, distant vision, uncluttered horizon, expansion and duration. The vertical axis, in this structuralist hierarchy, bespeaks loss, secrecy, darkness, decay, death, sedimentation and uncontrollability by sight, movement or touch. It contains and hides from view what cannot be visualised, observed or controlled by rational technological instrumental reason, strategy and tools. The challenges posed by the vertical axis have long been the subject of military technologies. But despite the historical (both ancient and modern) deployment of underground facilities for military purposes, solutions have often ended at the surface of the ground. The vertical axis also invites a challenge to visual prostheses to further outfit sight and render visible what has previously been invisible.

There are essentially two types of invisibility: one would be the invisible that can be rendered visible through technological intervention (e.g. x-ray, over-the-horizon viewing technologies, night vision or even the light that helps us avoid stubbing our toe on a table leg at night). This contingent invisibility operates with regard to the empirical realm of the visible, and thus is potentially visible. As potentially visible, this domain of visibility opens itself up to a range of representational manipulations and constitutes the visible and invisible as a continuum: vision and its horizon. The other kind of invisibility is of a more radical stripe for it is that which can never be rendered visible, the structural necessity of invisibility by which visibility is possible at all. It is the ground of possibility for visibility. Both military technology and experimental avant-garde aesthetics of the early part of the twentieth century address this second kind of invisibility in an attempt by the former to eliminate it, and by the latter to insist on its inviolability.[14] This tension about the invisibility that is the ground of possibility for the visible and for visibility tells a story that runs throughout the past 120 years as the drive by technoscience to extend the power of the senses, particularly vision, has been appropriated by military and corporate sectors in the name of defence, health and entertainment. Yet the ground of the image extrapolated from this tension necessarily lies below the surface and thus poses further challenges for various vested interests.

The tension between forms of invisibility and how to engage them depends upon additional structural relations, including surface and depth, as well as the visible or legible and what supports it. To engage aerial sightedness—or even vision in its most basic form—is to yield almost completely to the promise and problems posed by the surface. For the visible, like the tactile, can only engage surfaces. If something is visible, touchable, it is de facto a surface, and thus reliant upon some other entity, some other ground, not visible or graspable for its support. The ground of the image or the tain of the mirror, therefore, becomes something simultaneously necessary and uncontrollable. However, with instrumental justification and rationale in the ascendancy for the most powerful sectors of the majority of nation states, the most emergent and immediate dimensions of the object—those that are graspable by hand and eye—become dominant. Yet we know about the dangers of surface readings for literary study, religious hermeneutics, political analysis, knowledge accumulation, romantic love … anything that uses a literal or figurative surface-depth oppositional pattern. Aerial visual technologies and aesthetics—the view from above—are almost solely grounded, literally, in the terrestrial. They are rarely of the sky and almost exclusively from the sky for vertical, top-down capture and manipulation, and as a result remain stuck as surface readings.

Therein resides the problem posed by the elided relationship between image and ground. The limit is the ground: the ground of the image and the ground of the earth. This is where aerial technologies and aesthetics meet their limits. With the view from above, an image of mastery is derived, but only apparently so. The image, too, is 'inseparable from a hidden surface, from which it cannot, as it were, be peeled away: the dark side of the picture, its underside or backside'.[15] The backside of the image of aerial surveillance of the globe, the ground of the image, is the underground: the dark depths of unstable terra firma. So the aerial view always implies and depends on the subterranean invisible. Such implication and dependence necessarily evades the complete mapping and surveillance of the earth allowed now by satellite viewing technologies. The now-commonplace assumption that various militaries can see anything, anywhere, anytime is put asunder by the ground of what it sees, which is after all only the ground and the built environment atop it.

Mastery is inescapably haunted by that which eludes it. The backside or underside of what is visually apprehended—itself constituted by our apprehension of surfaces alone—means that the imagination begins acting up and wheedles its way to the fore. What lies beneath?, we ask. How can we see *that*? And once we do, how can we see what grounds even *that*? The drive for mastery generates such a set of questions that runs to infinity, rupturing its goal (i.e. mastery) in the process. This partially explains why the imagination has played such a pivotal role in conceptualising the underground—Plato's cave through Virgil's Aeneas to Orpheus to Pluto to Dante's Inferno to Poe's inner ocean to Jules Verne to H.G. Wells' many underground worlds to Dostoevsky's existential scribbler on through to Saddam's complex underground bunker network[16]—thus generating fear and fascination out of its elusiveness. A possible interim solution to this particular problem of visibility and invisibility, of viewing and imaging depth and not surface, can be found in a synaesthetic paradox—that is, through a confusion of the sensorium. Depth can be accessed by sound, revealing the limitations of sight while also providing it with a synaesthetic and prosthetic extension. Sound will let us see where vision stops.

*Geomancy and Artificial Lightning*

Underground is truly the final frontier.[17]

The archive is formed here in the other senses; an archive of the visible that remains secret, a visible secret, a secret of the visible. Supplemental, the archive of secret visuality appears in the other senses, everywhere but in visuality. (Merleau-Ponty calls this an 'echo'.)[18]

He was about to hurl his thunderbolts
At the whole world, but halted, fearing Heaven
Would burn from fire so vast, and pole to pole
Break out in flame and smoke, and he remembered
The fates had said that some day land and ocean,
The vault of Heaven, the whole world's mighty fortress,
Besieged by fire, would perish.[19]

The DARPA project called 'Transparent Earth' continues that agency's long-standing interest in 'mastering' or 'lording over' nature according to an article on the US Homeland Security newswire ('The Last Frontier'). From planet hacking to changing meteorological conditions (also known as 'enemy climate' or 'weather war') to experimenting with broadcasting frequency ranges in the ionosphere to create nuclear-sized explosions without radiation, DARPA has long sought technological solutions for the limits imposed by physics and nature in the successful operation of military strategy and engagement, including that oft-heard element of surprise (itself part of the DARPA mission statement). DARPA describes itself as an organisation that believes 'it is better to invent a head-mounted multispectral imaging device than curse the darkness', articulating perfectly its understanding of the relation between technology and physical limitations, especially when it comes to the division between the visible and the invisible.[20]

With 'Transparent Earth' the agency is attempting to convert its visual mastery of the earth's crust, its capacity to convert geography (geo-graphy, writing on the earth's surface) into a geology, a logos of the earth. The hermeneutics of aerial vision is turned into inscription, revealing the interpretive reality that each decoding is also an encoding. The geomancy of converting military geography into military geology entails the move from surface to deep reading, which is always preferred in analytic or interpretive situations. The shift to the emphasis on the underground brings a specific avant-garde movement in contemporary military technological research back to a pivotal moment in the history of the industrial scientific revolution with its emphasis on geology. Charles Lyell's groundbreaking *Principles of Geology*, published between 1830 and 1833, overturned long-held Christian dogma by helping to establish James Hutton's uniformitarianism in the popular imagination, inspiring Charles Darwin to make the same kind of mark in biology with similar theories, and sending thousands of Victorians sleuthing in shale in search of fossils. By proving that the earth was older than Church teaching and science influenced by religion claimed, and by finding evidence of change within and the loss of species, Lyell revolutionised the commonly held worldview from a static to a dynamic one. And all while also indicating the precarious nature of all living creatures on the face of the earth—a face that in the slow glacial movement of geological time would eventually be buried, leaving imprints of its former dwellers stamped in stone. His work, coupled with the simultaneous drive to fuel the Industrial Revolution with mineral resources, resulted in the import of mines and the study of the earth and thus the development of the extractive arts. The study of mines and mining became essential in the nineteenth century, and many tertiary institutions were opened to meet the rising need. Mines, of course, become the model for a host of underground structures, most especially fallout shelters for individual nuclear families or for large scale civil defence, as espoused by Herman Kahn and his RAND studies and immortalised in Stanley Kubrick's *Dr Strangelove* as a site of potential weakness in the US nuclear strategy, a 'mine gap'.

Using the ground as a mode of military defence has a fairly contemporary genealogy too, one that runs through the trench to the bunker. As the First World War honeycombed much of Europe's soil and inspired the post-war underground control centre called the Maginot Line, the use of reinforced concrete in these fortifications helped Hitler and his officers to envision and realise a massive underground command post that would use communications tele-technologies to conduct far-flung above-ground military activities from the relative safety of the underground, impervious to the newly sanitised mode of warfare called aerial combat. This kind of warfare was condemned as 'uncivilised' by the US and European nations when practised by the Japanese on China but was rehabilitated as essential for the Allied forces when they realised how effective firebombs in Tokyo could be. Further support of aerial warfare emerged with the strategic advantage brought about by 'dehousing the labor force' for the German war machine. Walt Disney helped with its propaganda feature *Victory through Airpower*—a film unavailable for decades until repackaged by the Disney Corporation along with its other Second World War propaganda films and cartoons in a box set offered for sale just after 9/11. The mastery of the sky, which translates into the apparent control of much of the ground from the air, meant defensive measures for underground installations would rise in tandem with and proportionate to the strength of airpower.

As is always the case with military-based research, DARPA makes dual-use claims about the work involved in trying to see, read and map the ground beneath the earth's surface by arguing that such knowledge could help generate tools for anticipating natural disasters in addition to 'detecting, targeting, and destroying hard and buried underground facility (UGF) targets,' with the latter presented almost as an afterthought. Toward that end the agency will spend $4 million to create real-time, 3-D maps of the physical and dynamic properties of the earth down to 5 kilometres, a small but important horizontal slice of the 3,500 mile stretch from crust to core. That 5-kilometer swath, though, is about as deep as most tunnelling or drilling projects can go, making it the most data- and information-rich portion invisible from the ground or the air (for military purposes but not necessarily for predicting earthquakes, one should note).[21] The entire project aims to be operational by 2015. The end result, so DARPA claims, will eclipse even the visual capacity from the air for mapping, tracking and sensing the above-ground natural and militarised world. The resultant imaging devices derived via sound will generate, so the grandiose claims run, perfect representational images of the uncanny landscape of the geophysical and human-created underground.

To generate these maps, a host of projective tools and developmental sensors will be deployed, including algorithms that estimate and predict tectonic shifts and other subterranean movements. Sensor technology will enable updates of the maps provided by the geoscientific algorithms and extant geophysical maps. New tools and techniques, though, play an integral role in bringing the 'Transparent Earth' to fruition. One of the more promising areas of research for rendering invisible the opacity of the earth has been the use of extremely low radio frequencies (ELF 3 Hz to 30Hz 100,000 km to 10,000km) and very low frequencies (VLF 3 Hz to 30Hz 100km to 10km). These frequencies create radio signals that can be transformed into a visual signal, much as radar or ultrasound technologies, rendering invisible sound waves visible as they move through subterranean spaces. Geophysical surveying has used naturally occurring ELF/VLF producers, such as lightning, for quite some time. Because they are naturally occurring, they are rather hard to come by. Now the US Department of Defense wishes to harness this strategy but to do so as and when they please: lightning on demand, as it were.[22]

This kind of research touches on the highly controversial work being conducted by the Department of Defense and many research universities under a project called High Frequency Active Auroral Research Program (HAARP). A lightning rod for conspiracy theorists because of its connections to weather warfare and 'geohacking', HAARP studies various forms of wave technology in strategic relation to the ionosphere for its potential with radio communication and surveillance (e.g. over-the-horizon radar) as well as exploring its potential as an offensive weapon.[23]

A small and seemingly more benign part of the strategy of the military with HAARP is a project to generate S-BUG, Sferics-Based Underground Geolocation (sferics is short for atmospherics—spherics), thus using artificial lightning to create a GPS for underground formations. However, the goal is grander than GPS, with the hope of a much more fully developed final collection of linked maps of the subterranean world than what we currently possess for the terrestrial one.

NIMBUS, a DARPA project to trigger and manipulate lightning, represents a development that will result in Zeus-like capacity with mythological lightning bolts being hurled across the aether at will largely to create VLF radio signals necessary for generating the sound waves required for mapping underground developments and structures. VLF radio signals were used by submarines to 'see' the bottoms of the oceans during the Second World War, and massive transmitters and receivers known as umbrella antennae dotted the landscape of Europe at the time. These antennae turn up in James Joyce's great novel about the senses, sensory information, technology and transformation, *Finnegans Wake*, while also providing the logo for the HAARP project: the spokes atop a long mast resembling an umbrella shed of its cloth, an umbrella's skeletal structure rendered visible through its surface, as it were, thus visually performing by analogy the surveillance work it carries out in the aural-optical domain.

Essentially, the goal of Transparent Earth is to create a very large tele-stethoscope that can make images, thus repeating the discourses of early medical science in early twenty-first-century military technological ones. The syntax of sight that Foucault argues in *The Birth of the Clinic* as emergent from late eighteenth- to early nineteenth-century medical practice focuses on the continued blurring of the once hard and fast boundary between the visible and invisible.[24] The medical gaze, aided and abetted by multi-sensory prostheses such as the stethoscope, opened up the body to the objective eye of the practitioner and brought forth its previously stored mysteries. At that particular juncture in the history of medical practice, the living body, the diseased body, the corpse, all opened up their dark tyrannical interiors to the libratory practices of the clinical space and the precise discursive practices that removed the magical sway that illness possessed when it operated under the signs of superstition and evil. The body became like a text, unfolding in its legibility and rendering the implicit as explicit to the trained eye of the corporeal hermeneutician. These are the same assumptions, goals and methods we find operative in the project called 'Transparent Earth'. Taking the layers and folds of the earth found in that pivotal natural science text, Lyell's history-changing *Principles of Geology*,[25] Transparent Earth melds Lyell's promises of reading the earth's interiors with those found in modern medical practice to survey the inner workings of the earth's core, and to do so, according to the current plan, nature must be mobilised by instrumental reason and for human ends.

The goal of generating and applying various attributes of lightning has been around for some time, as the studies by Dr P.L. Bellaschi in the November 1941 issue of *Popular Science* exemplify. The harnessing of natural phenomena has long captured the imagination of science and arts, especially that found in lightning for its uniquely charged qualities and effects on the atmosphere: as found in works running from mythology through Benjamin Franklin to Mary Shelley's doctor attempting to reanimate life, itself the stuff of Hollywood visual clichés of science gone mad with hubristic power. The apparently final stage in this trajectory would be the lightning unleashed over Hiroshima and Nagasaki that translated 'total war' into the constant state of surveillance, mobilisation and aggression that characterised the Cold War. The kind of surveillance potentially afforded to the military for underground mapping through the use of ELF/VLF radio frequencies, though, takes us to another cliché: that known as the *blitzkrieg*, which literally translates into English as 'lightning war' and which is meant to connote both power and speed. By the near-Promethean act of wresting lighting from nature to be used on demand, Transparent Earth hopes to render visible subterranean structures, to see with sound, and divine installations of weaponry or defence buried beneath the once opaque crust of earth. By yielding up the earth's Plutonic secrets and the human interventions contained therein, a geomancy of geology and geography is performed by those wizards of the air and earth known as the US military.

*Hiding Surfaces and the Autoscopy of Aerial Targeting*

From New York to California, in a total of 18 states, the U.S. is hard at work on the biggest, most complex and crucial military construction program in its peacetime history: the installation of attack-proof, underground launching sites for the nation's Atlas, Titan and Minuteman intercontinental ballistic missiles.

Never in military history has such a concentrated assemblage of destructive firepower been so completely masked in a setting of utter pastoral peace and tranquillity.[26]

And suppose further that the prison had an echo which came from the other side, would they not be sure to fancy when one of the passers-by spoke that the voice which they heard came from the passing shadow?[27]

We hear the wind blowing unimpeded, and we see the cinematic sweep of the horizon, great spaces of nothing and vast veldts stretching to the vanishing point. (A whole lot of nothing, as they say in Texas, often followed by 'It ain't the end of the world, but you can see it from here.') We see the massive plains of the Dakotas and those of Kazakhstan, the deserts of Iraq, the benign horizontality of these sites belie the lay of the land, housing as they do hardened silos with sleeping missiles or reinforced bunkers of weapons or wealth or both. The pastoral landscape yields to a giant tumult of natural forces mobilised and waiting to be unleashed, a large Turneresque landscape of boiling possibility and vaguely defined outlines. We know about the subsurface buildings, the underground architecture, the vast energies of burrowing and dwelling because the thin layer of ground on which we stand might support us, but it exposes us too.

We are laid bare on the earth's surface, and so thoughts turn to a genealogy of underground evasion, of defence and storage, protection and dwelling. Prehistory finds the Lascaux caves in France and the vast networks of caves in the American South West, followed by Hades and the Elysium Fields strewn with asphodel, up to today and the Taliban's caves and tunnels in Afghanistan and Pakistan. The caves become trenches become bunkers: First World War trenches and Maginot Line, Second World War and Hitler's bunkers, the Iraqi bunker complexes that so taunted both of the Bush administrations in the Gulf War and the Iraq War, and Mount Weather in Virginia, the underground base and living space for the executive branch of the US government built early in the Cold War in case a delayed blowback of the nuclear bombing of Japan should ever transpire, one rerouted through Moscow. Then there are tunnels: the elaborate and effective system devised by the Viet Cong that ironically echoed NORAD's massive tunnel in Cheyenne Mountain, as well as the improvised defence against the Blitz found in the London tube. The dark side of the underground can be found in the hardened silos—the plasticity and strength of concrete employed to

199

house, hide and protect ICBMs and people, the public and individual fallout shelters encouraged by Herman Kahn and the Hudson Institute as the only sound way to respond to a world laden with nuclear weaponry, as well as the massive Yucca Mountain storage site in Nevada holding the radioactive waste generated by the production of the very same weapons that also drive us underground. And of course, the ideals that each of these embody might find their origin in the cave of idealists known through Plato's parable discussed in the utopian vision of *The Republic*. All of these are hiding surfaces that are hiding other surfaces: to protect them from attack and surveillance.

Hiding surfaces pose the problem to be overcome by the various technologies mobilised and deployed for Transparent Earth. The phrase 'hiding surfaces' offers a nice middle voice articulation of the situation, for it ambiguously denotes surfaces that hide something or surfaces that are being hidden: surfaces that hide and surfaces that are hidden, surfaces hiding other surfaces, which is always the case with vision or touch. So sound enters the scene to find what other surfaces might surface if the earth could be sliced open, unfolded and explicated for those with the hermeneutical knowledge and skill to read what might be found there.

*Conclusion*

We return now, in conclusion, to Plato's cave and the epigraph from it that graces this section. In Plato's parable, those chained inside the cave mistake the images before their eyes for reality when they are in fact merely shades cast by a crude shadow puppet apparatus. The verity of the illusion, as the quotation explains, is rendered more fully by an aural mirror that reflects sounds as if coming from the images when they actually originate at the mouth of the cave and provide a kind of ventriloquistic tossing of the voice. We need to note that Plato's story centres on the fallibility of the senses, the mistaken impressions created by and as empirical evidence, and in this instance through a coordination of sound and image with regard to the position of the subjective recipient of the information. The Platonic conceit of the aural mirror as error, as further confusion of illusion for the real, is a cautionary tale, as are most parables, one with resonance (if we can pardon the term) for HAARP, NIMBUS and Transparent Earth. What if the immense technological prostheses producing synaesthetic images and maps of the underworld's forms, formations and installations prove illusory or simply mistaken? And what if they are in error, when the lightning flashes, the deep hermeneutics operate and the weapons follow hard on their heels in a closed system of identification, targeting and firing? The possibility emerges that the trick of the light as aural sound effect might well turn out to be just a trick of the light and a sound effect, and not a perfect image or map of subterranean structures.

And what then? Perhaps a return to a position of the sensing subject with the earth no longer liquid or transparent but newly reformed as solid beneath its feet, the objects of the world opaque once more and impervious to penetration, and no desire for them to be otherwise for the purposes of the extension and control of the domain perceived by this sovereign subject? No—it is likely that the standard answer we receive now is the one that will emerge: more technology to rectify the negative effects, problems and inadequacies generated and perpetuated by previous technological application. The very essence of the momentum of technicity takes over once again. Our grasp will continue to exceed our reach, not the other way around as commonly held, thus producing effects of which we know not.

The ubiquity of aerial views of landscapes or cities in cinema (whether as establishing opening shots or as part of the thriller or action genres) and photography exhibitions have rendered them commonplaces. Some kinds of aerial views, though, are rarer, indeed almost pathological. One of these is called 'autoscopy', viewing oneself, seeing oneself as a self viewing itself: both viewing subject and viewed object. Neurologists use the term to describe out-of-body or near-death experiences, and particularly in the latter cases when the body and the self are clearly *in extremis*, the perceiving subject sees his/herself from several feet in the air above the supine body. It is an aerial view of trauma or dementia. Some neurological studies link this phenomenon of seeing oneself in extrapersonal space as a pathological response to position, movement and completeness of the body, arguing that it results from a failure to understand and process proprioceptive, visual and tactile information. The effect is almost the neurological counterpart to ghosting for analogue broadcast television, but the experience of subjective viewing of the self changes that very experience. We are not seeing two of the same object as in televisual ghosting but rather seeing ourselves as an object from the position of our embodied subjectivity—like looking in a mirror but without the mirror or at a hologram projection of ourselves, or getting a glimpse of ourselves on a screen from a security camera or video camera display in an electronics store. However rare these neurological phenomena may be, the vastly successful eye-in-the-sky optoelectronic technologies used for global surveillance and targeting have rendered us all in a state of autoscopic *extremis*, able to see ourselves simultaneously as viewed and viewing subjects, embodied in both positions simultaneously in 'real time' in two distinct spatial positions. We can call this effect 'the autoscopy of aerial targeting'.

In this effect we project a viewing subject above, one that is not us but a simulation of us that allows us to see ourselves, and others, from above in such a powerfully mimetic manner that we can believe, as in the pathological state, that it actually is us viewing too: Plato's cave operating from above. In the process though we also view and target ourselves. We additionally target the earth's crust and now, with Transparent Earth, we target that which lies below the crust as if ground and underground were separate from us. We have reified a solipsistic loop of sensory projection and reception in which nothing exists outside the viewing subject, even when that viewing subject is also the object of the view. Yet this is a trick of optoelectronic teletechnologies, one that makes our astral out-of-body perceiving selves seem to be or feel to be our real selves, not understanding the effects of the actions as felt and experienced on the ground, which is where we actually dwell and as well as what we seemingly wish to render transparent. This perceiving hovering self is no longer a neurological anomaly or neo-necromantic epiphenomenon (like ectopolasm) but the consolidated result of massive spending, intensive research, military-driven geopolitical theorisation about and application of teletechnological prowess and synaesthetic manipulation. That is, it is us.

9

# AFP-731 OR THE OTHER NIGHT SKY

## AN ALLEGORY

*Trevor Paglen*

On a movable launch pad at Cape Canaveral sat the space shuttle Atlantis. It was night, and the shuttle's bone-white delta wings were lit with floodlights from below.[1] The date was 28 February 1990. It was a turbulent time: across the Atlantic Ocean the Cold War was thawing. The Central Committee of the Communist Party of the Soviet Union had pledged to give up its monopoly on power, and McDonald's had just opened its first restaurant in Moscow. Germany had agreed on a plan for reunification, and thousands had stormed East Berlin's Stasi offices looking for their files. In Lithuania, hundreds of thousands demonstrated for independence. In South Africa, Nelson Mandela had just been released from prison. For some, the world seemed to be opening up. But none of that extended to the Cape Canaveral launch pad that late February evening.

STS-36 was going to be an odd mission. NASA had issued no high-minded press releases about the wonders of space travel or the scientific instruments aboard Atlantis. There were no media kits for the aerospace journalists assigned to the story, and there was no talk of mission characteristics or payload. Few media outlets even bothered writing about it. The aerospace industry trade journal *Aviation Week and Space Technology* provided one of the few reports, and it was brief: 'A secret Pentagon shuttle mission set for Feb. 16 will carry a 37,300-lb. advanced reconnaissance satellite to be used by the Central Intelligence Agency and the National Security Agency. Designated AFP-731, it is a "combination" spacecraft carrying both digital imaging reconnaissance cameras and signal intelligence receivers.'[2]

It was not unusual for shuttle missions to receive so little press. After the Columbia's first few missions in the early 1980s, shuttle launches had become rather prosaic affairs, virtually unnoticed outside specialised aerospace journals such as *Aviation Week* and *Florida Today*, Cape Canaveral's hometown newspaper. Public interest in the shuttle was briefly revived in the aftermath of the 1986 Challenger explosion, but by 1990 shuttle missions had once again receded into the dim corners of the public imagination. It was supposed to be that way. Space shuttle missions were supposed to be reliable and routine. And to a great extent, they were supposed to be secret.

The space shuttle was designed in large part as a secret spacecraft. It conjoined two Cold War space races. On one hand was the public race to demonstrate state virility and generate nationalistic zeal by putting men in orbit, landing on the moon, and all the rest. But there was another space race, a 'black' space race conducted with similar vigour outside public view. The secret space race was about deploying ever-more powerful reconnaissance satellites, anti-satellite weapons and securing control over the strategic 'high ground' of orbital space. While John Glen, Allan Shepard and Neil Armstrong became national icons, stories from the black space race remain largely untold. The National Aeronautics and Space Administration (NASA) was the iconic organisation behind the 'white' space race, and the National Reconnaissance Office (NRO), the nation's 'other' space administration, led its murky counterpart. Although it enjoyed the largest budget of any agency in the intelligence community, the NRO was a black agency. In 1990, the NRO's very name was classified.[3]

First conceived in the 1960s, the space shuttle was meant to serve as a cost-efficient, reliable way to achieve low earth orbits. But the space

shuttle had a catch. To make up for the shuttle's development, maintenance and operational costs, the programme only made economic sense
if it put rockets out of business and enjoyed a monopoly on American
spaceflight. This meant that the shuttle would have two customers and
two varieties of missions: in addition to the public displays of American
space exploration, the shuttle would be tasked with the NRO's secret
forays into space. The shuttle was, in fact, a joint effort between NASA
and the NRO. Its payload bay was designed to house the NRO's colossal
reconnaissance satellites. For secret missions, the shuttle was supposed
to be able to land at Vandenberg Air Force Base on the California coast
after a single orbit.[4]

The relationship between NASA and the NRO was shaky at best. To
be sure, the space shuttle had some strong supporters within the 'Air
Force' (who, on paper, oversaw the Space shuttle's 'military requirements', as the words 'National Reconnaissance Office' were at the time
top secret), who imagined a space plane emblazoned with the letters
'USAF' on its fuselage. But, within the intelligence community, the
shuttle also had strong critics. To its detractors, the shuttle was known
as the 'turkey'. At classified briefings, its critics circulated cartoons
underlining the point.[5]

The NASA logo on the shuttle's fuselage was a kind of Barthesian
myth: the truth, but not the whole truth. On dozens of missions, the
NASA logo on the space shuttle's wings was a useful cover story.

*Cover Stories*

In this sense the space shuttle was nothing new. It had always been
impossible to hide titanic rocket launches from Americans living within
hundreds of miles from the launch sites at Vandenberg and the Cape.
Moreover, to Soviet observers, an unannounced launch might look like
the opening salvo of a nuclear war. Thus, beginning with the first classified satellite launches, 'black' space missions were conducted under the
guise of elaborate cover stories.

The legacy of cover stories began on 22 January 1958 when the
National Security Council issued Action Memorandum 1846. With this
document the Eisenhower administration, embarrassed by the launch of
Sputnik, made the development of a reconnaissance satellite the nation's
highest technical intelligence priority.[6]

FROM ABOVE: WAR, VIOLENCE AND VERTICALITY

Classified satellite development had been in the works for years. From 1955, the Air Force had been working on a satellite reconnaissance system code-named 'Pied Piper', but had made little headway. One of the many problems with Pied Piper was how to get reconnaissance images down from space. The Air Force wanted to relay a television signal from the satellite down to earth, broadcasting the reconnaissance 'take' in real time down to ground-based interpreters. But the idea was far ahead of its time—in August 1957, RCA (the company responsible for the system) informed the Air Force that this type of television signal would provide such poor resolution that it was not worth the effort to design and develop. The alternative was an ejectable film payload. When a reconnaissance satellite had exposed a requisite number of images, it would drop a film canister that could be recovered as it parachuted down to earth. To the Air Force, the idea of scurrying around trying to snatch film canisters mid-air was ridiculous. It chose not to pursue the idea.[7]

After Sputnik, Eisenhower publicly cancelled the Pied Piper programme, unleashing a torrent of anger from congressmen who interpreted the move as Eisenhower pinching pennies where the nation could least afford it. 'We of course couldn't tell anyone that the Air Force program was being replaced by a bigger one,' the CIA's Richard Bissell would later recall. The point of 'cancelling' the programmes was to hoodwink both the Soviet Union and the American media into thinking that the United States had given up on the idea of space-based reconnaissance. Of course, the Air Force programmes were not cancelled at all—the lead role in developing reconnaissance satellites was given to the CIA. The agency's new space reconnaissance programme would hide under the cover story of Discoverer.[8]

Ostensibly an Air Force project to conduct biomedical experiments in space, Discoverer's true purpose was to provide a public explanation for the new launch pad at Vandenberg Air Force Base on the California coastline and to hide—in plain sight—all of the attention-grabbing activities that go into putting a satellite into orbit. General Electric, which was building the camera capsule for Corona (the programme's real code name), went as far as publishing a pamphlet on the Discoverer programme, explaining how the satellite would ride an Agena rocket into space, and how its capsule containing 'scientific data' would be recovered. The space shuttle would continue the tradition of using 'scientific experiments' as cover stories, of wrapping secret missions in public images of civic science and national progress.

206

On 12 April 1981, the first shuttle mission (STS-1), lifted off from Cape Canaveral with all the fanfare associated with a great leap in the advancement of public science. An IMAX film called *Hail Columbia!* documented the spacecraft's debut and the rock band Rush put a song called 'Countdown' on their 1982 *Signals* album, whose liner notes contained a tribute to the first Columbia crew. A year later, the shuttle's 'other' patron would launch their first payload.

*STS-36*

STS-36 was going to be one of these 'other' shuttle missions. Its February 1990 launch window was classified; its weight was classified; its mission was classified; its payload was classified. That alone told aerospace-industry buffs that there was something special about STS-36. Another clue came from the résumé of the mission's pilot, a man named Colonel John H. Casper.

Casper, like scores of other astronauts going back to the Mercury programme, began his career in Vietnam-era fighter planes and later as an Air Force test pilot. Upon entering the world of flight tests, Casper was named 'Operations Officer and later Commander of the 6513th Test Squadron,' according to his NASA biography, 'where he conducted flight test programs to evaluate and develop tactical aircraft weapons systems.'[9] Translation: Casper flew secret airplanes.

Nicknamed the 'Red Hats', the 6513th Test Squadron was one of the Air Force's most unusual units. Their unit patch (see Image 14.3) sported a large brown bear wearing a red bowling hat climbing over the top of a globe. Above the bear, there was a collection of six red stars. The top of the patch read 'Red Hats'; at the bottom is the unit's slogan: 'More With Less.' To military insiders, the symbolism was unmistakable. The Red Hats flew a squadron of covertly acquired Soviet MiGs. The collection of stars on the patch referenced the unit's operating location. Six stars is the sum of five and one: Area 51, a 'black' flight test centre deep in the Nevada desert. The unit had been formed after the United States acquired several MiGs from Israel in the late 1960s. Israel provided the United States with two MiG-17Fs captured during the Six-Day War, and a MiG-21F provided by an Iraqi defector.[10]

After completing his assignments with the Red Hats, Casper went on to become the deputy chief of the Special Projects Office at the Air Force

Headquarters in the Pentagon, where he continued to work on 'black' projects. In 1984, NASA asked Casper to join. The following year, Casper became an astronaut. STS-36 would be Casper's first shuttle flight.

Among defence-industry analysts and aerospace journalists, common knowledge held that STS-36 would deploy another Keyhole-class reconnaissance satellite, a highly classified but nonetheless somewhat run-of-the-mill imaging craft. The Keyholes are essentially classified versions of the Hubble Space Telescope, but instead of pointing out towards the edge of the universe, they are pointed down at earth. The name 'Keyhole', however, is a bit of a misnomer: in 1990, the Air Force and National Reconnaissance Office had stopped using the code name to designate its imaging satellites, but in common parlance the name had persisted.[11] According to a leak in *Aviation Week and Space Technology*, STS-36 would launch a payload whose official designation would be AFP-731 (Air Force Project 731), a randomly chosen designation.[12]

Although the exact launch time was classified, we know that Space Shuttle Atlantis, commanded by John Casper, rumbled off the Cape Canaveral launch pad in the middle of the night at exactly 2:50:22 a.m. EST. Almost immediately, a strange sequence of events began to unfold. The first clue that something was amiss came from the characteristics of the STS-36 orbit. It went into a 62-degree inclination, the steepest inclination of any shuttle mission before or since. STS-36 went on to complete seventy-two orbits, and touched down on the dry lake at Edwards Air Force Base at 10:08 a.m. PST on 4 March 1990.[13]

Then something truly bizarre: twelve days after the Atlantis landed the Soviet news agency Nostovi reported that AFP-731 had blown up. The Soviet Union was not usually in the business of commenting on American space programmes, even failed ones. Moreover, the source of the information was obscure. Two days later, the *New York Times* picked up the story: 'A new American spy satellite has apparently malfunctioned after less than a month in orbit, and parts of it are expected to be destroyed re-entering the atmosphere within a month, United States and Soviet military space officials say.' *The Washington Post* quoted unnamed intelligence and congressional sources calling the reported explosion a 'serious setback' and a 'major concern'. An article in *Aviation Week* explained that the 'apparent failure of the US$500 million AFP-731 imaging reconnaissance satellite launched by the space shuttle Atlantis Feb. 28 is a serious setback in the U.S. strategic intelligence program,' and that 'the apparent failure of the AFP-731 imaging spacecraft was the third in a series of major Western space failures in the last month'.[14]

For its part, the Pentagon did little to confirm or deny the Russian report. President George H.W. Bush said nothing; the same held true for Dick Cheney, Colin Powell and the rest. There were no congressional investigations, no name calling or posturing in the halls of Washington. The political uproar in light of the many billions of dollars wasted on the AFP-731 project seemed strangely restrained. After the initial reports of the explosion, the matter simply went away.

It was as if AFP-731 had been nothing more than a ghost, seen like the residue from a bright light, fading from behind closed eyelids.

Enter Ted Molczan.

*The Other Night Sky*

In the threadbare apartment where he lives with his two cats, Rusty and Sparky, Ted Molczan tells me that he's always been a 'space nut'. His desk is neatly stacked with star atlases. Wil Tirion's *Uranometria 2000* sits atop the *Atlas Eclipticalis*, the Czech astronomer Antonín Bečvář's beautifully rendered 1958 guide to the heavens. Across the room is a bookshelf crammed with more space books: Patrick Moore's *Men of the Stars*, Robert Divine's *The Sputnik Challenge*, Desmond King-Hele's *A Tapestry of Orbits* and William Burrows' classic tome on secret satellites, *Deep Black*. Perched on a tripod next to the bookshelf is a pair of Celestron binoculars with lenses the size of two-litre Coke bottles. I've brought along my own small pair. In a good-natured way, Molczan tells me they might work at the opera.[15]

Ted Molczan is one of the world's leading figures of a very peculiar version of amateur astronomy. Most amateur astronomers content themselves to stare deep into the cosmic wonders sprinkled through the night sky: the Triangulum in the Orion Nebula, or the gaseous blue clouds surrounding the Pleiades, the globular clusters in Sagittarius, the Andromeda Galaxy near Cassiopeia or the great craters of Copernicus and Plato on the surface of the moon. At the high end of the hobby, amateur astronomers employ sophisticated automated systems to help university researchers search for extra-solar planets, collect gamma ray outbursts and generate observational data used by professional researchers. For decades, amateurs enjoyed a near monopoly on discovering new comets and tracking asteroids. In astronomy, the line separating back-yard amateurs from their cousins at major research universities can be blurry and indistinct.[16]

But Ted Molczan is a different breed of amateur astronomer. Molczan's specialty is a host of celestial objects even more obscure than the galactic cluster in Hercules or the dark nebula in Aquila. Using his pair of high-powered binoculars, a stopwatch and a self-fashioned computer program called 'Obsreduce', Molczan works with a handful of people around the world who observe and keep detailed records on nearly 200 classified American satellites in earth orbit. Molczan tries to understand a very different night sky than the starry night most of us see when we stare into the heavens on a clear night. Molczan maps the other night sky.

The other night sky is a landscape of fleeting reflections: of glints, glimpses, traces and flares; of unacknowledged moons and 'black' space-craft moving through the pre-dawn and early evening darkness, where the rising and setting sun lights up their stainless steel bodies, and they blink in and out of sight as they glide through the backdrop of a dark-ened sky hundreds of miles below. In most cases, the reflection is all we get. The other night sky does not want to be seen: even full-time defence-industry journalists and aerospace historians have a hard time knowing exactly what's what.

On the faintest end of the spectrum are the geosynchronous satel-lites—spacecraft like the Milstar constellation (originally for *Military Strategic and Tactical Relay*), perched 22,241 miles above the earth's surface so that they can 'cover' about half the planet. Milstar 5, for example, is parked on the equator over Eastern Africa, about halfway between Nairobi and Mogadishu, while Milstar 6 is perched on the other side of the globe near the Galapagos Islands off the coast of Ecua-dor. Then there are the Mercury, Mentor, Magnum and Advanced Orion eavesdropping satellites, purported to look like umbrellas the size of football fields, sitting in geosynchronous orbits to vacuum up com-munications, telemetry and electromagnetic signals emanating from below. Even among NRO and NSA insiders, the code names of these geosynchronous satellites are said in hushed tones. All but invisible to ground-based observers—and containing some of the most highly clas-sified systems and charged with collecting the planet's most sensitive data—they are earth's most secret moons.

The illuminated hulls of reconnaissance satellites follow the inverse-square law: objects closer to earth are exponentially brighter. Just visible to the unaided eye at an altitude of about 1,100 kilometres are the Naval Ocean Surveillance Satellites (code-named Parcae after Zeus's three daughters), whose mission is to track naval vessels by eavesdropping on shortwave and other transmissions. The NOSS satellites cruise across the night sky in formations of twos and threes, appearing as points of light moving across the sky in a triangular formation. In other words, they look exactly like late-generation UFOs, or delta-winged aircraft using a cloaking device. In fact, they are *so* easy to mistake for UFOs or black aircraft that UFO researchers have come to realise that a number of 'black triangle' sightings can be explained by the Parcae constellations.[17]

The kings of the night sky, however, are the imaging satellites. If you have ever looked up at the sky just after twilight or just before dawn and have seen a satellite moving across the sky, there is a good chance that you have seen one of these. They are some of the brightest objects in the sky. The size of school busses, and dwelling in precariously low orbits, their polished hulls light up like meteors when they reflect sunlight towards the earth below. The imaging satellites come in two basic 'flavours': those that use photographic imaging methods, known as Keyholes, and those that use something called 'Synthetic Aperture Radar', known by the code names Lacrosse and Onyx. And right about here, our knowledge of the other night sky's denizens starts to run out.

'One way to look at it,' Molczan told me about his unusual hobby, 'is that it's a form of science-based investigation, but more like detective work. I guess that it's like modern detective work, which is also based on science.' Molczan's pastime might sound unbelievably complicated, and to a certain extent it is, but in another sense, it is fairly straightforward if you know what you are doing. The hobby is possible, Ted explains, because no matter how many security classifications and code words the NRO uses to hide its secret satellites, the agency cannot classify Kepler's three laws of planetary motion. The tools of satellite observing are so simple that they seem almost anachronistic: a good pair of binoculars, some star charts and a stopwatch. Molczan does it all from the balcony of his apartment.

With these simple tools, Molczan can generate an incredible amount of detail about black satellites and secret moons. By synthesising his own observations with those of his fellow observers such as Russel Eberst in Scotland, Pierre Neidrick in France and Mike McCants in Austin, Molczan can predict where a spacecraft will be with astonishing accuracy. 'I have a fanatical devotion to accuracy,' he chuckles.

Molczan and the other observers' hobby shares origins with the various conflations of public science and underlying militarisms that have long characterised the space race. In the mid-1950s, it was clear that both the Soviet Union and the United States would begin populating earth orbit with artificial moons in the near future, but the United States lacked the capacity to track them. To solve the problem, the Smithsonian Astrophysical Observatory (SAO) created Operation Moonwatch, a national (and later international) programme to support satellite observing as a popular hobby. Inspired in part by the Ground Observer Corps, an early Cold War programme designed to teach ordinary citizens how to spot Soviet bombers and to act as a national 'early warning system' against potential Soviet attacks, the SAO encouraged legions of ordinary people to form local satellite-spotting clubs. While Moonwatch was billed as a large-scale citizen-science project, the military was an active sponsor and patron of the programme. Officials at the Air Force and Navy provided Moonwatch teams with telescopes and training support, and encouraged Moonwatch teams to set up shop on Air Force bases.[18]

Within a few years, Operation Moonwatch's popularity waned. The culture of McCarthyism, which motivated people to act as vigilant citizens on the lookout for Soviet machinations, began to fade. So too did the initial novelty of seeing artificial satellites in the night sky. While many Moonwatchers dropped out, those who stuck with the hobby became exceptionally good at it. By the time Moonwatch closed its doors for good in 1975, its remaining members were amateurs in name only. Observational data collected by the remaining Moonwatchers was often more accurate than data collected by their professional counterparts.

Throughout the Moonwatch programme, amateur observers enjoyed a symbiotic relationship with NASA. The space agency supplied amateurs with the lists of orbital elements it maintained for all the spacecraft it tracked. Amateurs, in turn, refined these orbital elements using data from their own observations. Even after Moonwatch ended, NASA's Goddard Space Flight Center continued doling out reams of orbital data to the amateur observers who had signed up for the service. Data from the Goddard Center contained elements for all the spacecraft it was tracking, including American reconnaissance satellites. Although the classified objects were not identified as such in the data set, it was relatively simple for amateurs to tell which elements corresponded to spy satellites. Until 1983.

That year, the Reagan administration abruptly stopped publishing orbital elements for American military and reconnaissance satellites. Ironically, this move gave amateur satellite observers a renewed sense of purpose. Russel Eberst, a former Moonwatcher based in Scotland, explained that the abrupt classification 'unwittingly set a challenge to the amateur network of observers'. The new game was 'to see if they could maintain reliable orbits for these "secret" objects'.[19] Amateur observers rose to the challenge: faxes, telephone calls and the mail were replaced by BBS boards, the precursor to the World Wide Web. Observations were shared, orbits refined. The amateurs developed a highly accurate guide to the blank spots in the official satellite catalogue. They monitored existing satellites for changes in behaviour, and updated their catalogue with every classified object the National Reconnaissance Office put into orbit.

When Ted Molczan (a self-taught observer) heard about the space shuttle STS-36 mission in 1990, and read that the shuttle would deploy something dubbed AFP-731, he assumed that the payload would be another Keyhole-class object, a standard optical reconnaissance satellite. Molczan also assumed that tracking the spacecraft would be a relatively straightforward problem, 'but being impatient and wanting as much data as I could get, knowing launch date and time, it was easy to determine that it wasn't going to be in range of most active observers. We were out of luck for weeks.' Because the skies above Toronto would be in the sun's shadow when Molczan expected the spacecraft to fly overhead, there would be no reflection. The payload would not be visible from his Toronto balcony. 'I needed to find observers in the North.' So, with a bit of 'inspired phone calling', Molczan recruited teams of agreeable amateur astronomers at Yellowknife in the Northwest Territories, at Whitehorse in the Yukon, and in Alaska, held a series of informal 'training sessions' to teach the astronomers the basics of satellite observing, then supplied the groups with 'look angles' where they could expect to see the classified spacecraft.

'Things went really well,' Molczan explained. After the space shuttle deployed the unidentified satellite, 'it went into [the] orbit that it was supposed to be in from *Aviation Week*. The whole idea of tracking it was to refine the estimated orbits that I'd predicted. That all worked like a charm.' The amateurs based in the Far North reported back that AFP-731 was nice and bright in the night sky—about Mag-1, the same brightness as Jupiter, and supplied Molczan with useful observations. In the UK, Russel Eberst made additional observations. 'A few days of this, and the guys up North asked if I had all the data I needed. Remember that it was about 40 degrees below [zero] outside.'

With the data in hand, Molczan decided that he had what he needed, even though it would be weeks before he would actually be able to see the object from his home. Then something unexpected happened: 'About a week goes by and a press release comes out from the Soviets saying that the satellite may have blown up.'

AFP-731 had disappeared. The last elements Molczan had for it looked like this:

USA 53 (AFP-731) 18.0 4.0 0.0 4.1
1 20516U 90019 B 90309.99079700 −.00002298 00000−0 −95528−3 0 03
2 20516 65.0200 194.0588 0009734 214.9671 144.9440 14.26241038 04

*Interlude*

On the twenty-third floor of his downtown Toronto high rise, Molczan is teaching me how to get the maximum accuracy from a satellite observation. We are sitting at his desk looking at a graphic on his computer screen: a line representing a predicted satellite pass bisects a pair of stars we plan to use as fixed reference points to measure the pass of a peculiar object launched in 1999 called USA 144Deb.[20]

'So it should be about 60 per cent down from this star to the next,' I say. 'Try to forget that number,' Molczan tells me. 'You don't want to bias your observation.' One of the dangers in observing satellites holds true for all strictly empirical work, he explains: you have to see what is there, not what you want or expect to see. To get accurate data, you have to be as objective and unbiased as possible. Trying to make oneself a 'reliable witness' requires a tremendous amount of self-discipline.

Minutes later, I am looking through Molczan's binoculars at the two stars we just saw on his computer screen. A point of light, shimmering like a diamond on blue-black velvet glides into my field of view and bisects the pair of stars. I click the button on Molczan's stopwatch as it passes. The USA 144Deb object appears to be exactly where Molczan predicted it would: about 60 per cent down from the top star to the bottom star. Then a moment of self-doubt: did I see it there because it actually was there, or because I expected it to be? I had not been able to forget the number. Molczan and I step back in the apartment. There is an hour until the next predicted satellite pass.

'I can't speak for everyone involved in this hobby,' Molczan once told me about his pastime, 'but for me, this is about democracy. There are elements out there who want to keep everything secret. I try to put pressure in the other direction. I try to put checks on that power. When people ask me about what gives me the right to make these decisions, I say "Citizenship in a democracy gives me the right to make these decisions." I don't break in, I don't steal stuff. I assert my right to study the things that are in orbit around the earth and study them with the belief that space belongs to all of us. I exercise my right to know what's there.'

The more we talked, the more I could not help but to start thinking of Molczan as a kind of latter-day Galileo, measuring reflections from secret moons in ways that recall Galileo's measurements of Jupiter's moons in *Sidereus Nuncius*, the book based on his initial experiments with the telescope. Galileo's observations helped open up the Pandora's Box of classical empiricism. His critics claimed that his self-fashioned telescope was feeding him illusions. The church told him to disavow the Copernican implications of his work—or else.

Molczan's work echoes the work of classical empiricists—the Keplers, Galileos and Newtons—and their contemporary descendants in the hard sciences. His work sits atop two epistemological assumptions and methods: (1) that abstracting space into something fixed and calculable can reveal (as opposed to produce) an underlying truth; (2) that reflections can serve as an index of verisimilitude. Both of these assumptions are, at best, quite dubious to contemporary critical thinkers. Molczan's predictions rely on the notion of space as something physical, fixed and calculable, the notion of absolute space most often associated with Newtonian physics and much maligned as a far too limiting account of space in contemporary spatial thought, and a powerful technology of domination.[21] Furthermore, Molczan's measurements of the other night sky stem from a second outmoded epistemological assumption: that signs are parts of things rather than representations of things; that signs do not necessarily conjure practical reality into existence, but can instead function as ciphers to decode underlying truths about reality as it is; that reflections exist prior to observation and naming. For Molczan, the truth is indeed 'out there' for anyone willing to look with open eyes and an open mind.

For theoretically inclined social scientists, humanities scholars, even a number of our colleagues in the hard sciences, Molczan's methodological assumptions might seem as epistemologically anachronistic as his pencil-and-stopwatch approach to collecting data. One of the critical tradition's most consistent projects is, of course, a thorough debunking of any easy link between reflection and correspondence. Critical scholars have come to see *vision*, and by extension, *truth*, like the NASA logo on the space shuttle or the cover story for a classified satellite. Vision and reflection easily function as myth and fetish: appearances easily mask far more than they reveal, naturalising the contingent, frangible and historical.[22] There is no underlying universal truth to reveal, and no rational subject or 'modest witness' to report disinterestedly whatever might be 'out there'.[23]

But the more I thought about it, the more profound Molczan's quip about democracy seemed. Almost 400 years earlier, the Enlightenment tradition had held out the promise of a world in which emperors and inquisitors could not dictate what would be true for all. The notion of democracy Molczan found by studying the other night sky was, of course, a liberal democracy, one rightly criticised for too easily producing the darkness of slave trades, satanic mills, genocides and arms races. But this tradition also animated countless other movements—from abolition to contemporary anti-torture campaigns—whose aims were to counter those dark spaces. Of course, the notion of critique is itself embedded in the Enlightenment tradition.

I wanted to believe in Molczan's methods.

I wanted to believe, because in a world of Abu Ghraibs, Guantanamo Bays, renditions, waterboardings, state secrets, wiretappings and black sites, it seemed as if the critical delinking of reflection and correspondence and the denigration of absolute space was no guarantee of a more equitable society. Wholeheartedly embracing contingency could also produce nightmares: a world in which torture was not torture, disappearances 'weren't happening', and the WikiLeaks cables were 'still classified' and therefore forbidden as evidence in an American court of law, even though they were out there for all to see. Not trusting one's own eyes can also have consequences: it can make it that much easier for authority to trump reason in dictating truth.

Molczan made me want to believe that truth was not always supposed to be the woman Nietzsche famously (and misogynistically) denounced as a temptress and a tease. Molczan made me want to believe that truth might be like a point of light in the evening sky, the sun's reflection against something authorities say is not there. Singular. Visible to anyone who bothered looking through a telescope. Insisting on the verisimilitude of reflections could be a radical gesture. I wanted to believe, as Winston Smith wrote in his secret notebook, that freedom can mean insisting on two plus two making four.[24]

But perhaps I was being seduced. Perhaps I was, ironically, doing the thing that a classical faith in empiricism forbids above all else: allowing myself to see something because I wanted to see it.

Soon, Molczan would come up against his own 'observer effect', the general notion that the act of observation has a tendency to change the thing being observed, an effect that undermines any easy distinction between observer and observed. Molczan would come to learn that the other night sky was also paying attention to him as he went out on his balcony each night when the clouds cleared enough to see the stars.

*14 March 1990*

Two days before AFP-731 appeared to explode, Teledyne Industries filed a rather unusual collection of documents with the United States Patent Office. Patent 5345238 described a solution to a problem that had plagued military and reconnaissance satellites throughout their entire history: they were too easy to track. 'Oftentimes space based weapons systems are looking out into the non-reflective background of outer space,' noted the patent authors. 'This makes the tracking of the target easier, because there is no background radiation or other noise background in the sensor's view. The satellite, which is a radiation source and a radiation reflector, is very evident in this radiation-free background.' Reflections, in other words, were a problem. They revealed the location of a satellite to anyone looking to track it. Moreover, once an observer or 'detection threat' acquired a few observations, she or he could predict a satellite's path with remarkable accuracy: 'Once a satellite or other space object is in orbit, they follow very precise orbital tracks. Therefore, once a satellite's position is accurately determined and tracked, predictions of the future location of this satellite are very accurate.' Teledyne Industries was proposing a solution to these pesky consequences of satellites' materiality. The patent was for a 'satellite signature suppression shield for camouflaging a satellite's location from ground based and airborne tracking and detection systems,' whose purpose was to 'suppress the laser, radar, visible and infrared signatures of satellites. ...'[25]

The Teledyne patent seemed to solve a problem that the National Reconnaissance Office had been worrying about for as long as it had been conducting classified space operations. As early as 1963, there were proposals within the NRO for a 'Covert Reconnaissance Satellite', an alternative to the bus-sized Keyhole imaging birds that shone so bright in the night sky. A proposed 'covert system' in a 1963 memo to the NRO Deputy of Technology would 'rely, above all, on concealment', in a dual sense: to facilitate bureaucratic secrecy, the covert system would have a 'separate and tight security system', and 'simplified check-out and handling procedures, requiring a minimum of personnel'. Physically hiding the covert system could be achieved through 'covert and at least portable launch and recovery, preferably mobile' and though a 'reduction of radar and optical cross-sections below the detection threshold'.[26] The National Reconnaissance Office had, in other words, long sought to develop a 'stealth satellite' that could appear and disappear at will, and that would thwart 'active and passive detection systems'.

The stealth system described in Teledyne's 1990 patent was in essence a giant, movable mirror designed to reflect the blackness of space towards predetermined 'threat detection sites' on earth. Its major limitation was that the threat detection sites had to be programmed into the satellite's control software. The mirror would not make a satellite invisible in general; the spacecraft could only hide from specifically programmed sites. The satellite suppression shield had to be 'oriented in the direction of the threat.'[27]

*October 1990*

Six months after AFP-731 'exploded' and disappeared, Ted Molczan came across something unusual in the sky. 'I get this message from Russel Eberst in Scotland and others,' said Molczan, 'and I realize we've seen precise unknowns on same nights.' The world's most accomplished spy-satellite observers had all seen an especially bright object. None of them could identify it. 'Russell refined the orbits, and we expected it to match something already up there,' Molczan explained. Figuring that the unknown object would end up being a wayward rocket body or a forgotten Soviet spacecraft, he checked the numbers against publicly available satellite catalogues and his own records of classified orbits. Molczan could not identify the spacecraft, so he turned to another set of records: objects that he had tracked and subsequently lost. Using the observations of the unknown object, Molczan precessed the orbits back in time. 'Lo and behold, they lined up on the seventh of March—the day the Russians said the satellite exploded.' The satellite 'had been exceedingly bright and was still bright, especially considering its height of about 800 kilometers'.

The 'explosion' of AFP-731 appeared to have been a ruse, something akin to an embattled submarine shooting oil, lifejackets and debris from its torpedo tubes to create the impression that it had been blown up. But Molczan's calculations showed that the bright object the amateurs found in October could only be AFP-731, alive and well. The Soviet press report could have been part of an international game of deception: perhaps the Soviets had not been fooled at all, but had wanted to indicate to the United States that it had.

Molczan and the other satellite observers continued tracking AFP-731 for several weeks, reporting their sightings and refining their data on a BBS board. Within a few weeks, a reporter named Todd Halvorson at *Florida Today* caught on to what was happening and wrote up an article about the sightings for Cape Canaveral's local paper. A month later, the *New York Times* added the titbit at the bottom of an article about a different shuttle mission: 'Amateur satellite trackers in Canada and Europe reported they had spotted the spy satellite in a higher orbit than anyone suspected—an indication that the craft was not only working but also highly maneuverable.'[28]

Autumn clouds obscured the night sky over Canada and Europe in the few days after Halvorson's article was released. And when the skies opened up, AFP-731 was once again nowhere to be found.

# SECTION THREE

# FROM CLOSE TO REMOTE

# 10

# THE PAIN OF LOVE

## THE INVENTION OF AERIAL SURVEILLANCE IN BRITISH IRAQ

*Priya Satia*

The cultural critique of aerial bombardment often begins with its chill-ingly dispassionate gaze: the sheer distance between observer and observed or between the trigger-puller and the prospective victim wreaks havoc with ordinary moral compunctions, turning the act of killing into a coldly technological affair, and thus enabling mass murder of an unprecedented order. Ironically, however, the technology is grounded in a surfeit of passionate emotions. It emerged in a particular historical context in which love was explicitly invoked as the justification for wreaking unearthly violence. The first bombers may have found relief in the distance between themselves and their victims on the ground; but the theorists who put them in the cockpit were motivated by a sense of emotional and cultural intimacy with and affection for their victims. I am talking about the inventors of the regime of aerial control in British Iraq after the First World War. Of course, this was not the first instance of bombardment from on high; but it was the first instance in which surveillance and punishment from above were intended as a permanent, everyday method of colonial administration.[1] It was thanks to the Iraqi laboratory that the RAF was able to justify its existence as an independent military service and air theorists were able to articulate their understanding of airpower's uses in conflict. For the designers of this scheme, Iraq's and Iraqis' particular qualities, about which they claimed exclusive and intimate knowledge, made such a scheme practicable. For them, Iraq was a site of unique sensory and emotional experience in which bombing fitted almost organically. It became part and parcel of the desert sublime of 'Arabia'. Moreover, British officials' avowed sympathy with the sensibility formed by this terrain anchored their faith in the local population's unique ability to tolerate a regime centred on aerial bombardment and surveillance. This British experiment with an empire in the sky has critically informed contemporary use of drones in Iraq, as we shall see in the conclusion of this chapter.

### The Desert Sublime

The ideas that shaped British aerial control in Iraq slowly crystallised on the ground in the years before and during the First World War. Intense British surveillance of the interior of the Ottoman Empire began at the turn of the century, when rivalry with Germany, nationalist movements within the Ottoman Empire and the intelligence failures of the South African War combined to recommend more energetic intelligence-gathering in the region. Britain's obligations to the Ottomans under their official alliance meant that military officers on leave, diplomats, gentlemanly scholars, archaeologists, journalists and other sorts of informal agents were the primary sources of intelligence.[2] But once on the ground, these agents complained of the great difficulty of gathering intelligence in a proverbially inscrutable land, 'peopled', as one put it, 'mainly by the spirits of the Arabian Nights, where little surprise would be occasioned in ... seeing a genie floating ... out of a magic bottle'. The naturalist and agent Douglas Carruthers felt 'suddenly transplanted to the ... moon'.[3] The fundamentals of cartography remained a challenge since in the apparently featureless, horizonless, protean and mirage-ridden desert, these dreamy and distracted agents often had great difficulty simply determining where they were.[4] Moreover, 'in keeping with the country', the local population was so prone to exaggeration that, as the agent Gerard Leachman put it, 'one cannot believe a word ... one hears'.[5] They ceased 'thinking geographically', concluding the region was 'very much the same everywhere'.[6] This was to be a unique intelligence project indeed.

Despite its geographical inscrutability, 'Arabia', as they referred to this geographic and cultural imaginary, did seem to possess certain virtues lost to an increasingly decadent and bourgeois Britain—as the difficulties of the South African War had made clear. Indeed, many of these informal agents had been drawn to the work as a means of venturing to an antique and romantic land under cover of patriotic duty. In Arabia, one wrote, 'one may step straight from this modern age of bustle and chicanery into an era of elemental conditions ... back into the pages of history to mediaeval times'.[7] Extending their romanticisation of the noble Arab to themselves, they saw the desert as a haven for individuals who prized 'boundless liberty', whether they had been born there or had fled civilisation's relentless smothering of their instincts to be there.[8] As one of the few places the Royal Geographical Society ranked 'Still Unknown', Arabia also offered the chance to revive the pioneer, heroic spirit of Victorian exploration.[9] The informality of intelligence arrangements, the agents' libertarianism and their nostalgic hankering after imperial fame conspired to keep intelligence in the Middle East outside the compass of the ethic of *discretion* that had begun to dominate the intelligence services. 'Arabia' was deemed a 'spy-space' in its very essence, a place where professional, methodological and ethical standards simply did not apply: 'Crossing the Mediterranean,' explained an agent, 'one entered a new realm of espionage ... full of Eastern ... cunning and subterfuge ... in which the spy no longer emerged bogey-like as in the West.'[10] It was the intelligence world of the emerging genre of spy fiction with whose development this intelligence project neatly dovetailed.[11] The sense that all was fair in Arabia would open that region up to possibilities that would not be contemplated elsewhere.

And that licentiousness could paradoxically be understood as a return to a truer grasp of reality. Deeply conscious of working in the region of the Bible and the Odyssey where espionage had always been an integral part of the epic struggle for knowledge of the self, these agents self-consciously followed their fictional contemporary, Rudyard Kipling's *Kim*, in seeing 'no contradiction between being a spy and being a spiritual disciple.'[12] To Edwardians, Arabia was a biblical homeland to which they could *return* in search of the 'perfection of mental content' that Arabs 'alone, even among Asiatics' seemed to possess.[13] In the 'infinitely mysterious' desert, faith, if not facts or visual data, seemed a reasonably practical objective.[14] This personal objective shaped their approach to their professional duties. Unlike previous generations of orientalists, they embraced an *anti*-empiricist, metaphysical epistemology based on notions of a shared past and racial affinities. If travel in blank Arabia numbed the senses, it did, as one traveller put it, allow one to '"see, hear, feel, *outside the senses*"'.[15] Many agents had pleaded with Whitehall for the job precisely for a chance to escape the existential dilemmas then being unleashed by Western science; the very forces fuelling the turn of the century mystical revival drove these adventurers to the ancient seats of occultism—indeed, many of them were involved in excavating those very sites (most famously, T.E. Lawrence).[16] The apparent limitations on empirical intelligence-gathering in Arabia tended to open the door to a breed of explorer-agent willing, like the contemporary philosophical and artistic avant-garde, to experiment with new theories of perception and more 'unscientific' ways of knowing.

As a basis of knowledge, faith could at once solve agents' intelligence-gathering difficulties and soothe their spiritual cravings. To them, Arabia, of all magical, mysterious places, was *the* place for miraculous conviction: in the words of the informal spy and infamous future secret diplomatist, Mark Sykes, 'the desert is of God and in the desert no man may deny Him'. This was the birthplace of the three monotheistic religions, all of which began with prophets who saw visions and heard voices.[17] Arab 'wisdom' was intuitive rather than intellectual, as beyond scientific check as all things Arabian: 'The European thinks, the Oriental only reflects,' theorised the journalist Meredith Townsend, 'and if left to himself the idea, turned over and over endlessly ... is part of the fibre of his mind.' This was as much a product of place as race: a traveller echoed, 'In ... desert countries ... the essential facts ... sink into you imperceptibly, until ... they are ... woven into the fibres of your nature.'[18]

So, at a moment when primitivism was in general seizing European imaginations,[19] agents' distraction from an empirical method of intelligence-gathering pointed a way forward. Rather than abandon the effort to grasp inscrutable Arabia, they invented an intelligence strategy that prioritised knowledge acquired through intuition over sense-data, stipulating lengthy immersion as the only effective preparation. By intuition, they meant the *acquired* ability to *think like an Arab*, an empathetic mimicry of the 'Arab mind', for, they held, 'Only by Orientals—*or by those whose long sojourn in the East has formed their minds after the Oriental pattern*—can the Orient be adequately described.' Agents like Norman Bray determined to 'merge ... in the Oriental as far as possible, [to] absorb his ideas, *see with his eyes, and hear with his ears*, to the fullest extent possible to one bred in British traditions'.[20]

Thus the intuitive ability to discern truth from the dross of Arabian deception was prized among these agents.[21] What made Arabist experts *expert* was their ability to see, like Arabs, beyond surface deceptions, to discern the real from the unreal, the mirage, the lie. Knowing Arabia was a matter of *genius*. 'Book knowledge' mattered little, for, Bray explained, 'we "sensed" the essence of a matter'. The gifted few seemed, to contemporaries, omniscient: 'Leachman ... knew everything.' Thus, at the outbreak of war, they exercised a remarkable influence over the tactics employed in the Middle East, not least the innovative use of airpower—it was there that 'the war proved that the air has capabilities of its own'.[22] And it was the agents' claim to a special knowledge about the region that made such confident innovation possible as Britain single-handedly occupied the three Ottoman provinces that make up present-day Iraq.

They were certain that the unmapped and unmappable desert would reveal its secrets to aircraft. The bird's-eye-view might, like intuition, offer vision beyond the mirages, sandstorms and horizonlessness that bedevilled two-dimensional observation. Early on, agents pined: 'Oh for some aeroplanes. If there was a country in the whole world eminently suited to these machines this one is: Flat flat as your hand.'[23] Aircraft could even enhance the expert's instinctual grasp. The Political Officer St John Philby (father of the more notorious Kim Philby) confessed that his 'altogether astonishing' first flight 'impressed itself deeply on my mind'; in a mere quarter-hour, the 'magnificent bird's-eye panorama ... doubled my knowledge of Mesopotamia'.[24] When aircraft began to arrive in 1916, military personnel affirmed their indispensability to 'obtaining quick and accurate information' in Iraq. Since 'in Mesopotamian battles, little can be trusted that is seen,' explained Brigadier General A.G. Wauchope, 'commanders are bound to rely on reports by aeroplane'.[25] In the desert, aircraft allegedly made it 'impossible for an enemy to alter his dispositions without discovery; the movement of a

Image 23: 30 Squadron above the Bazian Pass (undated).

few tents or shelters can be spotted at once, and there are no woods or buildings in which to hide his men'. In the Hejaz too, they were deemed 'the only means of overcoming the mirage' and the prevarications of natives; indeed, their information was 'the only sort that can be relied on'.[26] Aircraft, like the intuitive expert, could extract truth from an essentially deceptive land (in theory—in fact, there were real limits on using aircraft in the desert, as we shall see).

The agents helped the Royal Flying Corps develop aerial photography, which became central to its ability to improve geographical knowledge of the region. Gertrude Bell, having witnessed Lawrence's and Stewart New-combe's experiments in Cairo, inspired General Staff Intelligence (GSI) Basra to learn more about the technology. So, during a visit to Mesopotamia in 1916, Lawrence advised Basra on the uses of aerial photography. The hitherto ineffectual mapping section was soon able to turn out hourly editions of maps before attack, distributing them at the front by air. Contemporary and historical appraisals of wartime advances in aerial technology acknowledged aerial photography's precocious development in this theatre.[27]

Aircraft also promised to ease communication with tribes, officers and irregular troops marooned in the desert, since 'a good pilot could generally land by the unit itself, give them their accurate position and inform the commander of the situation personally'. With the desert functioning as an endless landing ground, pilots could act as *dei ex machina*, restoring a lost unit's bearings and ending its isolation.[28] It was not long before officials began to muse on the 'enormous political possibilities' of a technology that could so intensely impact morale and relations with distant tribes. It even provided access, of a kind, to otherwise forbidden sites. If British forces were not welcome in the Hejaz, the same objections did not exist to aircraft, which, Lawrence explained, Arabs found 'delightful'. Thus an air reconnaissance of Medina, where the Turks were holding out, was conducted in 1917. When the Iraqi tribes the British liberated got 'out of hand and require[d] a lesson', agents encouraged the use of an 'aerial raid with bombs and machine guns'. Many 'Politicals', a GSI intelligence officer noted sceptically, had come to regard aircraft 'as a panacea for all the ills to which tribal situations give rise'.[29]

Agents' ready resignation to such carnage might seem to fit awkwardly with their avowed affection for the region. However, it is important to recall that this generation of Arabists had not gone to the region in search of mere adventure but for the experience of the journey-quest. For this, they had to come back *having seen things*. They did not actually seek escape; they wanted to *enact* the role of the unknowing knight seeking escape but stumbling into great events that yielded real knowledge—following a romantic script of medieval adventure. Magnifying 'both defeat and victory', aircraft impressed Bedouin with British power, Bray explained to his superiors. Indeed, they magnified the British effort in Arabia to the epic proportions in which agents habitually conceived it; they were *ennobling*. Aircraft, despite and perhaps because of their lethal power, were their technological counterparts, 'knights of the air', vehicles on a hyper-quest. Then, too, this was a biblical land; from on high, the British pilot, a *deus* in *machina*, could enact the divine retribution that was fitting in such a land.[30] In short, it was because the intelligence community viewed Arabia as an inscrutable, delightful and otherworldly space that aircraft were used in these ways, at this time, in this region. Aircraft impinged on intelligence work in the Middle East in a way that they never did on the Western front.[31] They were associated with the intelligence community, who also favoured them as a means of personal transport. Indeed, aircraft were such an 'eccentric mode of transport for a high official' that the agents' (as much as the Arabs') oft-expressed 'delight' in them only fed the agents' reputation as unique, heroic, indeed superheroic, figures.[32] During the war itself, then, the emotional foundation for a peacetime aerial regime had already been laid: the desert sublime needed an aerial sublime.

## *The Aerial Sublime*

The British government turned naturally to its Arabist experts for advice when Iraq erupted in rebellion after the war. Life under protracted military occupation was not quite measuring up to the promise of 'liberation'; nor did post-war British policy, which, in a feeble attempt to fulfil the many diplomatic promises made during the war, attempted to dress Iraqi freedom in the new and relatively flimsy League of Nations garb of 'mandatory rule'. This mandate system assigned newly freed but allegedly immature countries to long-term, generally unsupervised tutelage under a more 'advanced' power. But Iraqis, and many others, suspected the term 'mandate' was but a semantic disguise for legalised imperialism; hence the massive rebellion of the summer of 1920,[33] just when the British were already facing mass nationalist resistance in Egypt, India, Ireland and elsewhere.

The militarily and financially overstretched imperial state fumbled for creative solutions to counter-insurgency. Early in 1921, Colonial Secretary Winston Churchill called a conference at Cairo, and there, with the help of the community of Arabist experts, it was decided to establish an Iraqi constitutional monarchy and to police Iraq from the air: rather than rely on expensive and unpopular troop deployments, the fledgling Royal Air Force would patrol the country, coordinating information from agents on the ground to bombard subversive villages and tribes. Air action was eventually used against Turkish and Najdi raiders (at a time when frontiers were very much a work in progress) as well as Kurdish and Arab rebels within Iraq itself.

As other scholars have noted, air control was cheap; hence, in large measure, its attractiveness. But reasons of cost and efficiency would have applied equally elsewhere, and indeed, after a brief incubatory period in

Iraq, air control was exported elsewhere in the empire, albeit in modified fashion.[34] So to understand the invention of this unprecedented scheme relying on a new technology whose uses had yet to be fully imagined, we need to look elsewhere. My argument is that the invention of air control depended on notions, dating from wartime, of Iraq's peculiar suitability to aerial surveillance. Lawrence, who was closely involved in the scheme's formulation, insisted at the outset that it was '*not* capable of universal application'.[35] Air control's theoretically panoptical power seemed capable of solving the peculiar information problems the intelligence community had come to associate with Iraq—the mirage, haze and lying natives. To the Arabist intelligence community, Iraq was destined for aerial control for aesthetic as much as practical reasons: the infrastructural austerity of air control seemed suited to a theoretically horizonless desert that allowed power to 'radiate' untrammelled 'in every part of the protectorate [*sic*]'. The region's actual topographical diversity—its mountains, labyrinthine marshes and varied deserts—when it was acknowledged, was held up as yet further proof of Iraq's suitability as a training ground for the RAF.[36] Then there was the much-touted 'natural fellow-feeling between ... nomad arabs and the Air Force ... both ... in conflict with the vast elemental forces of nature'.[37]

Finally, the many errors that plagued aerial control in practice were irrelevant in a region famed for its deceptions. Besides frequent reports of pilot disorientation, visibility problems and instances 'of quite inexplicable failures to identify ... whole sections of bedouin tribes on the move',[38] there were instances of aircraft bombing the wrong town,[39] and Bedouin finding cover in watercourses, hillocks and other features of the allegedly 'featureless' landscape.[40] But Arabia's timeless mystery provided a refuge from all this apparent inaccuracy. The civil commissioner and head of political intelligence, Arnold Wilson, explained that complaints about RAF observation failures were, like all information, necessarily exaggerated; after all, mirage prevented fair judgement of pilots from the ground. There was also little point worrying about casualties since assessing the effect of bombing operations was 'a matter of guesswork'.[41] Thus the air control experiment was pronounced entirely successful in 'this kind of turbulent country'.[42] In its Iraqi cocoon, the RAF was safe from criticism of its inaccuracy, protected by the notorious fallibility of all news emerging from 'Arabia'. Inscrutable Iraq provided the kind of technically forgiving space that air control needed to spread its new wings.

Image 24: 30 Squadron's DH9A in formation (undated).

This match made in heaven proved costly in Iraqi lives—a hundred casualties was not unusual in a single operation, not to mention those lost to starvation and the burning of villages. Whether attacking British communications or refusing to pay taxes at crushing rates, or harbouring rebels, many tribes and villages were ultimately bombed into submission.[43] In the wake of all this slaughter, critical voices emerged in Whitehall. The new War Secretary wrote witheringly, 'If the Arab population realize that the peaceful control of Mesopotamia depends on our intention of bombing women and children, I am very doubtful if we shall gain that acquiescence of the fathers and husbands of Mesopotamia as a whole to which the Secretary of State for the Colonies looks forward.'[44] This critique was amplified in the press and Parliament, where many had looked upon the Iraqi venture as outdated imperial foolishness from the very outset.[45]

Here, too, cultural notions came to the rescue. Besides inspiring the scheme, British representations of the emotional register of Arabia crucially defused criticism of its many practical failings, enabling it to last several decades. With haste, the Air Ministry responded to all the official displeasure, arguing in the main that air control was not unique in eliding the distinction between combatants and civilians. '[A]ll war is not only brutal but indiscriminate in its brutality,' they urged, pointing to the effects on civilians of naval bombardment, shelling, blockading, trampling by invading armies and so on; at least the lives of attackers were safer in air operations.[46] Speaking as unsentimental realists, the Air Staff even succeeded in convincing some 'of the great humanity of bombing', arguing that however 'appalling' and 'ghastly', it ultimately lowered even *enemy* casualties by forcing them to give up sooner in the face of 'continual unending interference with their normal lives'.[47] But behind this apparent stoicism lurked a particular passion about Arabia, as we shall see.

The Great War had certainly shifted notions about humanity and warfare. To many military thinkers, the moral imperative was to minimise casualties as a whole, rather than civilian deaths in particular, since modern combatants were merely civilians in uniform.[48] Nevertheless, the Air Staff did not really address the concerns of those critics who were equally offended by modern war's general brutality or of those who (rightly) considered aerial bombardment, in its all-seeing omnipotence, more lethal and terrible than older forms of barbarity.[49] But most saliently, for my purposes, their counter examples, naval bombardment, blockades and the like, were all *wartime* measures. The Air Staff paper was meant to discuss bombing as a *peacetime* security measure, a policing technique. What was permissible only in wartime in advanced countries turned out to be *always* permissible in Iraq. The air secretary acknowledged that there things happened 'which, if they had happened before the world war, would have been undoubtedly acts of war'.[50]

This military permissiveness was shaped by British Arabists' ideas about the kind of place they were dealing with: a mystical and romantic land existing somewhere beyond the pale of worldly and bourgeois convention, a place of perennial conflict and the violent emotions it unleashed. The RAF intelligence officer, John Glubb, later 'Glubb Pasha' of the Arab Legion, insisted, 'Life in the desert is a continuous guerilla warfare,' in which striking hard and fast was of the essence.[51] To the Bedouin, war was a 'romantic excitement' whose production of 'tragedies, bereavements, widows and orphans' was a 'normal way of life', 'natural and inevitable'. Their appetite for war was the source of their belief that they were 'elites of the human race'. In this view, it would almost be a cultural offence *not* to bomb them.[52] For their part, as the 'knights of the air', aircraft had brought chivalry, in the sense of honourable combat between elite warriors, back into an otherwise thoroughly grim and 'vulgarised' modern warfare. Arnold Wilson confirmed for the Air Ministry that the problem was only one of British public perception, that Iraqis were used to a state of constant warfare, expected justice without kid gloves, had no patience with sentimental distinctions between combatants and non-combatants, and viewed air action as entirely 'legitimate and proper'. 'The natives of a lot of these tribes love fighting for fighting's sake,' the former chief of the Air Staff, Hugh Trenchard, assured Parliament in 1930, 'They have no objection to being killed.'[53] In a place long romanticised as an oasis of a prelapsarian egalitarianism and liberty, defenders of air control could rest assured that Bedouin retained their dignity even under bombardment and were not miserable wretches deserving of a condescending pity.[54] After all, as the military theorist J.M. Spaight pointed out, chivalry was an influence quite distinct from 'the humanitarian one', which regarded with compassion 'those whom chivalry despised'.[55] Thus Iraqi women and children need not trouble the conscience, for, the British commander observed, '[Iraqi sheikhs] … do not seem to resent … that women and children are accidentally killed by bombs.' To them, Lawrence elaborated, women and children were 'negligible' casualties compared to those of 'really important men', conceding that this was 'too oriental a mood for us to feel very clearly'.[56] This was a population at once so orientally backward and so admirably manly and phlegmatic that, to a post-war imperium increasingly in thrall to cultural relativistic notions,[57] all principles of *ius in bello* were irrelevant. From the Colonial Office, ex-agents assured their colleagues in Iraq that 'Bombs dropped on men in the open seldom have much effect beyond fright,' and advised dropping the matter of results since observation of casualties was 'always misleading'.[58] Striking at an enemy practically begging for attack, and enjoying the bliss of wilful ignorance at the outcome, made air control sit more easily in the official mind. Only in Arabia, about which the British had long decided nothing could ever be really known and all could readily be tolerated, did such fecklessness make sense and thus make air control acceptable.

Besides its warrior ethos and mystery, officials drew on Arabia's biblical associations to explain air control's peculiar emotional purchase there. In 1932, when the inhumanity of air control was of pressing importance at the world disarmament conference in Geneva, the British high commissioner in Iraq argued that unlike the outrages inevitably committed by ground troops, 'bombing from the air is regarded almost as an act of God to which there is no effective reply but immediate submission'.[59] Lawrence similarly insisted that for Arabs, bombing was 'impersonally fateful'—'not punishment, but a misfortune from heaven striking the community'.[60] Arabia was a biblical place, and the people who lived there *knew* that; they expected periodic calamity and continual news of life and death. Bombardment was to them yet another kind of visitation. Air control played on their presumed fatalism, their faith in the incontrovertible 'will of God'. Such people could bear random acts of violence in a way that Europeans, coddled by secular notions of justice and human rights, could not.

Indeed, the problem with accuracy was partly dispensed with by the frank admission that 'terror' was the regime's underlying principle. Aircraft were really meant to be everywhere at once, 'conveying a silent warning'.[61] In an awesome terrain, awe was an entirely suitable emotion

Image 25: A 520lb bomb dropped on Sulaimaniyah in Kurdistan (1924).

Image 26: Another 230lb bomb dropped on Sulaimaniyah the following day (1924).

to conjure with. The terror principle rested on notions of the Arab propensity for exaggeration: where there was one plane, Arabs would spread news of dozens; a few casualties would instil fear of hundreds.[62] Air control would work like the classic panopticon, for, an official memo put it, 'from the ground every inhabitant of a village is under the impression that the occupant of an aeroplane is actually looking at *him* ... establishing the impression that all their movements are being watched and reported'.[63] If pilots could not be sure whether they were looking at 'warlike' or 'ordinary' tribes, Bedouin could not discriminate 'between bombing and reconnaissance expeditions'.[64] The foundation in terror was, rather paradoxically, held up as the regime's ingenious proof against inhumanity. In theory, invisible airpower would bloodlessly awe tribes into submission. Alternatively, interference with its victims' daily lives, through destruction of homes, villages, fuel, crops and livestock would produce the desired result.[65] Of course, the inhumanity of the system stemmed from its inability to distinguish between combatants and non-combatants in its violent destruction of their material wealth as much as their lives. Moreover, the idea that terror could be produced bloodlessly was entirely theoretical. Early RAF statements acknowledged that eliciting the 'moral effect' depended on demonstrations of exemplary violence, which could hardly be accomplished without loss of life.[66] But most importantly, official tolerance for such a level of fear and violence depended on their confidence that Arabs could take it.

As a biblical space, Arabia was a place of elemental emotions that transcended worldly morality. The Bedouin, Glubb said, possessed 'depths of hatred, reckless bloodshed ... lust of plunder of which our lukewarm natures seem no longer capable ... deeds of generosity worthy of fairy-tales and acts of treachery of extraordinary baseness'. Their 'love of dramatic actions' outweighed 'the dictates of reason or the material needs,' and even, the General Staff affirmed, overcame the 'inherent dislike of getting killed'.[67] In this last bastion of authentic experience, bombardment could be accommodated as yet another vitalising experience—shared equally by airmen. No group did more to fulfil this zealous vision of air control than Abdul Aziz ibn Saud's puritanical avant-garde forces who continually raided into Iraq from neighbouring Najd. Gertrude Bell, a powerful intelligence and administrative force in mandatory Iraq, was fiercely proud of 'our power to strike back' at the Ikhwan, who, 'with their horrible fanatical appeal to a medieval faith, rouse in me the *blackest hatred*'. All concerns about cruelty were moot in a region 'notorious for ... cruelty and ... inhuman injustices' perpetrated by these 'Die-Hards of the Moslem world.'[68] Bloodlust was the way of the place, and the mantra was 'when in Rome ...'

These clashes between good and evil transformed 'pacification' into a series of episodes of cosmic significance. During the rebellion, the Political Officer Captain Leachman wrote chillingly, 'I should like to see ... a regular slaughter of the Arabs in the disaffected areas.' His adoring biographer, fellow Political Officer Norman Bray, describes him living in constant fear of assassination, concluding, 'No wonder he ... revelled in dropping bombs on Arabs.' The transposition of real Arabia into the Arabia of myth, the consummate spy-space, made bombing palatable, even to those who believed they would revile it any other context. The vindication of air control did not rest merely on a simple racist dehumanisation of Arabs;[69] it grew out of long-circulating ideas about Arabia as a place somehow exempt from the this-worldliness that constrained human activity in other parts of the world. There, heroes could reach the most exalted heights and villains the profoundest depths; there, as in literature, agents could find escape from the pitiful reality of human suffering into an exalted sphere in which everything possessed a cosmic significance. There, where each soul was free to work out its cosmic destiny, violence was entirely *personal*: Leachman's assassin, Dhari, was the single exception to the general amnesty granted after the rebellion. He was not seen as a member of that uprising, but as someone who had violated the honour between two men; the Iraqi unrest was reconfigured as an episode of medieval battle, in which the mettle of chivalric men was tested and rewarded. In this 'supreme crisis,' Bray wrote, 'Every quality [Leachman] possessed, *even his faults*, served the cause of England.'[70]

Thus did the region's apparent emotional charge help British officials evade uncomfortable questions about the aerial regime's inhumanity. But the presence of British political officers on the ground claiming an intuitive understanding of the place also enabled the regime to project an actively humane image. If passion laid the groundwork for the dispassionate gaze, a claim to empathy facilitated the emotional estrangement that is essential to remote control.

From the outset, despite their fondness for the aerial regime, the intelligence community insisted on the enduring need for 'men who are specially gifted, who have got the feeling of the Middle East in their blood'.[71] Air patrolling was so new, the infrastructure of landing grounds so inchoate, and pilots so inexpert in deciphering the terrain that ground agents were deemed crucial supports to this system.[72] That desert signs could be only subjectively interpreted remained the intelligence orthodoxy. Aircraft relied on ground experts for political work, such as feeling out the intentions of local Arabs and ascertaining when the desired 'moral effect' had been achieved in order to avoid unduly prolonging operations.[73] The RAF's own group of special service officers worked closely with pilots and regular political officers, quickly absorbing the latter's tactics. Intuitive ability and canny knowledge of local custom were deemed indispensable to acquiring the information required for bombardment, given the 'peculiar mentality' of the tribesmen, 'who,' in Glubb's words, 'deemed it a duty to receive and to welcome a guest, although he was mapping their villages with a view to bombing them and told them so'.[74] The new agents continued to claim empathy with, even love for Arabs as the source of their genius; immersion enabled them to perform the near impossible task of understanding another race, explained one, allowing them to 'interpret … [the Arabs'] mind'.[75] The RAF trusted special service officers to 'sense impending events' accurately (if not to 'dig down to the facts').[76] Successful bombardment was often attributed to their genius, wireless technology allowing them to communicate swiftly with aircraft—from their mouths to God's ears.[77]

Defenders of air control pointed out that, far from rendering them aloof, aircraft enabled officials to roam freely, facilitating greater intimacy between administrators and Iraqis and preserving the tradition of British imperial benevolence.[78] Of course, agents' untrammelled mobility also ensured that the RAF could, the Air Staff explained, '[pick] out the right villages … to hit … when trouble comes'.[79] By this ironic logic, the RAF's successful persecution of a village testified to their *intimacy* with people on the ground, without which they would not have been able to strike it accurately. Indeed, the claim to empathy ultimately underwrote the entire air control system with its authoritative reassurances that bombardment was a tactic that would be respected and expected in this unique land. As late as 1957 RAF Marshal Sir John Slessor defended the regime by pointing to the fact that special service officers, who knew the place best and 'became so attached to their tribesmen that they sometimes almost "went native",' were rarely critical of air control. Well into the eighties, Glubb continued to insist, 'The basis of our desert control was not force but persuasion and love.' In 1989, a military historian also vindicated the regime by citing Glubb—since, 'No European was ever closer and more sympathetic to the Arabs than [he].'[80]

At the end of the day, the claim to empathy was built, literally, on sand. From its Edwardian invention as an intelligence epistemology, the agents' cultivation of empathy had signalled not the recognition of a common humanity but an effort to transform the self to cope in what they saw as a radically different physical and moral universe. After the war, aspiring agents, inspired partly by the legends surrounding their predecessors, continued to venture to Arabia to escape the bonds of *too much* civilisation, to recover a noble, free spirit lost to 'utilitarian' England. Their effort to gather intelligence in the Middle East began with the same baptismal sensations of moving in a fictional, unreal, biblical, enchanted and uncanny space. They too found in the desert sublime a remembrance of God, a rekindling of faith. Their travel in the desert was still understood as an escape into the blue, a truant fulfilment of patriotic duty. Glubb knew that 'in the desert I was alone. The government was indifferent.' To enter Arabia was still to exit the *customary* world, since, as he put it, 'The desert is a world in itself …'[81]

The 'extraordinary and romantic' world of the RAF in Iraq compounded this feeling of being in a world apart. Its tenuous links to 'civilisation' through a miraculous wireless infrastructure, and rumours of Lawrence's presence in the ranks, only fed the Arabian mystique. If flight over the austere biblical terrain reached new heights of sublimity and divinity, it also produced 'quite a bad effect upon one's nerves,' a feeling that 'the end of the world had really come,' according to one RAF official. For new pilots, this 'sense of being lost at sea' was deemed a crucial 'mental factor'. Pilots could identify 'that air of quiet weariness which comes to those who have been in the desert too long,' and fell prey to a 'nameless terror' that made them go mad over time.[82] This was not a place for empathy, but for total psychic breakdown; without some kind of bracing, Britons risked losing their minds. Emulation of Arabs was intended to enable their survival in this extraterrestrial space, but did not produce true compassion for the Arab victims of the surreal world of bombardment they actually created by pulling the strings of fate from the sky.

Ordering bombers was thus entirely consonant with the sensibility of the Arabist agent enchanted with notions of Arabian liberty. The agents loved Arabia for its otherworldliness, and it was that quality that also made it fit to bear the unearthly destruction wreaked by bombers. Britons considered the moral world of Arabia distinct from their own.[83] From the outset, the intelligence project in Arabia had been infused with a philosophical spirit, which did not depart it at this stage. The search for passion, and passion itself, produced among certain critically positioned British Arabists a willingness to impose a level of violence and pain in Arabia that, they knew, the rest of the world could not take. Airpower was not, for them, defined by its dispassionate gaze, but by a miraculous quality, a sense that it was not fully of this world.

This Arabian window of emotional acceptability opened the door to wider and eventually more dispassionate uses of aerial bombardment: in 1921 the Air Staff deemed it better, for fear of allegations of 'barbarity', 'to preserve appearances … [and] avoid emphasizing the truth that air warfare has made [distinctions between military and non-military targets] obsolete and impossible. It may be some time until another war occurs and meanwhile the public may become educated as to the meaning of air power.'[84] Iraq offered the Air Staff a means of selling the new warfare by exhibiting it in a famously romantic and chivalric place where, everyone knew, the bourgeois rules lately exposed by the war as utterly bankrupt did not apply anyway. It worked: eventually British bombs fell all over the world, including Europe. The gruesome relish evident in a 1924 report by the officer commanding Squadron 45 in Iraq is striking in this regard:

[T]he Arab and Kurd ... now know what real bombing means, in casualties and damage; they now know that within 45 minutes a full sized village ... can be practically wiped out and a third of its inhabitants killed or injured by four or five machines which offer them no real target, no opportunity for glory as warriors, no effective means of escape.

This officer later achieved distinction, and, as one scholar puts it, 'in the ruins of this dying village one can dimly perceive the horrific firestorms of Hamburg and Dresden,' for the officer was none other than Arthur Harris, head of Bomber Command in the next world war.[85] During that war, it was largely under Harris's influence that Prime Minister Churchill warded off periodic pangs of conscience about bombing Germany with faith in the 'higher poetic justice,' that 'those who have loosed these horrors upon mankind will now in their homes and persons feel the shattering stroke of retribution'.[86] It was the Ikhwan all over again, and Europe itself had become the scene of a clash between good and evil.

### The Legacy of Love

As the affairs of the Second World War dramatically evidence, at some point the notion that Iraq was peculiarly suited to aerial control vanished. Indeed, in the 1930s, when aerial control was being actively exported to other parts of the empire, experts were disparaging as 'absurd' the contention that 'some peculiar quality about the country ... has enabled aircraft to achieve in Iraq what they could not achieve anywhere else'.[87] I have tried in this chapter to expose the power of that original contention about Iraq's peculiar qualities and its ability to sustain faith in the scheme throughout the interwar period despite often vocal scepticism of its efficiency and humanity. Certain cultural notions about Arabia, inspired by the search for a particular kind of affective experience, shaped the practical organisation of surveillance in the Middle East—and its violent excesses. The 'idea of Arabia' circulated by the intelligence community during the early decades of the twentieth century enabled the British official mind to, with mostly clear conscience, indeed with confidence in a consistent paternalism, invent and implement the world's first air control regime. Criticism was never totally silenced,[88] but enough people were convinced, even impressed, for the regime to remain viable well beyond Iraq's formal independence in 1932. The country was swiftly reoccupied during the Second World War, and it was only the revolution of 1958 that at last sent the Royal Air Force scuttling home. This story forces us to re-examine received wisdom about the relatively benign nature of the interwar British state[89] and begin to understand how British officials reconciled genuine ethical scruples with the actual violence of imperial policing in the Middle East. Dispassion mattered less, in this instance, than passion; love more than hate. Intimacy is no guarantee of humanity, or even true empathy.

Perhaps more saliently, for my purposes, the myth of air control's success and humanity has critically informed American tactics in the region since 2003. This is partly the result of institutional collaboration between the two countries during the Second World War and the Cold War, when the United States took over as the masterminds of covert activity in the region—indeed, a mere two years after the British departure in 1958, the CIA attempted to assassinate the new Iraqi republic's head of state. It is also partly because military historians, citing the orientalist experts who created the regime, as we have seen above, have confirmed their view of the region. They in turn have been cited by USAF academics in their studies of airpower in counterinsurgency.[90] Even historians critical of aerial bombardment have proven susceptible to the notion that aerial control actually *worked* against the desert's 'clearly defined, completely visible targets'.[91] This was despite the numerous errors aircraft made and the unexpected resourcefulness of desert Arabs. Media and cultural representations of Iraq continue to portray a site of unique sensory experience centred on the desert sublime and urban inscrutability. The presumption of special insight into a special people remains, too: thus, in the very first year of the most recent Iraq war, an American captain defended harsh security measures in Iraq, 'You have to understand the Arab mind. ... The only thing they understand is force—force, pride and saving face.'[92] These continuities meant that, as early as 2003, the Pentagon began to dream of replacing troops in Iraq with airpower that could 'strike everywhere—and at once'.[93] In 2008, Defence Secretary Robert Gates launched a secret programme that put hundreds of unmanned surveillance and attack aircraft in orbit over Iraq and Afghanistan and later Pakistan too. After President Barack Obama came into office, drone use increased dramatically. He and Gates became convinced that constant, ubiquitous drone surveillance coupled with airstrikes triggered remotely will solve our tactical problems in these regions.

In June 2009 the UN Human Rights Council condemned the US failure to count and disclose, much less prevent, civilian casualties from drones. The US government still refuses to share even the most basic information about drone use and attacks. It is difficult to take the military's certainty of drone effectiveness on trust, when it neither counts nor identifies those killed. Quite the opposite—official secrecy fuels rumours of the worst, and periodic scandals prove that often the worst is true. People who live in these countries are cynical (and wise) enough to assume from past experience that the secrecy covers up facts too grisly for public airing. Memory of the past crucially shapes the military and political dynamics of any aerial strategy in the region. Colonel David Kilcullen, a counterinsurgency expert and former adviser to General Petraeus, has publicly attested that people in the region see the drones as 'neocolonial'. He calls their hit rate immoral. Lord Bingham, a retired senior British judge, compares them to cluster bombs and land mines, weapons that have been deemed too cruel for use. While military experts warn of the impossibility of usefully analysing the enormous amount of data the drones collect, Iraqis proved adept at using cheap software to hack into the video feeds from the drones.

But generalised terror, rather than precise strikes against insurgents, remains the unspoken tactical foundation of today's plan. Enthusiasts praise not only the data but the uneasiness produced by the hunter–killer machines' 'persistent stare capability'. Oddly, they believe this imprisoning gaze will help win Iraqi hearts and minds, as if 'bad guys' can be surgically removed and are not every day produced by precisely such impositions. News reports have testified to Iraqi civilians' fear of the drones—thanks largely to their inevitable imprecision. So, if love underwrote the invention of aerial control in Iraq, and terror was its desired outcome, those emotions have been only partially revolutionised in more recent iterations of the technology's use. The love has vanished, but the terror remains, as well as the reasoning that terror somehow saves lives (regardless of its political costs). But as we ponder the coldness of the drone operator and the impossibility of empathy through remote control, it is worth recalling the powerful passions that put a joystick in those unassuming young hands. Perhaps the simplest summation of this story is that chilling piece of old wisdom: the road to hell is paved with good intentions.

There is a third emotional state at stake here, too: indifference. Air control was a mechanism of control for a region in which more overt colonial rule was a political impossibility, since, as the Air Ministry theorised, 'in countries of this sort ... the impersonal drone of an aeroplane ... is not so obtrusive as ... soldiers'.[94] But the discretion was not only for the sake of Iraq's citizens. It was as much for the sake of British public opinion: indeed, the scheme's cheapness was explicitly intended to elude the democratic check of taxpayers.[95] Air control allowed covert pursuit of empire in an increasingly anti-imperial world; the secrecy surrounding the regime guaranteed the indifference of the British public. Similarly, today, the reign of drones enabled the narrative of US troop withdrawals to appease American public opinion. The plan was to relocate 'withdrawn' forces to forward operating bases where they could discreetly hunker down for a long *entr'acte* off-stage in expensive, built-to-last facilities. They would be renamed 'trainers' and 'advisers', making the American 'invisible'. Obama made it clear that 'departure' would not mean withdrawal of advisers, special forces, drones, and helicopters. Withdrawal would work the same way as the 1932 British grant of independence, which was confessedly nominal; the Air Staff made it clear that the change would be 'more apparent than real'[96] since Iraq was 'an oriental country where intrigue is rife'. Making it just that themselves, they privately conceded, 'we really have no defence'.[97] In the event, Obama did not succeed as well as he had hoped in maintaining a rump American force on the ground because of Iraqi concerns about sovereignty—history had taught Iraqis a bitter lesson. Hence has Iraqi airspace provided a corridor between Iran and Syria and Israel.

The American public knows little of the debacle of negotiations surrounding these endgame details—just as we were told little about the hundreds of thousands of Iraqis killed during the eight-year war there. Surveillance drones remain in the country officially for the protection of

American personnel, but Iraqi officials consider them and the outsized American embassy to be affronts to Iraqi sovereignty. Meanwhile, discreet imperial aerial control remains the American *modus operandi* elsewhere in the war on terror, stoking Iraqi suspicions that they too will be subjected to fresh strikes under some pretext. As Iraqis continue to suspect a hidden hand guiding their country's fortunes, the discreet American presence significantly compromises the legitimacy of the ostensibly sovereign Iraqi government. The Iraqi government looks feeble, corrupt, and ineffectual and the country's claims to sovereignty laughable when the sky is full of American drones with no sign of an Iraqi air force. In such a context, security there will remain a mirage and Americans remain susceptible to a paranoid groupthink about Iraqi politics, making real withdrawal impossible.

# 11

# TARGETING AFFECTIVE LIFE FROM ABOVE

## MORALE AND AIRPOWER

*Ben Anderson*

## *Introduction: Violence and Targeting Life*

The beginning of the US-led 'Operation Iraqi Freedom' in 2003 was marked by 'Shock and Awe', an explicitly performative display of violence. Shock and Awe was simultaneously an 'effects-based' aerial bombing campaign and a perception management operation where images of the event became weapons. As a way of warfighting, it mixed the punctual interruption of shock with the rapt attention of awe. As such, Shock and Awe resonated with everything from the Kantian sublime to electric shock treatment. The 2003 bombing campaign provides one example of the focus for this chapter—the relation between affects and the targeting of life from above. The target for the effects of 'shock' and 'awe' were collective affects interchangeably named as 'morale', 'will to fight' and 'popular support'. A 1996 strategy document—where the term Shock and Awe is first used—explicitly names the injuries that Shock and Awe aims to bring into being. These are forms of affective harm and damage that disrupt habitual perception and are named vaguely but ominously as 'comatose and glazed expressions', 'feeling of impotence' and 'fear of his [*sic*] own vulnerability and our own invincibility'.[1]

The status given to targeting what I will refer to as morale (given that 'popular support', 'will to fight' and 'morale' are used interchangeably) in the doctrine of Shock and Awe is not distinctive to the 'war on terror'. Of the innumerable affects of war, morale is unique: both because it has long been part of war in the form of 'esprit de corps' or 'intense fellow feeling',[2] and because it has subsequently accumulated a promise in Western military thinking. Morale and other amorphous collective affects such as 'will to fight' are elusive targets which, if destroyed, promise to guarantee that war ends in victory. Hope has, therefore, long been invested in targeting morale—the hope of victory by damaging and destroying an enemy.[3] Of course, since the German Zeppelin and Gotha bombing raids over England during the First World War, the affective effects of airpower have both enchanted and terrified entire populations. The hope invested in airpower is inseparable from this fascination with the destructive power of falling bombs and flying machines. As Sven Lindqvist, Peter Adey and others have shown, the vertical geographies of bombing from the sky are inseparable from a history of the intentional production of terror from above, and a history of attempts to measure and bear witness to that terror.[4] Shock and Awe is but one name given to what became known after the First World War as the 'morale effect' or 'terror effect' of bombing from the air. In this chapter I address, then, a simple question: given its long and varied status as part of war, how is morale assembled as a target in the contemporary forms of military violence that make up the US-led 'war on terror'? Put differently: under what conditions has targeting morale accumulated a violent hope—the hope of victory over an effaced, or annulled, Other-designated 'enemy'?

The chapter addresses these questions through an analysis of the targeting of collective affects in network-centric 'effects-based' airpower doctrine (in relation to aerial bombing doctrine, and more recent counterinsurgency doctrine). I argue that a hope has been invested in targeting morale in the context of changing forms of war and warfare. The reason? Because collective affects are taken to be both the cause and consequence of how the enemy, understood as a network, holds together. This version of the *form* of morale and other collective affects has, I will show, a troubling, and perhaps even terrifying, consequence—all of life is targeted and subject to intervention through the creation of named biopolitical 'effects'. 'Shock' and 'awe' are but the most high-profile examples of 'effects' which promise to create life and bring death by targeting affective life from above.

*Anticipating Violence and Networked Affects*

In the context of the multiple relations with life that generate and sustain the 'war on terror', targeting morale is closely entwined in the resurgence of sovereign power associated with airpower under conditions of occupation. The catastrophic role of airpower is, of course, sadly evident in the ruined civilian geographies of Iraq and Afghanistan. My aim in this chapter is not, however, to bear witness to the visceral affective geographies that follow these and other examples of violence, although this work is clearly necessary and important. My focus is, rather, on how collective affects become the 'object' of forms of power through processes of targeting. Or, to put it differently, my interest here is on how collective affects are known, rendered actionable and made subject to intervention. The first stage in such targeting is to name the target. As previously mentioned, the term morale is used interchangeably with other names for affects. On the one hand, the generic affective state of the 'enemy' is targeted (variously named as 'will and cohesion', 'opponent's psychological will', 'popular support' or the 'will to fight'). On the other hand, specific affects that are the property of the individual or collective bodies of combatants and civilians are targeted (e.g. 'hope', 'confidence' or 'optimism').[5] The first effect of these acts of naming is to make distinctions within affective and emotive life. Affectivity is on the road to being controllable and manageable once housed within the stable unity of a name. Naming also functions to individuate within lines of descent. In this case, and as mentioned above, morale carries with it the long history of targeting morale and the will to fight. Naming is, of course, referential. The name 'morale' refers to the present existence of something real that supposedly pre-exists the act of naming. At the same time, naming can be evocative.[6] Naming an affect sets forth chains of association that evoke embodied, sensory, life.

My focus is, therefore, on how a 'version' of these named collective affects has been cultivated in the intersections between think tanks, the US military and military journals.[7] What does it mean, then, for collective affects to be targeted in military thinking? Versions of morale, popular support and will-to-fight are cultivated through bombing surveys, journal articles, reports, official doctrine and other texts; all are explicitly future orientated. Even as they draw on past examples, the texts of military thinking address future encounters with an imagined 'enemy'. Military thinking is, in other words, anticipatory. But in a very specific way. Military thinking anticipates better ways of destroying or damaging through dialogue with the actions of a negated other (i.e. an 'enemy') in the context of future events ('battles' or 'wars'). It aims to be purely operational. But the operationality of its 'concepts' is perpetually deferred. There is no guarantee that such ways of injuring will ever be put into practice. The referent object for military thinking is always a war or battle to come. Military thinking on targeting collective affects is, in short, animated by promises and threats. It aims to seize possession of a future. Perhaps in naming targets—and in all the other names given to targets, affective or not—we therefore have an example of what Denise Riley terms 'wishful naming'. By which she means a naming that would ward off danger and avert crises. A naming that would promise immunity to those who are able to destroy or damage the target.[8]

Let's take one example of such 'wishful naming' from the milieu of thinking that pre-dates the current 'war on terror' but nevertheless surrounds Shock and Awe. The example has been cited in recent discussions of how to target 'insurgents' or 'terrorists' in counterinsurgency operations.[9] A 1999 special issue of the US Air Force journal *Air Power*, which focuses on 'Targeting Morale', names the 'effect' to be created over the bodies of soldiers in the 'battlespace' as a combination of six 'stressors'. One of these stressors—'helplessness'—is described in the following way:

The feeling of not being able to fight back is a major combat stressor. This often stems from a belief that the enemy's weapons are superior and one has no defense. This often leads to feelings of impotence and lack of control. These feelings often lead to panic.[10]

Taken together, 'helplessness' and the other stressors will produce a collective affect of 'hopelessness' that would disrupt morale and thus enable military victory. The promise attached to targeting collective affects occurs, however, under particular conditions of violence. The 'effect' of 'hopelessness' promises to damage and destroy in the context of the spatial-temporal dispersion and indeterminacy that characterises a network. A set of affective effects—the other stressors are 'claustrophobia', 'noise', 'ignorance', 'isolation' and 'fatigue'—are understood as properties of individuals *and* networked phenomena held in relation with other elements in a battlespace.

It is necessary, therefore, to place the targeting of named collective affects in the context of the 'network' as the anticipated space of violence for the US military. Network-centric warfare responds to a set of changes to the conditions of state violence and war in a post-Cold War context. These pre-date the present 'war on terror' but have crystallised and intensified since 9/11. They include the proliferation of asymmetric struggles, low intensity conflicts and the reduced frequency of state-vs-state wars in the context of the hegemonic position of the US military. Although its precise meaning is contested, network-centric warfare can be briefly summarised around two changes in warfare and warfighting.[11] First, it names a change in the spatial formation of military technologies, doctrines and organisational forms from one organised around the mass state-vs-state mobilisation of standing armies in 'total war', to one organised around the targeting of elusive and unstable 'networks' of state and non-state actors—networks that are characterised by their spatial/temporal dispersion and indeterminacy. Second, it names the process of warfighting becoming networked in response to those changes in the enemy, and in conceptualisations of the enemy.[12] Notwithstanding differences in the form of networks, such as 'chain or linear' or 'hub, star, or wheel' networks,[13] the presumption is of the relationality of the entities that make up the battlespace.[14]

So the network is now the designated space of violence, while the enemy is understood as a network sited within a broader series of networks. It is within this context that morale is established as a certain sort of target and a violent hope has been invested in its targeting.

*Targeting 'Morale' and De-Contextualising/Disappearing*

The simple act of targeting, in which a target is seen and taken aim at, concludes a complex operation. As Samuel Weber argues, targets only appear or materialise in relation to a specific tactical or strategic problem.[15] Morale, popular support and will-to-fight emerge in 'effects-based airpower' as targets in relation to a certain problem. A problem that follows from conceptualising the enemy as a complex network sited within a complex network. This can be summarised in a simple question: how do networks 'hold together'? Or, more precisely, how, in the context of the spatial/temporal dispersion that characterises the form of the network, do heterogeneous elements gain and retain a certain consistency? The threat is that a *decisive* target that disrupts how an enemy 'holds together' will not appear because of the radical relationality of the elements within the network. The problem of the failure of a target to appear is compounded when discussing blurring or swarming—the forms of movement taken to be proper to the logic of networks[16]—and how they disrupt such a search for the centre of a network. If the enemy is a network—sited within a network that includes military actions—where, for example, is the enemy's 'centre of gravity'? Where are the enemy's 'vital' nodes? What are the 'critical' relations?[17]

It is in relation to the strategic problem of how to target a network—and the accordant threat that a network cannot be targeted—that morale and other collective affects accumulate a promise as a *certain type of target*. A target that can be found throughout the enemy network because collective affects are central to how that network develops and continues. Morale, will-to-resist and popular support are given quasi-spiritual qualities as the source of a networked enemy's ability to continue war. The act of naming these pre-determined targets offers, then, a 'certain island of predictability' in the 'ocean of uncertainty' that is network-centric war.[18] Naming morale as a target promises to '[d]ispose of the future as if it were the present'[19] by anticipating how a target network or system will no longer hold together.

Before collective affects can be subject to intervention, they must first be known and rendered actionable. Most explicit discussion of this issue has centred on how to target morale, where a heterogeneous set of forms of knowledge are drawn on. Consider, for example, an argument in the *Air and Space Power Journal* for the use of airpower in counterinsurgency operations.[20] In a discussion of the psychological 'effects' of airpower, morale is targeted by reference to anecdotes from Afghanis who have experienced aerial bombing, the documented experience of Second World War soldiers and aphorisms from strategic thinkers and military leaders. Military thinking on airpower more broadly cultivates a version of morale through reference to numerous forms of evidence. These do not add up to a single paradigm through which 'morale' is definitively named and known. Quite the contrary. Beyond a consensus about what morale does—it sustains the ability to continue with war despite present suffering—there is no consensus as to what it is. Versions of what morale is proliferate across: classic strategic texts (such as Sun Tzu and Clause-witz) in which morale is a universal form of 'fellow feeling' intertwined with 'will'; social science and humanities work on the experience of bombing, where morale is akin to a shared capacity to sustain belief in the future in the face of present suffering;[21] and anecdotes, lore and tacit military knowledge where morale is a possession of subjects, and becomes akin to a shared feeling of being alive based on a sense of belonging to a collective. In addition, other work on targeting morale makes use of the biopolitical techniques of surveillance and measurement that monitored 'morale' as a property of domestic populations in 'Total War'.[22]

What matters about this unruly proliferation of truth procedures is not the content of what morale is—whether it includes within it optimism, confidence, hope and so on—but the consequences for the *form* of morale. These are twofold. On the one hand, morale is not transparent to techniques of knowledge. Across these sources, morale accumulates an aura of indeterminacy and elusiveness. It is described, for example, as part of the 'fog of war' or as existing in 'shades of gray' as 'an elusive but critical factor' in war.[23] Indeterminacy, and elusiveness, haunt the various attempts to capture 'soft factors' such as morale through techniques of measurement and calculation (see the US Department of Defense's current JWARS simulation or Perception Assessment Matrices).[24] On the other hand, the indeterminacy of 'morale' as a target is multiplied when supplemented by other questions about where affectivity is located—that is whether morale is a property of a population or individuals or a property of mind or body.[25]

If a target must be made to appear as such—made present to take on a promissory status through de-contextualising—targets also disappear. As McKenzie Wark argues, targeting is a fight to 'hit' the target before it is no longer present.[26] The absence of a target, therefore, normally denotes the failure or limit of targeting. This is not the case when targeting morale. Indeed the opposite is the case. It is precisely the movement of morale from presence to absence that is central to its promise. Morale promises as a target because it cannot be reduced to the status of a self-contained object. Morale is only knowable in its effects, specifically how an enemy network continues or will continue the strategic, tactical and logistical activities that make up war. In short, morale cannot be known in and of itself. Techniques of measure and calculation must know a range of factors seemingly unrelated to morale in order to know morale. The disappearance of morale as a direct target has been met with the constant deferral of morale on to a range of other targets throughout a network.

Let's illustrate this process whereby, on the one hand, morale as a target disappears while, on the other, what is to be targeted multiplies. This move can be seen in Colonel John Warden III's influential 'five ring theory' of airpower, which was central to the air campaign in 'Operation Iraqi Freedom' (especially in the initial 'decapitation strike'). In a 1996 paper, he argues that the focus of military planners should be on the capabilities of 'physical' systems because they are '[i]n theory, perfectly knowable and predictable', whereas capabilities such as 'morale' are '[b]eyond the realm of the predictable in a particular situation because humans are so different from each other'.[27] But Warden stresses that morale must, however, still be targeted as it is central to the 'national will to fight'. It just cannot be targeted directly. Morale is to be targeted through action on the various 'rings' of power—fielded military, population, infrastructure, system essentials and leadership.

Although morale cannot be targeted 'directly', it can be targeted 'indirectly' because other targets 'hold together' or 'manifest' morale in the context of complex networks. For Colonel P. Meilinger, to give another example, 'national will' is similarly an 'amorphous and largely unquantifiable factor' in 'all war' that cannot be targeted 'directly' ('national will' is used interchangeably with 'morale').[28] Consequently, the challenge is:

[t]o determine how to shatter or at least crack that collective will. Because it is an aggregate of so many factors and because it has no physical form, attacking national will directly is seldom possible. Rather, one must target the manifestations of that will.[29]

It is possible, therefore, to target morale 'indirectly' with the aim of damaging it by acting on its 'manifestations'. Meilinger takes these to be 'military capabilities' but, exemplifying the expansion of war to all of life in network-centric warfare, he includes within 'military capability' '[t]he sum of the physical attributes of power: land, natural resources, population, money, industry, government, armed forces, transportations and communications networks, and so forth'.[30] Morale becomes immanent to all of life. But because it is difficult to destroy capability entirely:

The key lies in selectively piercing this hard shell of military capability in one or several places, thereby exposing the soft core. Through these openings, one can puncture, prod, shape and influence the national will. In most cases, will collapses under such pressure before capability has been exhausted.[31]

In the context of the disappearance of morale as a 'direct' target that can be de-contextualised, and thereafter made a fixed presence, 'effects-based' warfare becomes a means of 'exposing' morale by targeting its constitutive network relations. Morale is deferred on to an inexhaustible set of 'manifestations'. The erasure of morale as a fixed object is integral to sustaining the promise of targeting 'morale'.

Any act of targeting normally relies on an attempt to bracket space and time as media of alteration and alterity. The aim is to locate, and fix, a target through an act of de-contextualisation.[32] Morale, however, cannot be known 'in itself'. It can only be known through its varied and varying manifestations. 'Effects-based' targeting from above thus involves what I would call the virtualisation of morale-as-target. This is virtualisation in Levy's precise sense of 'an "elevation to potentiality" of the entity under consideration'. Such an 'elevation' involves a process whereby the target (in this case morale) comes to exist as a real virtuality found across life. So that '[i]nstead of being defined principally by its actuality (as a "solution"), the entity henceforth finds its essential consistency in a problematic field'.[33] The outcome is that morale cannot be targeted as such. It is not an object with fixed coordinates. A version of morale is assembled that makes morale equivalent to the secret source of an enemy's ability to sustain the activities of war. But it can be targeted 'indirectly' by acting over those 'manifestations'. The result is that all of life is potentially targeted. This is the only way to expose a 'morale' or 'national will' that is both everywhere and nowhere. The result is the creation of a 'target-rich' environment—*life*.[34]

*Providential and Catastrophic 'Effects'*

Morale (and popular support or will to resist) disappear as fixed targets. But it is precisely for this reason that a hope has come to be invested in their targeting. By becoming immanent to a network, collective affects are elevated to the cause and effect of a network holding together. Targeting them from above is, in short, the answer to the strategic problem of how to damage or destroy a network defined by its spatial-temporal dispersion. The very act of targeting in 'effects-based targeting', that is the process of de-contextualising, is therefore an originary act of violence. What is to be targeted multiplies even as the cutting of network relations—the act of de-contextualisation—fails. As such, power exercised in a hierarchical form as domination is integral to 'effects-based' targeting *as a process of targeting*. Or, put differently, being exceptional to the network is the condition for violence, where violence is understood in its widest and most elemental sense of a force that is exerted by one thing over another.

How, then, does airpower promise to act over targets that disappear? In Shock and Awe it was through destroying an enemy's ability to perceive, including 'decapitation strikes' and, we should note, the extensive use of psy-ops such as leaflet drops and broadcasting to persuade soldiers not to fight.[35] Airpower doctrine has, however, shifted in the context of a turn in the US military to counterinsurgency operations.[36] Here the object for military action is the 'popular support' of a population alongside the 'morale' of an enemy. In terms of form and content, morale and popular support are equivalent. First, 'popular support' is also understood to exist only in its variable manifestations across all aspects of life. Second, 'popular support' is given the same quasi-spiritual characteristics as morale, involving a combination of belonging, pride, confidence and belief in a future.[37] The difference is in the form of military action. Counterinsurgency doctrine is based on the premise that the military must 'win' the 'popular support' of the population, as well as destroy the 'morale' of the enemy. While counterinsurgency doctrine and shock and awe may initially seem to be very different, one destroys, the other cultivates, both share an emphasis on airpower as creating life. This introduces a new class of actor in which the hope of victory in networked war is placed: an actor that exceeds the relations and nodes that make up a constituted network—'effects'. It is through 'effects' that power can become part of the targeted network and disrupt it. In order to act over life, airpower must become immersed in every aspect of life. It must create kinetic and non-kinetic 'effects' that become immanent to the networks targeted: 'effects' that follow precision-guided strikes, or repeated bombing raids, and can be called catastrophic in that they are designed to damage and destroy morale; or 'effects' that follow leaflet or food drops, and can be called providential in that they are designed to provide repair and improvement and thus win popular support. So, in order to expropriate a network, the military (or its effects) must become part of that network. In the case of Rapid Dominance, the effects of 'shock' and 'awe', for example, are designed to 'paralyse' and 'confuse' by acting before perception, thus '[c]ollapsing the will to resist'.[38] Whereas in counterinsurgency, by contrast, the effects of 'confidence' that follow leaflet drops or the use of airpower in infrastructure repair are designed to repair and improve life, thus building 'popular support'.[39]

In order to destroy or damage the enemy as network or system, force must be concentrated at one particular node or relation through the creation of events and/or conditions. From within this context, 'effects-based' operations are defined as a means of acting on the multiple elements that 'hold-together' a network.[40] They are:

[c]onceived and planned in a systems framework that considers the full range of direct, indirect, and cascading effects, which may—with different degrees of probability—be achieved by the application of military, diplomatic, psychological, and economic instruments.[41]

Because popular support or morale are scored across life, the 'effects' of airpower must also occur across life. A networked 'battlespace' must be engaged with militarily, politically, economically and psychologically. The expanded environment of military action is named as the 'total situation' of Shock and Awe or the 'full spectrum' of counterinsurgency.[42]

Reflecting on the origins of the word 'effect' in the Latin *ex-facere*, meaning 'to make something out of something else',[43] is instructive. The etymology of 'effect' hints to how 'effects-based' targeting promises to damage life by becoming part of the conditions of life of an enemy. First, the term 'effect' gives the sense that an 'effect' makes something out of something else (i.e. that it expropriates one thing in the production of another). Second, that this is a pre-determined movement towards a deliberate or intentional goal. We can think of 'effects-based operations' as a form of violence that aims to make explicit the implicit conditions of life in order to create life and bring death.[44] The key, then, is that morale, will to resist and popular support are not acted on directly. Instead force is concentrated on vulnerable nodes, relations and conditions. These are at one and the same time manifestations of those virtual targets *and* constitutive of them.

In advocating a certain sort of intervention over the 'full spectrum' or 'total situation' of life, 'effects-based' airpower involves what, after Foucault, can be termed 'environmental type interventions' that aim to modulate the relations between the (natural and artificial) elements that make up a network or system.[45] Let's look in more detail at examples of this process of creating catastrophic and providential 'effects' to show how they anticipate a form of power that aims to create life through military action.

One oft-cited example in military thinking is the use of specific air-based weapons—the B-52 bomber, AC-130 gunships or unmanned aerial vehicles—to create 'psychological effects' of hopelessness or helplessness over certain types of life; those designated outside of the population and named as 'extremists' or 'terrorists'. These 'effects' are designed to become part of how a network breaks, changes and takes on a new damaged or destroyed form. Here airpower destroys the ability to partake in war (as 'shock' and 'awe' are designed to do). Writing on the use of airpower in Afghanistan, one theorist asserts that the B-52 bomber creates 'hopelessness':

Death per se does not extinguish the will to fight in such opponents [discussing the Taliban]; rather, it is the hopelessness that arises from the inevitability of death from a source they cannot fight. Sheer impotence in the face of superior weaponry and the denial of a meaningful death will crush war-fighting instincts.[46]

In this case hope is invested in airpower as a particular technique of war that brings meaningless death—and thus hopelessness—by removing the capacity for action. The intentional 'effect' is an impactive experience that shatters normalised relations, cannot be rendered ordinary or inconsequential, and therefore produces an environment of suffering without end, with the supposed result that the enemy loses the will to fight (described in terms of demoralisation) and the network is either damaged or destroyed.

Airpower in counterinsurgency doctrine also promises to create providential 'effects' that become part of a network and thus establish an environment that may sustain certain sorts of life.[47] The US Army/Marine Corps counterinsurgency (COIN) field manual stresses that airpower should be used with 'exceptional care' because of the unpredictable effects of excessive violence in 'new wars' that mix war and peace, and population and insurgent.[48] The field manual therefore advocates calibrating any action for its prospective effect on winning the population's 'popular support' and destroying the enemy's 'morale'. While the field manual stresses the need to use airpower to destroy life stripped of a specific form of life ('terrorists' or 'extremists'), it also contains warnings about the use of airpower on the grounds that: 'An airstrike can cause collateral damage that turns people against the host-nation (HN) government and provides insurgents with a major propaganda victory.'[49] The rationale for suspending a violence that is always potentially available is not primarily ethical/moral, or based on the need for adherence to the rule of law. Whether to use airpower or not is instead based on a pragmatic–speculative calculus of the relation between airpower and popular support/morale—how will 'effects' produce an environment given that insurgents blend and blur with the population, with the result that '[i]nappropriate or indiscriminate use of air strikes can erode popular support and fuel insurgent propaganda'?[50]

Violence is defined in COIN airpower as a creative act that produces life. It is also a risk factor. The wrong 'effects' may produce the life that threatens—insurgency—and they may produce that life from the very collective—the population—that the insurgent must be cast out from.[51] What is therefore advocated in addition to destroying enemy morale is a 'positive' modulation of the popular support of the population. Somewhat absurdly, the role of airpower is extended to include actions that repair, improve and transform life. These include the use of airpower to support essential services (such as food drops), and the revival of colonial-era perception management operations, including the use of aircraft to disseminate information in the form of leaflets and broadcasting.[52] Take the example below from the US *Armed Forces Journal*, where a huge range of 'non-lethal' airpower effects are described and advocated:

Initially, air power can be used to foster a sense of accomplishment by teaching the target population how to effectively employ air power in its own defense ... Additionally, air power can be used to assist with new elections. As was done by the Marine Corps in Nicaragua, air power can be used to transport ballots and election monitors to protect the integrity of the process. Concurrent with these uses of air power, airlift also can be employed to transport the necessary supplies to rebuild communities.[53]

In this case—in comparison to the creation of a 'meaningless death'— 'effects' are designed to act throughout life and positively repair and improve a valued life.

The two examples may seem very different: one revels in the creation of meaningless death, the other advocates rebuilding communities. However, in both the catastrophic and providential use of airpower, the aim is to create life. The former in terms of an insurgent life that air-power strips of meaning in advance of death, the latter in terms of a ruined, damaged life that airpower helps repair and improve. The former relation with life is, we should remember, only suspended in the latter. The providential can always become catastrophic. Both rely on the violence of de-contextualising discussed at the start of this section. But they also do something more. Power derives from becoming immanent to life. Airpower aims to become part of networks by intervening strategically in how bodies come in and out of formation (rather than exclusively destroying life). Naming an 'effect' is, then, an attempt to pre-determine the formation of life. Acting over ambiguous, indeterminate, collective affects is the means to create an environment that will support certain types of valued life and bring death to the enemy.

*Targeting Networks*

Targeting a network from above involves being simultaneously inside and outside the targeted network.[54] Outside in order to name 'effects', inside because those 'effects' must become part of how a network (de)forms. This topology is enabled by elevating the object of airpower—morale, popular support or will to resist—to the status of a real virtuality. Morale escapes being fully knowable, with the result that it can be rendered actionable. Indeed, the elevation of a target to a virtuality justifies the extension of targeting to the entire biopolitical sphere of life. It also sustains the hope that creating catastrophic or providential 'effects' promises victory.

Acting over amorphous targets such as 'morale', 'will to resist' or 'popular support' involves, therefore, the co-functioning of two different forms of power—power-as-domination and vital power-to. These coexist with the resurgence of sovereign power associated with acting over the 'total situation' or 'full spectrum' of life in spaces of military occupation. This conclusion suggests the need for attention to the various relations between types of airpower and specific modalities of power. Work on airpower must take care not to presume the particular form of power or, following on, how airpower relates to life as targeted. We make a mistake if we presume that techniques of power are always orientated to a stable, de-contextualised, target. In the case of morale or popular support, power has no object. Instead it virtualises, endlessly proliferating what should be targeted, and hopes, promising to create life and bring death through catastrophic and providential 'effects' delivered from above.

# 12

# ECOLOGIES OF THE WAYWARD DRONE

*Jordan Crandall*

*Introduction*

On a clear evening in December 2010, as the sun was setting over the Texas horizon, a Mexican drone entered US airspace and crashed into a backyard in El Paso.[1] In another time and place, a drone falling from the sky could have elicited a high degree of alarm. In this particular border neighbourhood, however, it is not much of an event. Flanked by an army of surveillance cameras, floodlights, thermal imaging systems, inspection apparatuses, ground sensors and mobile surveillance units—fortifications that, together with an enormous barricade of cement, steel and barbed wire, define the US border with Mexico—the region is home to a cavalcade of mysterious machines that populate the skies: reconnaissance aircraft, relentlessly prowling for illegal activity, extending ground patrols into the air.[2] As the doomed drone lay tangled in the desert scrub brush and softly blowing sand, its processing ability weakened and its connective capability disabled, the home-owner calmly picked up his phone. He did not call emergency services. At the onset of this particular catastrophe, he did what any vigilant citizen in this part of the world would do: he called US Customs and Border Protection.

# FROM ABOVE: WAR, VIOLENCE AND VERTICALITY

Drones—also known as Unmanned Aerial Vehicles (UAVs)—are prone to system failures and pilot mistakes. Bad weather can bring them down; relatively small and vulnerable, they can be felled by something as simple as a gust of wind. The source of this particular El Paso crash was revealed to be a mechanical malfunction. It caused operators, who always operate at a distance, to lose control of the pilotless plane. As is often the case with unmanned vehicles, it was not clear who those operators were. A Mexican attorney general spokeswoman denied her country's involvement with the drone, but later that same day another Mexican official said it was being operated by the Ministry of Public Security and was following a target at the time of the incident. A spokesman for the Mexican embassy in the United States said that the drone, while belonging to Mexico, was part of an operation in coordination with the US government.[3]

The impact had opened up more than just a small hollow in the sand. It disrupted and opened the rituals of neighbours, the connectivities of machines, the routines of public agents and the choruses of desert cicadas. It destabilised the coherency of the crashed drone itself, which, far from sitting intact, was now distributed into the routines and spaces of the various agencies that were engaged in parsing its failure, sustaining its role or coordinating its return. At the onset of its weakened capacities, phones were dialled, conversations started. A collection of material and discursive components, it was now available for reassembly. At the international level, the accident brought into play the governmental agencies concerned with the maintenance of relations between the two countries, along with the global Israeli company Aeronautics Defense Systems, agent of the drone's manufacture, all of whom sought to maintain the perception that the drone 'works', whether in terms of mechanical infrastructure, data or public relations. At the national level, it brought into play the investigative agencies of the National Transportation Safety Board, concerned with the regulation of the skies in which UAVs travel, and the Department of Homeland Security, chief enforcing agents in the area of border protection and regulation.[4] These arbiters of the safe and legal passage of people, traffic and goods harnessed the drone as a case study, distributing its components within an investigative landscape of national security discourse, drug smuggling, gang violence, public health and private commerce. At the city level, it brought into play the El Paso Fire Department, attender to the emergency and modulator of its material risk, along with the local Police Department, dispatched by a US Customs and Border Protection agency whose presence in the region, along with its brother agencies in immigration and drug enforcement, is considerable.

The return of the UAV was a humble affair. The US Border Patrol, in the capacity of several agents and a van, pulled up to an international bridge in a cloud of smoke and dust. They stepped out of the vehicle, extracted the drone from the rear and handed it back to Mexican officials. Perhaps a ceremony of some kind was involved, but the handing over of the drone was, in terms of physical exertion, fairly easy, since it was in the 'Mini' class—an Orbiter Mini UAV—with a total wingspan of about 7 feet and capable of being disassembled into a backpack.

The recovered drone, relatively intact, might have been reassembled quickly. According to the manufacturer's specifications, this takes about ten minutes.[5] Yet the UAV was dismantled by an array of forces that violated its coherency in a deeper, more long-lasting sense. Between the drone's destruction in a backyard and its delivery at a bridge, the component agencies necessary to operate and maintain it became newly revealed. Dislodged from their mainframe and rendered vulnerable, these component agencies, however operational, institutional or discursive, become newly active in their negotiations and attachments. Phone calls are made, conversations started, extensions orchestrated. At the same time as they are distributed, however, they are consolidated—resolved to a territorial or ontological specificity. Escaping abstraction, they become embroiled in a geopolitics that may have been overlooked or erased. In the onset and aftermath of the catastrophe, the coherency and centrality of the drone is destabilised, its deceptive unity revealed. It cannot be reassembled in quite the same way.

## Space for Drones?

UAVs made their appearance on the world stage just after 11 September 2001, in the wars in Iraq and Afghanistan, although their technological legacy can be traced back to the Second World War and the US war in Vietnam.[6] The United States now relies on them heavily, most notably for surveillance and bombing missions across the Middle East and North Africa. Primarily due to their perceived success in these military operations, their potential has come to be widely recognised in many sectors of the US government, and pressure has been growing to allow them into domestic skies. The Department of Defense and the Department of Homeland Security have besieged the Federal Aviation Administration (FAA) with requests for the flying rights of a range of pilotless planes into civilian airspace for the purpose of domestic security operations, law enforcement and disaster relief. So far, they have obtained FAA permission to operate unmanned planes along limited zones, including the south-west border from California to Texas.[7] In the case of large-scale catastrophes, drones can be operated nationwide in the search for survivors.

While many countries across Asia and Latin America—Israel, Japan, South Korea, Brazil and Mexico, to name a few—already allow UAVs for domestic use, and while the EU is planning to have them integrated into civilian airspace by 2015, the process of developing regulations in the United States has been slow and fraught with complication. FAA officials are concerned that, in domestic skies, there is a greater danger of collisions with smaller aircraft than in the war zones in which UAVs have been tested. The fact that UAVs come in such a wide range of scales—the Global Hawk is as large as a small airliner, while the hand-launched Raven is just 38 inches long and weighs 4 pounds—makes matters worse. The FAA are worried that drones might plough into airliners, cargo planes and corporate jets at high altitudes, or plunge into low-flying helicopters and hot air balloons. The rapid growth of unmanned planes of all sizes not only threatens safety in the skies but on the ground. With UAVs coming as small as the 13-inch Wasp, it is easy to imagine a tiny drone, malfunctioning or wavering off-course, crashing not into a border-town backyard but through a living room window.

One of the FAA's key concerns is that remote operators can lose communications with the aircraft. In the world of drones, loss of communication with the aircraft can lead to loss of control. Many UAVs, when they lose a connection to ground stations, are programmed to fly off to a safety zone and try to regain contact. But often this does not work. The plane goes renegade, disappears or plummets to earth. Loss of communication and control can occur from a systems failure, a software glitch or, as in the case of the Mexican drone crash in El Paso, a mechanical malfunction. The drone can also be cut off by an atmospheric disturbance, a hostile interception from the ground or an enemy hack. In one way or another, human error often plays a role, whether in the form of a faulty program, mechanical oversight or coordination mistake.

Image 27: Global Hawk UAV taxies ready for take-off (photo by the author).

## Drone Desire

The demand for unmanned vehicles is not limited to the military, home-land security, and law enforcement. Civilians, too, want their drones! Tornado researchers want to send them into storms to gather data. Energy companies crave their use for geological surveying and pipeline monitoring. Security companies want to send them up for new surveil-lance applications. Commercial upstarts yearn to service them and train their operators.

Of the vigilante groups who now fly drones along the US–Mexico border, the most visible and technologically advanced is the American Border Patrol.[8] Its UAV, a ramshackle plane called the Border Hawk, is operated from a ground control station on a private ranch situated on the south-eastern Arizona borderlands. Endeavouring to provide public access to transmissions that are usually shrouded in secrecy, the group streams its drone video footage on the web. Their plane's inaugural flight took place over the San Pedro River, a popular site of cross-border activ-ity. To ensure that the drone proved its effectiveness in spotting actual, living people, volunteers from the American Border Patrol masqueraded as illegal border crossers. Jutting to and fro, stealthily wending their way across the harsh desert borderland, conscious of the view from above, they mimicked the very people they aimed to target, adopting their renegade behaviour in a caricature of criminality. The complex pleasures of crossing over in this manner, through appearance, disposition and demeanour, are well known to the deviant manoeuvring to 'pass', with whatever degree of conformity and sacrifice this might entail. These pleasures are often undetectable to those who man the optics: visual mastery is privileged over ground-level display, at the expense of any awareness of the correspondences of self that the targeter finds reflected, extended and propagated in the scope.

As glimpsed in the amateur officiality of their nomenclature, groups like these straddle the line between governmental and non-governmental agency. Aiming to assist the US Border Patrol in the apprehension of illegal immigrants, they see themselves as providing a valuable public service, filling in the gaps among the limited number of Border Patrol agents that are available to patrol the entire 2,000-mile stretch. At the same time, they regard themselves as government watchdogs. Suspicious of their state apparatus and disillusioned by the ideologies of their generation, these groups, dominated by retired military and security men, patrol the border as if in search of something far more than illegal activity: the recuperable myth of white male privilege. Situated far from the contemporary sci-fi imaginary, they seem to embody, instead, the genre of the Western—the pre-technological harbinger of its cyber-frontierism. Drifting about the desolate landscape, drones at the ready, they guard their version of the American Dream.

Allowing unmanned aircraft into domestic space heightens a number of civil liberties concerns. It expands the government's ability to carry out surveillance on its citizens—adding to its already substantial patrol arsenal of sensors, night vision scopes, video surveillance systems, directional listening devices and data mining systems. The cameras on drones like the Predator can read a licence plate from 2 miles up; the electro-optical sensor systems of the Global Hawk can identify an object the size of a milk carton from an altitude of 60,000 feet. And while domestic drones are not presently armed, they can easily be outfitted with weaponry—as they were after the 11 September 2001 attacks, when Predators were quickly armed with Hellfire missiles (fired, frequently, at the wrong targets). Drone strikes often slip into the cracks between regulatory domains; their responsible parties, often combinations of actors working across the boundaries of national governments and domestic agencies, are difficult to pinpoint. Among the hundreds of deaths—some say thousands—that America's drone strikes abroad have caused there is little accountability.

Many of these concerns are superseded by the drone's allure. Even when considering social costs and ethics, the use of drones is widely supported by the general public. Guarding the border is understood to be paramount to US national security, and the practicalities of domestic security loom large. Politicians do not want to risk appearing 'soft' on border security. They argue that UAVs could operate as 'force multipliers', allowing the Border Patrol to deploy fewer agents and improve coverage along remote and sparsely patrolled sections. The synthetic-aperture radar, infrared sensors and electro-optical cameras on a UAV like the Global Hawk can provide the capability to survey over 60,000 square miles a day. According to Homeland Security, UAVs have proved their effectiveness, helping to intercept thousands of illegal immigrants and pounds of drugs.

In a more general sense, it is widely understood that unmanned systems, for both military and domestic security operations—considering, for the moment, that this distinction still stands—are the way of the future. The Department of Defense has invested aggressively in their development and use, and by Congressional mandate this investment must continue to increase. The perceived advantages are many. As with many robotic systems, drones are unhampered by the physiological and psychological limitations of humans; they can easily take on jobs that are dirty, dangerous or dull. They can stay aloft and loiter for prolonged periods of time, persisting on targets over ten times longer than piloted aircraft, at far less cost. The human risk factor, at least on the US side, is vastly reduced. As a general rule, drones do not result in the injuries and deaths of their crews.

But they do crash. They crash frequently—many more times than manned aircraft. They crash not only into American border regions and backyards but into global hotbeds of military activity. They have slammed into Sunni political headquarters in Mosul, Iraq; nose-dived into the runway at Parc Aberporth in Wales;[9] struck power lines and cut off power in Alberta, Canada; vanished into Pakistan's tribal region in

North Waziristan; plummeted into uninhabited terrain near Ghanzi, Afghanistan, and the Indian–Pakistani border; collapsed into the Gaza Strip; plunged into the Mojave Desert; disappeared into Turkey's desolate Mardin province; cannonballed into the coast of Spain; ditched into the Iraqi countryside; and rolled with scrub brush across the rough desert terrain near Indian Springs, Nevada. The Italian Air Force even discovered one of its downed drones floating along the surface of the Adriatic Sea, its body glistening in the sunlight like the bleached skin of a whale.[10]

If a demo reel of Oscar-worthy drone crash moments were assembled—perhaps in order to pitch the drone for a starring role in the ubiquitous action-adventure movie—it would be composed of clips like these. In true commercial fashion, it would seek to harness the drone's menace and its allure, its potent combination of desire and threat. Like any good object of desire, it would give us what we want and what we fear. As a conduit of identification and affect, it would allow us to extend ourselves, in all our sensory acuity, into a landscape devoid of everyday political rationales and ethical or moral judgements: to plunge headlong into the melee.

*Salvage Operations*

When UAVs crash they provide a bounty of potentially valuable information and parts. Their databases are rendered vulnerable to access, their components susceptible to retooling—absorbed into affiliations that can enhance the warfare capacity of foes. In order to prevent enemies from obtaining sensitive materials, almost every drone crash involves an intensive recovery operation. It can be difficult to secure the wreckage. When, in 2009, a Canadian UAV crashed around 3 kilometres from the US military base at Ma'Sum Ghar in western Kandahar province, American military forces were too late: within twenty-two minutes, the drone had been fully stripped and hauled away by locals. If a recovery is not possible in time, a drone may be destroyed by its own government: British Special Forces once bombed a Reaper that had crashed in Afghanistan in order to prevent its parts falling into the hands of the Taliban. Smaller drones like the Raven often simply disappear into the hands of enemies, as they have frequently in Iraq.[11]

Similarly, in 2011, a US Predator crashes in Jahayn, a remote village in Yemen. Local residents, who frequently complain of the noise that the widely used drones make as they relentlessly circulate in the skies over-head—some say it sounds like a lawnmower—most likely meet this downed drone with some degree of relief. Discovering the wreckage, they call the police. The felled drone is recovered, hauled away from the oil-stained sand. As the convoy heads back, however, it is intercepted by gunmen. The armed rebels, reported to belong to al-Qaeda, hijack the plane. The Yemeni Ministry of Defence dismisses these reports as base-less rumours. How solid is its claim? According to diplomatic cables released by WikiLeaks, the Yemeni government deliberately covered up the crash of a previous American drone, the Scan Eagle, claiming that the aircraft, which washed up on the coast of Hadramout, was an Ira-nian spy plane.[12]

Unlike the smooth coordination of agencies involved in the El Paso incident, which resulted in the crashed UAV's return to Mexico, there is no handing over of the drone by the Yemeni government. It scatters into the routines and spaces of renegade agencies. Its parts, however material or discursive, are absorbed into other systems of meaning and affect, however straight or wayward, countered or modulated, amplified or diminished. The dispersion, more the rule than the exception, is always accompanied by gathering, a consolidation. As the drone's material parts, each endowed with a distinct spatial boundary, are assembled in a coherent, stabilised form, its discursive components are often consoli-dated into a linear narrative—outfitted with a beginning and an end. Which story to believe and invest in? One will most likely prevail: a descriptive phrase, like a material part, seeks consistency, endurance and relevance, of which those that work best for the task at hand, or become most useful, achieve a higher degree.

At the onset of the El Paso crash, Mexican officials were pressed to speak. They dodged the inquiries by citing national security concerns—replying, as most government officials do, that all information related to unmanned aircraft systems is classified and restricted. The dodging is typical. Governments will disclose the nature and quantity of their UAV operations and arsenals only when hard-pressed, and only when drones drift—or rather, plunge—into the public sphere, often in the form of an accident. The CIA has a highly active but covert drone programme—its bat-winged Sentinel stealth drone played a role in Osama bin Laden's capture—but while drone crashes are publicly acknowledged by agencies like the US Air Force, its accident figures are never released. Crashes in the 'Mini' and 'Micro' classes are seldom if ever reported by anyone.

Even when UAV failures are acknowledged, the technical details are often obscured in bureaucratic manoeuvrings. The National Transportation Safety Board, which is responsible for investigating the cause of the El Paso incident, note that a typical investigation can take almost a year. Even when finally released, the accident reports of military institutions can be difficult to decipher. Inquiries regarding drone crashes in the testing and marketing stages fall into the cracks between private companies and client governments. When pressed about the details of a catastrophe, manufacturing companies often reply that they were ordered by governmental officials not to discuss the details. In spite of these manoeuvrings, the crash, as an *event*, cannot be contained, and this is precisely the source of its compelling power. Destabilised, its parts scattered, it cannot be reassembled, however hastily, in quite the same way—in spite of the considerable rhetorical power that might be mobilised to accomplish this feat. Fault lines appear, allowing new discursive openings.

Stories develop coherence, weaving together disparate parts into a whole, yet they also create separation where there was none. Conversations gather around the *event* as it reverberates through its discursive agents, whether official agencies through their portals or gathered friends at a social settings. As there are entire websites devoted to the drone's fetishisation as an object, there is a growing body of interest in its destruction and disappearance: drone crash lore. Stories are woven around downed drones and their sites, however accurate they might be or outrageously fabricated they might seem. As drones are outgrowths of the histories of UFOs and robotics, as they have been integrally tied to warfare, war technology and anxieties of invasion, however real or fictional, at least since the mid-twentieth century, the inevitable corporate and national spin that is woven around the accident and its aftermath is often, as with mid-century UFO crashes, seen as a cover-up or conspiracy. As with many news reports, intentions are interpolated in ways that conform to one's own beliefs, and in a world of viral media even reports that seem ridiculous are given legs. At one time three US drones were reported to have been deliberately flown into the dome atop the Iranian nuclear reactor at Bushehr. The reports of this event, in common with others relating to many other drone events, can still be found online with a simple search. Such stories propagate with little or no verification, especially as they activate the imaginary, affirm ideological orientation and offer easy munitions in wars of attention.[13]

Conversations intersect with or spin off into others, amplifying or diminishing in scale and intensity as they become harnessed to personal concerns, anxieties and desires, aligned with group imaginaries and ethical codes, and enabled by communications platforms. They might involve the particularities of technology and impact site, the vagaries of luck and community governance, or the generalities of warfare, nationhood, freedom and oppression. They might stabilise into stories, some propagating and enduring, some vanishing by sunrise. They might create new conflicts or fuel existing ones, produce new images and dreams, rearrange or reinforce existing routines. They might obscure specific details, overlook obvious connections, or forge entirely new ones.

As these conversational actors magnify or wane, speed up or decelerate, and accumulate relevance, influence and intimacy, so, too, does the material *event* that they draw from—the material occurrence with which they have entered into affiliation. Agential networks and events are intricately tied together and mutually influencing. At the most basic level, even causality and temporality are up for negotiation. In this way the narratives that are woven around the drone's fate—circulated around crash sites, dinner tables, cookouts, online forums and board rooms—have a vitality. They are social actors that negotiate realities even as they are negotiated by them. Yet the fate of the drone's carcass is but one narrativised outcome of a much larger and more vital function that the catastrophe performs. The crash is important because it destabilises the coherency of the drone and embroils it in a politics that was heretofore invisible or diminished.

## Drone Geographies

The El Paso crash, in its dispersions, helped reveal the specific details of Mexico's UAV programme. Until then, it was not publicly known that Mexico was using drones along the border. To maintain this level of focus on the drone is to amplify its history, its manufacture—how it came to exist in a specific built form. Mexico operates a fleet of drones purchased from three Israeli companies—Aeronautics Defense Systems, Israel Aerospace Industries and Elbit Systems. The Heron UAVs that it has purchased from Israel Aerospace Industries have also been sold to Canada, Turkey and Ecuador, where the company now has a branch office, in addition to its offices in Brazil, Colombia and Chile. Elbit

Systems, in addition to selling its UAVs to Mexico, has also sold them to the United States, the UK, Singapore, Croatia, Georgia and Brazil, where it has a subsidiary, AEL Sistemas. Drones are marketed to these countries for a variety of purposes including jamming signals, locating enemy satellite dishes, spotting drug plantations or cartel hideouts, or monitoring police forces for corruption. The specificity of drone manufacture, when pursued, opens out into multiple corridors, each of which can be followed to reveal others: networks of affiliation that operate at a number of scales, magnitudes and degrees of stability, from research to assembly to testing and marketing. Zooming out to the largest consolidating scale, the production and consumption of UAVs is a global phenomenon, with about sixty manufacturers operating in at least forty-eight countries. The US military is the single largest consumer. Along with their manufacture and selling, drone operation is also a transnational endeavour: the Turkish Herons that Canada owns, for example, are flown and operated in Afghanistan by the Australian Air Force. So, too, with drone training and logistics.

The geographical specificity of the material *event* reveals the distinct spatial politics in which these distributed drone economies are embroiled—economies that, in their vast scale and speed, and in their considerable rhetorical arsenals, blur impact at ground level. The site tells its own story. The Mexican Orbiter drone crash in Texas occurred in a specific spot on the earth, its collision etched on to the ground of a unique El Paso backyard, its temporal streams collapsed into a singular date and standardised time. The impact occurred on 14 December 2010 around 6:25 p.m. The site is located on Craddock Avenue near South Yarborough Drive, in the city's Lower Valley neighbourhood. It is just over the border from Ciudad Juárez, one of Mexico's key epicentres of violence.[14] The drone could have easily landed there, amid the very region that it was clearly intended to monitor—a region where rival drug cartels battle for control over smuggling and drug trafficking routes, their caches filled with American weaponry, and thousands of killings occur each year.

The contrasts between these two sites could not be more extreme. The operation of American and Mexican drones along this stretch of border—often only glimpsed in terms of their failures—reveals its specific geographical, social and political climates, as it is sliced through with a border barricade, surrounded with a surveillance apparatus, and embroiled in discourses around domestic security and drug use. For politicians, ever more intricately connected to the global economies of drone manufacture, sales and operation, the 'force multiplying' factor of UAVs saves lives, increases manpower, improves coverage, enhances relations and reduces crime. It draws a harder line in the soft desert borderland.

However hardened, the line is breached ever more intricately by the trade in weapons, people and narcotics, much of it driven by US demand. The material specificity of the *event* reveals its technological infrastructures, however actualised or latent. The GPS signals necessary to steer drones and locate their targets are weak and easily interfered with, as are the electromagnetic waves generated by radio signals, computers, electronic equipment and various other machines. Spurious, unidentified signals can cause engines to shut down mysteriously (as they have with Bell Helicopter's Eagle Eye UAV—a signal whose mysterious source has never been identified).

Perhaps the Mexican Orbiter went off course into Texas because it was hacked. The drone's devices and communications are vulnerable because many of its software and electronics components are 'off the shelf', riding on existing structures. About 95 per cent of the military's communications travel over commercial telecom networks, including satellite systems. As drones populate the imaginary through games, sci-fi literature, television and film, so, too, does the hacker ethic, often embodied in the agency of the hero. Bringing down a drone can engender as much affective thrill as launching one: the *jouissance* of the crash reverberates across the body that it helps render social, firing up electro-chemical connections and igniting its sensorium. In keeping with the ethos of DIY cultures, the 'stupidity' or simplicity of home-grown solutions is prized: a quotidian 'know-how' that resists the dictates of commercial knowledge production regimes. The underdog sweeps in to score—the resurgent rebel fighter who, in the nick of time, shoots down the vessel of the oppressor.

As the material specificity of the *event* reveals its deeper technological substratum, the renegade force that is the drone's undoing may be buried within the machine, in a site where human agency is much harder to locate. A US Air Force Global Hawk, the largest unmanned plane in the military's arsenal, was once brought down by a small, renegade part. An investigation of the crash revealed that the plane's rudder had become loosened during a previous mission.[15] It was not detected on routine maintenance checks. During the fatal flight, it began flapping uncontrollably. Though comparatively small, this particular part plays a crucial role, and its flailing was persistent enough to destabilise the mammoth plane and send it plummeting to earth. Failures of the Global Hawk are not uncommon. The Air Force had lost two of them just before the excessively flapping rudder catastrophe. The first crash was due to a simple input error: the plane was programmed to taxi at 155 miles per hour. The second was due to operators inadvertently engaging a self-destruct code while the plane was aloft.

Inquiries into the maintenance of the rudder, the programming of the mission and the writing of the code reveal the drone's concealed infrastructures—its systems of operation, logistics and maintenance. For most American drone operations abroad, ground crews service the mechanical systems of the planes at regional bases in war zones, as flight crews operate them thousands of miles away, out of converted trailers at air force bases in Nevada (and soon to be joined by bases in Arizona, California and Texas). The material realities and infrastructures of these bases, along with their geographical and institutional embeddedness, play a large role, as do the highly specific, routine practices they register

Image 28: Problem Technologies (author's photo).

and call forth. The single-wide trailers out of which drones are operated are oddly humble, given the considerable expense of the UAV programme. They sit isolated amid the vast desert landscape, parked among the scrub brush. If not for the barbed wire around their peripheries, one could mistake them for the typical mobile home encamped at any trailer park in the American south-west, with enormous satellite dishes and cables that link them to the world beyond.[16]

A typical UAV requires a flight team of four. A pilot manoeuvres the plane and is the chief person responsible. A sensor operator manipulates cameras and sensing devices. A mission monitor receives requests from 'customers' and sends them required images or information. A flight engineer monitors the status of the aircraft. While the ground crews wrestle with the vagaries of small parts, the flight engineer monitors their operational states in the form of technical data arrayed on one of the crew's displays. Another display contains navigational data: GPS signals and other locational data downloaded via satellite transmission and translated as coordinates on a GIS, for use in directing the plane and positioning targets. Another display contains image data: the drone's view from above.

Data flows through satellite transmissions that link the assembled flight crew to the plane, as data flows through radio transmission or otherwise connect the flight crew to intelligence teams and arrangements of commanders and troops on the ground. These links and flows are determined through existing connections, platforms and procedural agencies, yet at the same time they also help instantiate them. Transmitted signals are modulated and rendered discrete as code, in concert with the actors—programs, hardware, organisations, personnel—that rely on them. As they flow through actors, they are filtered, constrained, related and interpreted, and in the context of this activity, actors acquire rhythm and articulation. They configure, and are configured, through limitation and correspondence.

Most of the algorithmic and machinic operations necessary to operate the plane and negotiate its trajectories across geographical, national and institutional territories are those that bypass the corporeal agency of personnel. The plane flies as an affiliation of maintained and monitored states through the activity of a multiplicity of actors, however human, mechanical, informational, environmental or institutional. These actors operate at various scales and levels of complexity, whether at the level of hardware, software, image, data, controls, flight or ground crews, or at the scale of logistical support, service or operator and maintenance training.

Through it all, the rudders remain stable. The transmissions are cleared, the connections maintained. Collective intelligence and skill emerges for operation. Hardware, personnel and supplies are integrated into tactical formations. Communication protocols and pathways fit together in stable systems. Ideas fit together in doctrines. These actors stabilise and cohere because of the procedural structures, standards and programs of the higher-order affiliations into which they fit—affiliations that might exist at the level of algorithm, hardware or logistics, or at the local, the regional or the national.

The component actors within these affiliations are relatively discrete and stabilised. Yet they are active: they band and disband, accumulate and release, extend and consolidate. Some links are weak and some more solid. A dispatch is simple while a doctrine is complex. Even internally, composites that would seem to be solid and enduring are embroiled in bandwidth battles and inter-service rivalries. All must be actively maintained, with varying levels of frequency and force.

Since affiliations, and their component actors, come to perform in certain ways, at various scales and speeds, with varying degrees of reliability, influence and intimacy, the next step in the salvage operation is to explore *how* they come to perform—the relational structures and organising principles through which they are coordinated and combined together at various scales, magnitudes, speeds and levels of complexity, and the mechanisms through which this is sustained. The next step involves opening up the possibility that these components can be hacked, retooled, reprogrammed—appropriated into new patterns of use. The operation is not primarily reductive and critical but affirmative and constructive: the production and mapping of new ontological platforms, epistemological itineraries and political possibilities.

## Ontologies of the Drone

Consider the rudder, excessively flailing on the luckless Global Hawk. The function it performed was relatively simple. When understood from the overall scale of the mission, it did poorly. When understood from the scale of the actuator, it did just fine. As with all actuators, its job was to initiate physical response or intervene within the environment for a purpose, whether this might involve moving, positioning, regulating, pumping or filtering. Other component actors of the drone may not have any actuation capacity. They may be only engaged with the processing of data, whether for communication, identification or locationing. They may only have the capacity of sensing—detecting and responding to a change in state, within a medium that might be mechanical, electrical, magnetic or chemical. In a general sense, the UAV can be understood as a collection of sensing, processing and actuating functions. Here the alternative descriptor of the UAS (Unmanned Aircraft System), frequently used in the military's long-term projections, is most appropriate. The drone is a distributed cognitive and actuating system that can sense the environment, create a context for that information, communicate internally among components, draw inferences from the data and initiate physical response.

The burgeoning global UAS market relies heavily on companies and organisations that provide service, logistics and training for the unmanned vehicles that proliferate across the skies. Since access to national airspace is largely restricted, much of the training is done on simulations. The interfaces of these simulations are familiar to any aficionado of games, roleplaying environments and high-tech adventure films. Like the control panels of actual flight crews, sitting in their Nevada desert trailers, they bear the traces of the commercial game formats from which they are derived. One can detect the influence of Xbox controls, used by the Army, and the engines of games like Halo, upon which Raytheon's UAV control system is based. Even simulations like the Marine Corps' Virtual Battlespace 2 are based on commercial game engines that are boldly reflected in their titles.[17]

Like the actual drones of which they are a component, the coherency and discreteness of these interfaces and applications dissolve upon scrutiny, scattering into arrays of component actors that are shared by other affiliations. These component actors—visual and rhythmic motifs, behavioural conventions, perspectival formats, software codes, tags, controllers, users, corporate procedures, game architectures, rules—circulate and bond across multiple domains of experience, traversing the divides between corporation and government, combat and entertainment, simulation and reality. The particular applications in which they accumulate, largely developed by the commercial game industry and influenced by commercial formats of cognitive and affective engagement, are made to excite the gamer, with characters that run faster and jump higher than is humanly possible, and explosions and flames that burn more fiercely than normal, in ways that would surely disappoint the seasoned player. The problem is not that people, environments and behaviours are un-lifelike, but that they are more than lifelike and must be downscaled, along with the expectations of their human accomplices, to calibrate with the velocities, magnitudes and textures of the real world.

Simulations often require nothing more than a joystick and personal computer—a laptop can run all of the vehicle dynamics, including the sensors. Rendered portable, the same high-end environments that are found in stationary systems can be brought home for practice or taken directly into the field. Further narrowing the gap between rehearsal and mission, some simulations are plugged directly into the ground control stations that are used to manipulate real UAVs, allowing for training and operation to be done together, with operators toggling between simulation and actuality within a functional crew station.

The integration between gaming, simulation and mission happens not only at the scale of the crew station but at the level of the command structure. Simulations like Virtual Battlespace 2 allow data that is gathered by UAV sensors within the gaming environment to be fed directly to command and control systems for a commander's strategic planning. It is said to provide a more comprehensive view of the battlefield, with real participants vying with simulated ones for evaluation, engagement and participative hands on training, in networks that amplify access to knowledge, situational awareness and collaborative endeavour. It also provides analysts with simulated back-end processing of the raw data collected by the sensors. Ground-base operations, inter-service and multinational training events, and game-based training situations together become essential precursors to deployment, increasingly integrated into command and control systems and actual operations in real time.

Ground control stations, training simulations and video games occupy a common economic, affective and cognitive terrain: sites of data rendered actionable. Together they constitute an interlocking, visual, rhythmic and orienting complex, harnessing the imaginary, which conditions orientation in the world. Like the material realities and infrastructures of the bases and training facilities within which it unfolds, however virtually, geographically or institutionally embedded, the enacted routines of this complex play a large material and materialising role. Their transmitted signals, electrical and vibratory, are modulated and rendered discrete as coded meaning, in concert with the software and personnel that channel them. The relations and modulated transmissions are mutually generative: they configure agency, and are configured by agency, through their limitations and correspondences in the enacted routines of practice. They course through their attendant actors, as these actors perform within the dynamics of the various situations that arise, in various degrees of attunement to the shared priorities that they may reveal: priorities acted upon and inhabited, in various degrees of frequency, scale and magnitude, in stance and position.

As the material agency of trained crew members coalesces with an affiliation of maintained states, in alliance with a multiplicity of actors—however human, mechanical, informational, environmental, institutional—so, too, does the material agency of the drone that is flown. Through practices, they are maintained in continuities, cooperating and competing for endurance, in whatever degree of simulation or actuality. They do not always conform or affect one another in linearly causal terms. The affiliations of the pilot and plane are connected, in resemblance and limitation, to a degree and scale that they can be stabilised and sustained: they exist in the world, inform and influence one another, with some degree of reliability, relevance and intimacy.

Across these dynamic, entraining affiliations, functional organisations of knowledge and skill—capacities enacted and roles played in the organisation of the system—are redistributed and re-constrained, along with positions, categories and divisions of labour. As agencies circulate and bond across multiple domains of experience, traversing the divides between corporation and government, combat and entertainment, research and commerce, affiliations composed of unlikely bedfellows are brought together through economic need. If the 'unmanning' of systems moves soldiers off the battlefields, it brings technology companies directly into them, in search of ground-level feedback for updating existing products and developing new ones, in an increasingly competitive global industry. The redistribution of manpower in the 'unmanning'— the shift from soldiers in battlefields and fighter planes to those in double-wide desert trailers and high-tech command centres—challenges the stances, positions and qualifications that have defined previous generations. The values, tastes and dispositions of unmanned warfare do not always align with the gendered roles, imaginaries and concepts of adequacy that were present in the noble, heroic ideals of the past. The US Air Force now trains more drone pilots than fighter and bomber pilots combined. The 'top gun' archetype is on the wane. Yet, as past ideals of heroic masculinity are threatened, new ones are created, embroiled in new forms of agility, knowledge and displays of prowess.[18] They might be manifest in the subtle alignments of the body, its pacing, expression, stylistic action, inclination or mood. Warrior archetypes migrate into alternative geometries of privilege, however gendered—myths of male identity wrestled with in the reinventing, rather than resuscitating, of a fading ideal.

*Amplifying Expertise*

Imagine the scene—it is a crisp October day in Afghanistan, in the midst of Operation Enduring Freedom. A US Predator has just taxied and departed from Kandahar Air Field for a routine reconnaissance mission. The plane is assigned to an Expeditionary Wing at Creech Air Force Base in Nevada and operated by crew members at March Air Reserve Base in California. Suddenly, during the flight, the crew receives a direct task order from the Combined Forces Air Component Commander: they are to provide immediate air support to US and Afghan ground forces that are under siege. The enemy fighters, numbering about 300, appear to be carrying out a large, coordinated attack. Given the intensity of the battle on the ground below, the circumstances of the attack and the immediate and critical need for support—US soldiers were being killed—the Predator crew is consumed with a high-degree of urgency. Their attention fully focused on the battle, their awareness of the bigger picture diminishes. The pilot's distraction leads to a fatal mistake: he fails to see that the UAV is headed toward a looming, 17,000-foot mountain. The drone smashes into it, abstracted into a cloud of black smoke, its parts scattering into the desolate terrain below.

Human attention can be too tightly focused along one zone of experience, to the exclusion of a wider expanse of contextual information—what the military calls 'situational awareness'. It can also be too scattered: not focused enough on anything. In order for it to be effective, a dynamic between stabilisation and destabilisation must be actively maintained. Yet however vigilant it might be, human attention is imperfect and undependable—ill-equipped to keep up with the demands placed upon it. As drones gain the ability to 'dwell and stare'—recording activities on the ground over much longer timeframes—the vast amounts of data they absorb can easily outrun the capacities of personnel. On a single day the Air Force must process nearly 1,500 hours of full motion video and another 1,500 still images. Cameras and sensors become ever more sophisticated, yet they are of limited value unless they can be accompanied by improved human intelligence and skill. The task of interpreting what the UAV is seeing falls partly into the hands of the flight crew, who always has access to the aircraft's live video feed; they may also be joined by an image expert trained precisely for this purpose. The video is also sent to image analysts at other bases—analysis and dissemination sites like the Joint Base Langley-Eustis in Virginia, inside of whose cavernous rooms image analysts sit, filtering vast streams of data arrayed on constellations of monitors. They, too, are hard-pressed. Staring for hours on end, nearly inert at their chairs, they try to ferret out the singular target—the single, telling deviance in the normalised flow. Armed with the skill of extracting relevant data from image flows and information arrays, they attempt to organise that data into patterns of affiliation from which further extrapolations can be made.

The UAS, as an affiliation of components and systems, relies on analysis and dissemination sites like these. They are vital platforms of the drone in its shared perceptual and analytical capacities, its sensing, processing, communicating and actuating functions—nodes through which its data is streamed, formatted, tagged and rendered searchable across networks of datasets. The platforms and nodes of these affiliations are many, from personnel bases in Nevada to storage facilities in Iowa—repurposed shipping containers within which arrays of servers, tasked solely to house the video data generated by Air Force drones, quietly hum, as unremarkable as the double-wide trailers in which their operating crews sit. As the image data is organised and stored, it becomes the primary site through which correlations can be made and inferences drawn. Databases, activated through search algorithms, become the primary repository of knowledge. A backlog of replayable events is generated, seen from above: a searchable, historical record of a region's activity, as viewed from the UAV.

This movement that is detected, however geographically understood, is not necessarily causal and continuous: a history inscribed upon ground or air. It is rather a result of calculations calibrated across datasets: correlations among relatively stabilised and standardised elements that do not move across space so much as flicker or fluctuate within it. Movement is less a continuous transfer—over ground, land or spatial volume—than a configurative interpolation. It is a trajectory assembled in retrospect, piecemeal as a correspondence of points: a behavioural composite into which movement and intention are inferred.

A typical drone requires nineteen analysts. A single drone outfitted with 'Gorgon Stare' technology—which can capture live video of an entire city—requires 2,000 analysts.[19] This advanced video capture system, paralleling the teeming array of snakes emerging from the head of the mythical creature referenced in its title, has a spherical array of nine cameras—five electro-optical and four infrared—emerging from the underbelly of its platform. Its software compiles the various camera views into a broad, continuous mosaic, of which personnel on the ground or elsewhere can simultaneously grab slices—its analytical requirements farmed-out, in real time, to a network of analysts and computing platforms.

Like the winged fusion of human, beast and machine of its namesake, with claws of steel, bulbous head and large unblinking eyes, the gorgon drone is equipped with a deadly, commanding stare: it looks at you, but you cannot look back, lest you be turned to stone. Although menacing in its demeanour—fangs bared and nostrils flared as it readies to inhale and consume—it is understood to have protective qualities for its deployers. Heir to the symbolic apparatuses of myth, its figures are everywhere present on objects, documents, ideologies and units of value exchange, manifesting hopes of warding off evil. Its potential for domestic protection is not lost on the US Department of Defense and Homeland Security, who are exploring its panoptic potential for border security: Gorgons now joining the cavalcades of machine beasts flying high over the desert borderland, with no illegal activity going unchecked, be it immigration, drug trafficking or the very flows of terror.

The technology is not without its substantial problems. While it may 'see' over a wider swath of territory, it does not necessarily understand the significance of what moves within it. The challenge remains that of tracking vehicles, objects and humans on the ground with a higher degree of precision, in ways that lessen the demands on human personnel. The challenge with UAS in general is to amplify the overall intelli-

Image 29: UAV Dystopias (author's photo).

gence of the system—heightening the level of skill and expertise that the affiliation can engender. This often takes the form of enhancing the capacity of tracking and search algorithms, along with the network processing capability required to parse and coordinate the data. It involves increasing the ability of UAVs to sense, reason, learn and make decisions, and to collaborate and communicate, with a minimal degree of direct human involvement.

In popular terminology, this is called automation or autonomy. When approached within the terms of the 'salvage operation', however, where agency is situated in shared composites of intelligence and skill—affiliations among all manner of actors operating at various scales, magnitudes and degrees of complexity, whether at the level of hardware, software, flight crews or institutions—these discourses of autonomy are resisted. The unmanned system does not eliminate the human so much as redistribute the agencies of warfare. The capacities of sensing, dispatching, analysing and alerting—the intelligence and skill required to interpret and store information and act on the results—are shared by an affiliation of actors, however algorithmic, organic or systemic. The focus is on their performative practices within the functional organisation of the system. It is a matter of how they are maintained as dynamically stable entities—sustained, naturalised and rendered discrete—and the programs through which this is accomplished.

Image analytics software is used for the recognition of objects, vehicles and people. Even the most rudimentary drones, such as the Mexican Orbiter in El Paso, have some form of algorithmic tracking, even if it is only basic motion detection. The software takes input from cameras, then recognises and identifies the objects in each frame to learn what activity normally takes place within the area under observation. With the norm firmly in place, the software then aims to detect activity that deviates from it: the exceptional occurrence that stands out from the domain of the ordinary. Algorithms screen out non-critical movement activity and foreground the critical, in order to maximise the attention spans of observing experts. Once alerted to aberrant movement activity, its nature and intent is to be inferred by these experts, who then decide what action to take.

The norm is based on the practised rhythms and regularities embedded in everyday space—the patterns inscribed in the timespaces and infrastructures of observed populations through travel routines, social habits, building configurations and communication forms, as these are aligned with the rhythms and regularities of the observing institutions of the UAS, cohering through flight patterns, transport timetables, interchanges, regulations, monitoring systems, base locales and operating habits. It is a calibration among systems and data derived from both the object of observation and the observing institution itself—the stabilisation of dynamic fields that have been limited and brought into correspondence with each other. The UAS learns the everyday norm as it co-constitutes it. Embedded regularities coalesce in the 'air' and on the 'ground': embodied practices respond to the regularities that they help sculpt, in ways that can further stabilise or destabilise their defining platforms and programs.

Advanced-stage UAVs now incorporate cognitive architectures and machine learning capabilities that allow them to recognise and identify objects with a more complex and integrated capacity of expertise. The parameters for the algorithms to recognise behaviour or objects need not be set in advance. A learning engine gathers information about dominant object content—tracking, for each object, features like size, colour, reflectivity, sheen, shape and level of autonomy—and forges object classifications without any pre-programmed definitions or specifications. The software analyses the scene to learn and identify normal and anomalous behaviours by way of a constant study of the types of objects that exhibit those behaviours. It learns from experience, internally adapting to changes in the observed environment, detecting and classifying activity that was not previously defined or anticipated.

13

# SATELLITE IMAGES, SECURITY AND THE
# GEOPOLITICAL IMAGINATION

*David Campbell*

In the early hours of 6 September 2007 local residents on the Syrian–Turkish border reportedly heard five jet fighters overhead. The following day a Syrian government official told the Reuters news agency that Israeli aircraft had bombed an empty area in the country's eastern desert before being engaged by air defences, but Israel refused to comment officially on the incident, imposing a news blackout on the story.[1]

So began weeks of international speculation about what had taken place. US newspaper accounts questioned whether the still unconfirmed flight of Israeli warplanes over Syria had been an exercise in intimidation, intelligence-gathering or the targeting of conventional weapons caches bound for Hezbollah in Lebanon. However, within two weeks the media had settled on the claim that the target of Israel's strike was a nuclear facility on the Euphrates River, some 90 miles from the border with Iraq. According to these accounts, Israel had supplied the United States with dramatic evidence of this nuclear site, principally in the form of photographs and satellite imagery.

Syria countered these claims by taking journalists on a tour of the alleged site in early October 2007. Showing the visitors an agricultural research centre, government officials denied any attack had taken place, despite the fact that, one week earlier, the Syrian president had told the BBC an abandoned military site had been targeted, something the Israeli military censor subsequently confirmed. So what did take place in Syria's eastern desert in early September 2007? The public release of satellite images by a private American group in late October 2007 started to make the picture clearer.

The Institute for Science and International Security (ISIS)—headed by former weapons inspector David Albright, and dedicated to 'employing science in the pursuit of peace'—obtained commercial satellite imagery from the American-based satellite operator DigitalGlobe. ISIS argued that the images, taken six weeks apart, revealed a nuclear reactor under construction and then a bomb-damaged site being concealed by earth movement.[2]

The international media lapped up these images, and stories based on them ran globally. And yet, for all their surface veracity, these satellite images recalled earlier concerns about the amount of trust that should be placed in such annotated pictures. When the then US Secretary of State Colin Powell went before the UN Security Council in February 2003 to offer evidence of weapons of mass destruction (WMD) to back the planned invasion of Iraq, for example, the bedrock of his argument was a series of satellite images said to show the weapons programmes the Iraqis were allegedly hiding. Powell called his slideshow 'an accumulation of facts' from solid intelligence, and David Albright told CNN at the time that he found Powell's presentation 'compelling' evidence of the Iraqi failure to comply with the UN's demands for disarmament.[3]

Despite such certitude, the US occupation of Iraq failed to disclose any evidence of Iraqi WMD. After the war the official Iraq Survey Group concluded that all Iraq's WMD were effectively destroyed in the Gulf War of 1991, and the country's nuclear ambitions had ended in the same year. If Syria had embarked on a nuclear programme close to the Iraqi border then America's preoccupation with Saddam Hussein's non-existent WMD programmes was even more misplaced. And that is what another commercial satellite photo seemed to show. The day after the ISIS report deployed the DigitalGlobe images, the US company GeoEye released a 16 September 2003 image that showed the Syrian site in

Image 30: Slide from Secretary of State Colin Powell's address to the U.N. Security Council on the issue of Iraq and disarmament (5 February 2003).

much the same state as the August 2007 photograph.[4] The Syrian reactor story was thus becoming a battle of images.

How have satellite images become powerful resources for claims about international security? Why are satellite images regarded as compelling evidence? Part of the answer is historical, for since flight has been possible, military authorities have developed technologies for remote sensing so that data about distant or obscure things can be captured by contemporary imaging technology. Cameras were attached to tethered balloons as early as 1858 in order to produce aerial views. The First World War was the first conflict in which aircraft-borne cameras enabled widespread photoreconnaissance, and this technology progressed as both photography and flight became more sophisticated. In the Second World War the Allied Central Interpretation Unit in Buckinghamshire was the headquarters for the Royal Air Force's photographic intelligence. Handling 25,000 negatives and 60,000 prints of these images per day, the unit was staffed with personnel using primitive stereoscopes to find evidence of enemy military placements in order to coordinate future attacks. Analysts could only see what they were tasked to find, however. This meant, for example, that although Auschwitz had been photographed more than thirty times by Allied aircraft before Soviet forces liberated it in January 1945, nobody had connected the images to the Final Solution until the former CIA photo interpreters Dino Brugioni and Robert Poirer retrospectively analysed the intelligence files in 1979.[5]

The Cold War was the context for an expansion of photoreconnaissance. The difficulties for intelligence posed by a closed Soviet Union meant that Western agencies sought more advanced means to spy on their opponent's military forces. At a 1955 summit, President Eisenhower proposed that the United States and the USSR enter into an 'open skies' agreement that would permit military overflights of each other's territory in order to provide photographic reassurance that neither was planning to attack the other. The Soviets refused and the Americans developed advanced aircraft like the U-2 that enabled long-range and high-altitude photoreconnaissance. From a height of 70,000 feet its cameras could record images with resolutions as low as 2.5 feet. The U-2 famously documented the Cuban Missile Crisis in 1962—photographs showed Soviet vessels carrying nuclear missiles to Cuba and the construction of launch pads on the island—providing the Kennedy administration with visual evidence to mobilise its allies and to force Khrushchev to back down. Through moments like this, photoreconnaissance—which continues using unmanned aerial vehicles like the 'Global Hawk'—established a record as a revelatory force that could overcome political obstacles to transparency and knowledge.

Photoreconnaissance is but one possible remote sensing technology among many others. It depends on what Jeffrey Richelson calls 'limited black and white visible-light photography' that cannot record things if they take place at night, in the shadow of solid objects, or under the protection of cloud cover.[6] However, panchromatic technologies cover only one small aspect of the electromagnetic spectrum from which images can be produced. The history of remote sensing in the period since the Second World War, then, has been a history of new technologies designed to increase the resolution of images, the speed of their delivery and the variety of sensors through which they are recorded. The quest has been for what Lisa Parks terms 'tele-vision'—the best means of seeing what cannot be seen because of either distance or the limitations of the eye.[7] As a result, satellite reconnaissance using earth-orbiting platforms fitted with infrared sensors and imaging radars as well as visible light cameras greatly increase what can be recorded regardless of conditions. Russia, France, China and the United States are the major nations that have developed these capacities, but the American programme has been by far the largest and best known.

Satellite reconnaissance began in the Eisenhower administration as a way to achieve open skies unilaterally, and this capacity remained an official secret until 1978. Beginning with the CORONA programme in 1960, satellites carried Keyhole (KH-1) film cameras with a resolution of between 25 and 40 feet. They ejected canisters that had to be collected from the air by special aircraft before the film could be processed and interpreted. This meant that although satellite observation could

Image 31: A degraded KH-11 photograph of the Zhawar Kili Base Camp, Afghanistan, which housed training facilities for Osama bin Laden. The photograph was used by the Secretary of Defence to brief reporters on the 1998 cruise missile attack on the facility. This was the first official release of contemporary classified satellite imagery by the US government (DoD, 1998).

proceed without any constraints from another country, the detail of the images and the time it took to get them was some way behind what photoreconnaissance aircraft could achieve. By the end of the CORONA programme in 1972, however, resolutions of 5–6 feet were being achieved, and by the time the more advanced Keyhole optical systems were operational, digital signals transmitted by relay satellite ground stations were producing photographs as low as 6-inch resolution in near real time. Given that a satellite image, taken at an altitude of hundreds and sometimes thousands of miles, is equivalent to a conventional photograph taken from an aircraft flying at only 3,000 feet, this precision is a considerable technical achievement.

The technological developments that have increased both the accuracy and speed of images have given satellite reconnaissance an aura of unsurpassed objectivity. With their ability to thwart environmental and political obstacles, the notion that space-based cameras can capture images anywhere, anytime, thereby revealing what others do not want disclosed, extends the cultural understanding of mechanically generated objectivity (literally) to new heights. As advanced optical systems orbiting anywhere up to 22,000 miles above the earth, satellite reconnaissance embodies Denis Cosgrove's idea of the god-like 'Apollonian eye' surveying all below from a remote vantage point.[8] This omniscient perspective establishes the earth as a domain over which the technological masters can exercise global control. Given that the best satellite resources are controlled by the US government or American-based corporations, the power to see globally is unequally spread. As the Syrian reactor story shows, the international media readily consume satellite images—something that additionally bolsters the images' claim to objectivity—but pays little heed to the fact they are distributing a necessarily partial perspective.

Although the sophisticated technology, orbital position and media appetite for satellite images contribute to the heightened aura of objectivity surrounding these pictures, none of these features can conceal the fact that they are anything but self-evident. When Colin Powell introduced satellite photos of Iraq as part of his Security Council presentation, he made a revealing observation. 'The photos that I am about to show you,' Powell observed, 'are sometimes hard for the average person to interpret, hard for me. The painstaking work of photo analysis takes experts with years and years of experience, pouring for hours and hours over light tables. But as I show you these images, I will try to capture and explain what they mean, what they indicate to our imagery specialists.'[9] Given the rise of digital technologies, Powell's description of photo analysis as taking place at a light table is somewhat anachronistic, but the importance of analysis in the production of satellite imagery cannot be underestimated.

Satellite images do not arrive fully formed as photographs. The raw data has to be computer processed before it can be viewed in a meaningful form. One outcome is an 'orthophoto' whereby distortion from camera perspective is removed, another are stereo pictures to create three-dimensional scenes. This is done in outfits such as the CIA's 'National Photographic Interpretation Center' where more than 1,200 staff work on images through 'photogrammetry' (using pictures to extract the three-dimensional coordinates of what is photographed) and 'photo interpretation' (which draws on previous cases to establish the nature of the objects photographed).[10] Because satellite images do not speak for themselves, the end result is that they appear in public festooned with arrows, captions and claims designed to anchor what is otherwise—increasing resolutions notwithstanding—a blurry and imprecise picture. Moreover, even though satellite images can be produced in real time, their tense is one of latency. Satellites are constantly scanning the earth's surface but their data is most often archived rather than analysed. As Lisa Parks writes in *Cultures in Orbit*, 'satellite image data only becomes a document of the "real" and an index of the "historical" if there is reason to suspect it has relevance to current affairs'.[11] Much like CCTV footage, then, satellite data is dormant and only becomes an image after the fact, when it is rendered, analysed and circulated in response to an event.

These issues of analysis and timing are clear in the case of the Syrian reactor story. Not only did the satellite imagery require captioning, subsequent efforts by the US intelligence community to secure the meaning of the story drew upon other photographic forms to anchor the images. In April 2008 two CIA officers briefed the media in support of the argument that North Korea was working on the Syrian reactor. The transcript has been published on GlobalSecurity.org, and the associated video presentation has been posted on global news sites.[12] While the satellite images are prominent, they are linked to dozens of hand-held, ground-level photographs of the reactor under construction. Dating from 2002, and obtained by Israeli intelligence, these snapshots helped generate a 'photography-based computer model' of the site. Central to this analysis are repeated comparisons with satellite images of North Korea's nuclear facilities at Yongbyon (themselves published on the BBC News site alongside the Syrian story). Here then we have all the components that generate the conventional but strong understanding of satellite images as objective—their technical complexity, association with other forms of photography, rendering into a computer model, links to prior historical cases, textual annotation and propagation by media outlets around the world. Israel has still not officially acknowledged its raid on the Syrian site, but with these images freely circulating such an acknowledgement would be redundant.

A notable feature of the Syrian reactor story has been the role played by commercially produced satellite images. Whilst programmes like LANDSAT have provided broad but coarse multispectral imagery of the earth's surface since 1972, high-resolution satellite imagery remained a prerogative of the state until commercial operators emerged in the late 1990s. The first image from a US-owned commercial satellite was released in October 1999, and since then companies like DigitalGlobe and GeoEye have expanded their technical capacities.[13]

These operators can provide satellite imagery for a variety of purposes, and various non-governmental organisations have purchased them to highlight issues of humanitarian concern. For example, the American Association for the Advancement of Science's 'Geospatial Technologies and Human Rights' programme enabled Amnesty International to get imagery that exposed the government-organised campaign of house demolitions in Zimbabwe. The director of Amnesty's Africa programme described the images as 'irrefutable evidence' of injustice, demonstrating that non-state actors are equally attracted to the aura of objectivity that such images continue to carry.[14]

While commercial operators have challenged state control, US-based firms are subject to 'shutter control'. The government is permitted to terminate the right to take images that could be detrimental to national security, although this is yet to be directly exercised. Instead, licensing restrictions limit the circulation of satellite imagery. Pictures with 1-metre resolution or less cannot be freely sold while images of Israel with 2-metre resolution or less are prohibited by Congress. There is also 'censorship by contract' whereby the government limits the public flow of satellite images by purchasing all the output from commercial operators for a specified time. During the 2001 US attack on Afghanistan the Pentagon spent $2 million per month to gain exclusive rights over Space Imaging's pictures of that country. This policy was modified during the US invasion of Iraq because the Bush Administration made commercial operators the US government's principal supplier of satellite imagery, something that netted Space Imaging a $120 million contract, and left the secret National Reconnaissance Office's space assets free for specific missions.[15]

Satellite imagery is an information resource that helps construct our global geopolitical landscape. Once the tool of a few dominant states, it is now playing a wider role, even though its meaning cannot be predetermined. Google Earth (which gives the appearance of real time but uses DigitalGlobe images between one and three years old) can either mobilise activists alleging genocide in Darfur, or it can be used by Hamas to target their rocket attacks from the Gaza Strip. In this context, we

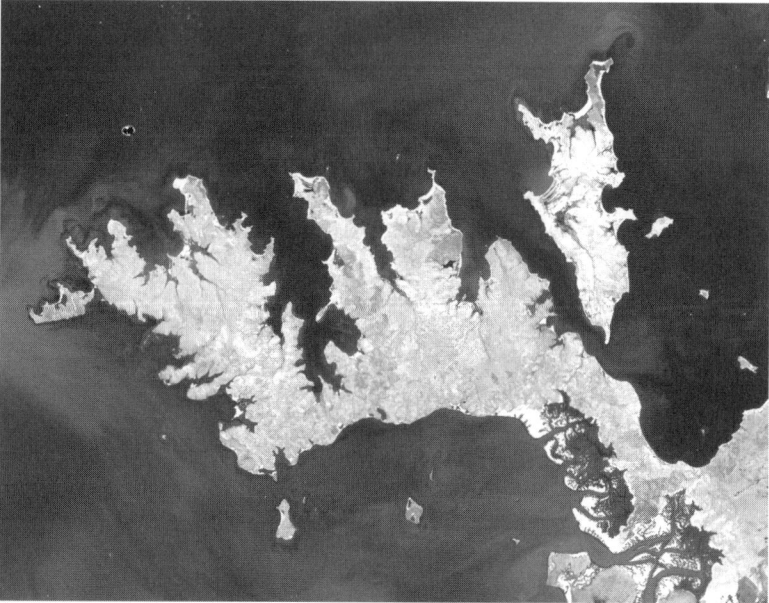

Image 32: Landsat 5 Image of Garig Gunak Barlu National Park in Australia's Northern Territory, 30 July 2006.

need to follow Chad Harris's assessment that aerial and satellite images should be understood and interpreted as artefacts of our broader visual culture. We need to read them critically like any photograph, exploring their production, interpretation and circulation in terms of the political work these supposedly objective regimes of vision support.

# NOTES

## INTRODUCTION: VISUAL CULTURE AND VERTICALITY

1. Eyal Weizman 'The Politics of Verticality', Open Democracy, 1 May 2002 (www.opendemocracy.net/conflict-politicsverticality/article_810.jsp; accessed Aug. 2009).
2. Stuart Elden, 'The Space of the World', in El Hadi Jazairy (ed.), *New Geographies 4: Scales of the Earth*, Cambridge, MA: Harvard University Press, 2011, p. 28.
3. Stephen Graham, 'Vertical Geopolitics: Baghdad and After', *Antipode*, 36 (2004), pp. 12–23.
4. Robert Wohl, *A Passion for Wings: Aviation in the Western Imagination 1908–1918*, New Haven: Yale University Press, 1996.
5. Ann Douglas, *Terrible Honesty: Mongrel Manhattan in the 1920s*, New York: Farrar, Straus and Giroux, 1996.
6. Priya Satia, *Spies in Arabia: The Great War and the Cultural Foundations of Britain's Covert Empire in the Middle East*, Oxford: Oxford University Press, 2008.
7. Marshall Berman, *All That is Solid Melts into Air: The Experience of Modernity*, New York: Verso, 1988.
8. David Omissi, *Air Power and Colonial Control: The Royal Air Force 1919–1939*, Manchester: Manchester University Press, 1990; Mark Neocleous, 'Air Power as Police Power', *Environment and Planning D: Society and Space*, forthcoming, 2013.
9. Stephen Graham, *Cities Under Siege*, London: Verso, 2010; Mike Davis, *Planet of Slums*, New York: Verso, 2002; Derek Gregory, *The Colonial Present*, New York: Blackwell, 2004; Peter Adey, *Aerial Life: Spaces, Mobilities, Affects*, London: Wiley-Blackwell, 2010.
10. Denis Cosgrove and William L. Fox, *Photography and Flight*, London: Reaktion, 2010.

11. Peter Sloterdijk, *Terror from the Air*, New York: Semiotext(e), 2009.

12. Paul N. Edwards, *The Closed World: Computers and the Politics of Discourse in Cold War America*, Cambridge, MA: MIT Press, 1997.

13. Richard Holmes, *Age of Wonders*, London: Harper Press, 2008.

14. Tyler Wall and Torin Monahan, 'Surveillance and Violence from Afar: The Politics of Drones and Liminal Security-Scapes', *Theoretical Criminology*, 15 (Aug. 2011), pp. 239–54.

15. Santanu Das, *Touch and Intimacy in First World War Literature*, Cambridge: Cambridge University Press, 2006.

16. Sven Lindqvist, *A History of Bombing*, New York: Granta Books, 2002.

17. Tami Davis Biddle, *Rhetoric and Reality in Air Warfare: The Evolution of British and American Ideas about Strategic Bombing, 1914–1945*, Princeton, NJ: Princeton University Press, 2004.

18. Peter Singer, *Wired for War: The Robotics Revolution and Conflict in the 21st Century*, London: Penguin, 2011.

19. John Gibson, *The Perfect War: Technowar in Vietnam*, New York: Avalon, 2000.

20. Judith Butler, *Precarious Life: The Power of Mourning and Violence*, New York: Verso, 2006.

21. Peter Adey, 'Holding Still: The Private Life of an Air Raid', *M/C Journal*, 12, 1 (2009).

22. Paul K. Saint-Amour, 'Modernist Reconnaissance', *Modernism/Modernity*, 10, 2 (2003), pp. 349–80.

23. Terrance Finnegan, *Shooting the Front: Allied Air Reconnaissance in the First World War*, London: The History Press, 2011.

24. Maja Zehfuss, 'Targeting: Precision and the Production of Ethics', *European Journal of International Relations*, 17, 3 (2011), pp. 543–56.

25. Bregjie van Eekelen, *Shock on Awe: War on Words*, New York: New Pacific Press, 2004.

26. James Der Derian, *Virtuous War: Mapping the Military-Industrial-Media-Entertainment Complex*, New York: Routledge, 2002; David Alberts, John Gartska and Frederick Stein, *Network Centric Warfare: Developing and Leveraging Information Superiority*, 2nd edn, Washington, DC: National Defense University Press, 2000.

27. Zehfuss, 'Targeting: Precision and the Production of Ethics'.

28. Derek P. McCormack, 'Aerostatic Spacing: On Things Becoming Lighter than Air', *Transactions of the Institute of British Geographers*, 34, 1 (2009).

29. Greg Miller, 'Under Obama, an Emerging Global Apparatus for Drone Killing', *The Washington Post*, 2011; Derek Gregory, 'Supplying War in Afghanistan: The Frictions of Distance', Open Democracy, 2012; Craig Jones, 'NYT Death Reports & $150 Collateral Lives', War, Law & Space, 2012 (http://warlawspace.com/2012/12/30/nyt-death-reports-150-collateral-lives/).

30. Peter Forbes, *Dazzled and Deceived: Mimicry and Camouflage*, New Haven, CT: Yale University Press, 2009.

31. Kenneth Hewitt, 'Place Annihilation: Area Bombing and the Fate of Cities', *Annals of the Association of American Geographers*, 73 (1983), pp. 257–84.

32. James C. Scott, *Seeing like a State*, New Haven: Yale University Press, 2000.

33. Davide Deriu, 'The Ascent of the Modern Planeur: Aerial Images and Urban Imaginary in the 1920s', in C. Emden, C. Keen and D. Midgley, *Imagining the City. Vol. 1: The Art of Urban Living*, Oxford: Peter Lang, 2006, pp. 189–211.

34. See Graham, 'Vertical Geopolitics: Baghdad and After', pp. 12–23.

35. Chris Perkins and Martin Dodge 'Satellite Imagery and the Spectacle of Secret Spaces', *Geoforum*, 40, 4 (2009), pp. 546–60.

36. See Ben Rich and Leo Janos, *Skunk Works*, New York: Back Bay Books, 1995.

37. David Pascoe, *Airspaces*, London: Reaktion, 2001; David Pascoe, *Aircraft*, London: Reaktion, 2002; Denis Cosgrove, *Apollo's Eye*, San Diego: University of California Press, 2001; Bronislaw Szerszynski, *Nature, Technology and the Sacred*, London: Blackwell, 2008.

38. Patrick, M. Cronin, *The Impenetrable Fog of War*, Westport: Praeger, 2008.

39. Gavin Bridge, 'The Hole World: Scales and Spaces of Extraction', *New Geographies*, 2 (2009), pp. 43–8.

40. James Verinni, 'The Tunnels of Gaza', *National Geographic*, Dec. 2012; Philippe Le Billion, *Geopolitics of Resource Wars*, London: Routledge, 2005.

41. Robert Wohl, *The Spectacle of Flight: Aviation in the Western Imagination 1920–1950*, New Haven: Yale University Press, 2005.

42. Michael Paris, *From the Wright Brothers to 'Top Gun': Aviation and Popular Cinema*, Manchester: Manchester University Press, 1995; Robert Wohl, *A Passion for Wings: Flight and the Western Imagination, 1908–1918*, New York: Yale University Press, 1996.

43. Kitty Hauser, *Bloody Old Britain: O.G.S. Crawford and the Archaeology of Modern Life*, London: Oxford University Press, 2008; John Morrow Jr, *The Great War in the Air*, Washington, DC: Smithsonian Institution Press, 1993.

44. Paul Virilio, *War and Cinema: The Logistics of Perception*, London: Verso, 1989.

45. Reminded of course of Donna Haraway's classic exposition of the 'god trick'.

46. Susan R. Grayzel, *At Home and Under Fire: Air Raids and Culture in Britain from the Great War to the Blitz*, New York: Cambridge University Press, 2012.

47. See Trevor Paglen and A.C. Thompson, *Torture Taxi: On the Trail of the CIA's Rendition Flights*, New York: Hoboken, 2006.

48. Stephen Kern, *The Culture of Time and Space 1880–1918*, London: Harvard University Press, 1983; David Harvey, *The Condition of Post-Modernity*, Cambridge, MA: Harvard University Press, 1989.

49. Eyal Weizman, *Hollow Land: Israel's Architecture of Occupation*, London: Verso, 2007.

50. Judith Butler, 'Contingent Foundations', in Seyla Benhabib et al., *Feminist Contentions: A Philosophical Exchange*, New York: Routledge, 1995; Philip Knightley, *The First Casualty*, New York: Harcourt, 2004.

51. Gregory, *Colonial Present*, p. 54.

52. David Omissi, *Air Power and Colonial Control: The Royal Air Force 1919–1939*, Manchester: Manchester University Press, 1990.

53. Tim Blackmore, *War X: Human Extensions in Battlespace*, Toronto: University of Toronto Press, 2005.

54. Lisa Parks, *Cultures in Orbit: Satellites and the Televisual*, Durham: Duke University Press, 2005.

55. John Andreas Olsen (ed.), *A History of Aerial Warfare*, Washington, DC: Potomac Books Inc., 2009.

56. Alison J Williams, 'Enabling Persistent Presence? Performing the Embodied Geopolitics of the Unmanned Aerial Vehicle Assemblage', *Political Geography*, 30 (2011), pp. 381–90.

57. Cosgrove, *Apollo's Eye*.

58. Stuart Banner, *Who Owns the Sky? The Struggle to Control Airspace from the Wright Brothers On*, Cambridge, MA: Harvard University Press, 2008.

59. Taylor Downing, *Spies in the Sky: The Secret Battle for Aerial Intelligence during World War II*, New York: Little and Brown, 2011.

## 1. THE BALLOON PROSPECT: AEROSTATIC OBSERVATION AND THE EMERGENCE OF MILITARISED AEROMOBILITY

1. Paul Virilio, *War and Cinema: The Logistics of Perception*, London: Verso, 1989 [1984], p. 6.

2. Peter Adey, *Aerial Life: Spaces, Mobilities, Affects*, Malden: Wiley-Blackwell, 2010.

3. Saulo Cwerner, 'Introducing Aeromobilities', in S. Cwerner, S. Kesselring and J. Urry (eds), *Aeromobilities*, London: Routledge, 2009.

4. Ibid., p. 2; Mark Whitehead, *State, Science and the Skies: Governmentalities of the British Atmosphere*, Oxford: Wiley-Blackwell, 2009; David Edgerton, *Warfare State: Britain, 1920–1970*, Cambridge: Cambridge University Press, 2006. See also essays in D. Cowen and E. Gilbert (eds), *War, Citizenship, Territory*, New York: Routledge, 2008.

5. Michel Foucault, *'Society Must Be Defended': Lectures at the Collège De France, 1975–1976*, New York: Picador, 2003; Julian Reid, *The Biopolitics of the War on Terror: Life Struggles, Liberal Modernity, and the Defence of Logistical Societies*, Manchester: Manchester University Press, 2006.

6. Col. Terrence J. Finnegan, *Shooting the Front: Allied Aerial Reconnaissance and Photographic Interpretation on the Western Front—World War I*, Washington, DC: National Defense Intelligence College Press, 2006.

7. Richard P. Hallion, *Taking Flight: Inventing the Aerial Age from Antiquity through the First World War*, New York: Oxford University Press, 2003, p. 366; Tami Davis Biddle, *Rhetoric and Reality in Air Warfare: The Evolution of British and American Ideas About Strategic Bombing, 1914–1945*, Princeton: Princeton University Press, 2002, p. 19.

8. Giulio Douhet, *The Command of the Air*, New York: Coward-McCann, 1942; William Mitchell, *Winged Defense: The Development and Possibilities of Modern Air Power—Economic and Military*, New York: G.P. Putnam, 1925; Conrad C. Crane, *Bombs, Cities, and Civilians: American Airpower Strategy in World War II*, Lawrence: University of Kansas Press, 1993.

9. David Omissi reminds us that, even before the First World War, in 1911, Italy used airplanes for aerial observation and to drop bombs over Libya while similar tactics were used by both sides in the 1913–15 Mexican civil war. David E. Omissi, *Air Power and Colonial Control: The Royal Air Force, 1919–1939*, Manchester: Manchester University Press, 1990, p. 5. See also Beau Grosscup, *Strategic Terror: The Politics and Ethics of Aerial Bombardment*, London: Zed Books, 2006; Priya Satia, *Spies in Arabia: The Great War and the Cultural Foundations of Britain's Covert Empire in the Middle East*, New York: Oxford University Press, 2008; James S. Corum and Wray R. Johnson, *Airpower in Small Wars: Fighting Insurgents and Terrorists*, Wichita: University of Kansas Press, 2003; Alison Williams, 'Hakumat al Tayarrat: The Role of Air Power in the Enforcement of Iraq's Boundaries', *Geopolitics*, 12 (2007), pp. 505–28.

10. George W. Goddard, *Overview: A Lifelong Adventure in Aerial Photography*, Garden City: Doubleday, 1969; Stephen L. McFarland, *America's Pursuit of Precision Bombing, 1910–1945*, Washington, DC: Smithsonian Institution Press, 1995. It should also be noted that between 1937 and 1941 China and Japan fought each other by air as well as by other means. All major industrial countries invested in airpower during this crucial period.

11. Kevin D. Haggerty, 'Visible War: Surveillance, Speed, and Information War', in K.D. Haggerty and R.V. Ericson (eds), *The New Politics of Surveillance and Visibility*, Toronto: University of Toronto Press, 2006, pp. 250–78; Michael J. Shapiro, *Violent Cartographies: Mapping Cultures of War*, Minneapolis: University of Minnesota Press, 1997.

12. Beaumont Newhall, *Airborne Camera: The World from the Air and Outer Space*, New York: Hastings House, 1969; Col. Roy M. Stanley II, *World War II Photo Intelligence*, New York: Charles Scribner's Sons, 1981.

13. In an innovative multimedia essay, Jennifer Terry has argued that the 'visual codes' pertaining to the most recent US-led wars in Iraq and Afghanistan operate 'in the grain of other popular entertainments that may indeed serve to make the violence of war something that we have already incorporated into daily life and bodily practice, but of whose many effects we remain disturbingly unaware'.

Jennifer Terry, 'Author's Statement', in 'Killer Entertainments', *Vectors: Journal of Culture and Technology in a Dynamic Vernacular*, 3, 1 (2007) (http://www.vectorsjournal.org/index.php?page=7&projectId=86).

14. Chris Hedges, *War Is a Force That Gives Us Meaning*, New York: Anchor Books, 2003; Rachel Woodward, *Military Geographies*, Malden: Blackwell, 2004; Nick Turse, *The Complex: How the Military Invades Our Everyday Lives*, New York: Metropolitan Books, 2008; and Rey Chow, *The Age of the World Target: Self-Referentiality in War, Theory, and Comparative Work*, Durham: Duke University Press, 2006.

15. Nigel Thrift, 'Immaculate Warfare? The Spatial Politics of Extreme Violence', in D. Gregory and A. Pred (eds), *Violent Geographies: Fear, Terror, and Political Violence*, New York: Routledge, 2007, pp. 273–94.

16. John Law and Ruth Benschop, 'Resisting Pictures: Representation, Distribution and Ontological Politics', in K. Hetherington and R. Munro (eds), *Ideas of Difference*, Oxford: Blackwell, 1997, pp. 158–82.

17. Virilio, *War and Cinema*, p. 3.

18. For example, Charles Waldheim, 'Aerial Representation and the Recovery of Landscape', in J. Corner (ed.), *Recovering Landscape: Essays in Contemporary Landscape Architecture*, New York: Princeton Architectural Press, 1999, pp. 121–39; Lucy Budd, 'The View from the Air: The Cultural Geographies of Flight', in P. Vannini (ed.), *The Cultures of Alternative Mobilities: Routes Less Travelled*, Burlington: Ashgate, 2009, pp. 71–90; Karen Piper, *Cartographic Fictions: Maps, Race, and Identity*, New Brunswick: Rutgers University Press, 2002; Davide Dériu, 'Between Veiling and Unveiling: Modern Camouflage and the City as a Theatre of War', in M. Funck and R. Chickering (eds), *Endangered Cities: Military Power and Urban Societies in the Era of the World Wars*, Boston: Brill, 2004, pp. 15–34; Matthew Zook et al., 'New Digital Geographies: Information, Communication, and Place', in S.D. Brunn et al. (eds), *Geography and Technology*, Dordrecht: Kluwers, 2005, pp. 155–76. I made a similar argument in earlier work, cf. 'Mobility and War: The Cosmic View of U.S. Air Power', *Environment and Planning A*, 38, 2 (2006), pp. 395–407.

19. Virilio, *War and Cinema*, p. 3.

20. Denis Cosgrove, *Apollo's Eye: A Cartographic Genealogy of the Earth in the Western Imagination*, Baltimore: The Johns Hopkins University Press, 2001, p. xi.

21. Astrit Schmidt-Burckhardt, 'The All-Seer: God's Eye as Proto-Surveillance', in T.Y. Levin, U. Frohne and P. Weibel (eds), *Ctrl [Space]: Rhetorics of Surveillance from Bentham to Big Brother*, Cambridge: MIT Press, 2002, p. 23; Albert M. Potts, *The World's Eye*, Lexington: University Press of Kentucky, 1982, pp. 68–78; and E.H. Gombrich, 'The Dream of Reason: Symbolism in the French Revolution', *The British Journal for Eighteenth-Century Studies*, 2 (1979), p. 200.

22. Schmidt-Burckhardt, 'The All-Seer', p. 24.

23. Michel Foucault, *Discipline and Punish: The Birth of the Prison*, New York: Vintage Books, 1979, p. 199.

24. Ibid., p. 201.

25. Kevin D. Haggerty, 'Tear Down the Walls: On Demolishing the Panopticon', in D. Lyon (ed.), *Theorizing Surveillance: The Panopticon and Beyond*, Cullompton: Willan, 2006, p. 23.

26. Michel Foucault, 'The Eye of Power', in C. Gordon (ed.), *Power/Knowledge: Selected Interviews and Other Writings, 1972–1977*, New York: Pantheon, 1977, p. 148.

27. Martin Jay, *Downcast Eyes: The Denigration of Vision in Twentieth-Century French Thought*, Berkeley: University of California Press, 1994, p. 97.

28. Chow, *The Age of the World Target*, p. 31.

29. David N. Livingstone and Charles W.J. Withers (eds), *Geography and Enlightenment*, Chicago: University of Chicago Press, 1999; Barbara Maria Stafford, *Voyage Into Substance: Art, Science, Nature and the Illustrated Travel Account, 1760–1840*, Cambridge: MIT Press, 1984.

30. In addition to Virilio's influential work, see Nicholas Mirzoeff, *Watching Babylon: The War in Iraq and Global Visual Culture*, New York: Routledge, 2005; Manuel De Landa, *War in the Age of Intelligent Machines*, New York: Zone, 1991; James Der Derian, *Virtuous Wars: Mapping the Military-Industrial-Media-Entertainment Network*, Boulder: Westview, 2001; Chris Hables Gray, *Postmodern War: The New Politics of Conflict*, New York: Guilford Press, 1997; Paul Edwards, *The Closed World: Computers and the Politics of Discourse in Cold War America*, Cambridge: MIT Press, 1997; and Stephen Graham, *Cities Under Siege: The New Military Urbanism*, London: Verso, 2010.

31. Virilio, *War and Cinema*, p. 11.

32. Ibid., p. 4; p. 66.

33. Jamil Smith, 'The Coming Drone Wars', The Maddow Blog, Tuesday, 5 July 2011 (http://maddowblog.msnbc.msn.com/_news/2011/07/05/7018408-the-coming-drone-wars; accessed 6 July 2011); Elisabeth Bumiller and Thom Shanker, 'War Evolves with Drones, Some Tiny as Bugs', *The New York Times*, 19 June 2011 (http://www.nytimes.com/2011/06/20/world/20drones.html?_r=1&nl=todaysheadlines&emc=tha2; accessed 6 July 2011). See also Michael Ignatieff, *Virtual War: Kosovo and Beyond*, New York: Picador, 2000.

34. Giuliana Bruno, *Atlas of Emotion: Journeys in Art, Architecture, and Film*, New York: Verso, 2002, p. 16.

35. For a recent, thoughtful exploration of the limits of the visual, see Derek P. McCormack, 'Aerostatic Spacing: On Things Becoming Lighter than Air', *Transactions of the Institute of British Geographers*, 34 (2009), pp. 25–41.

36. 'Extract of a Report of Fourcroy to the Convention Nationale (1795)', cited in Christopher Hatton Turner, *Astra Castra: Experiments and Adventures in the Atmosphere*, London: Chapman and Hall, 1865, p. 283.

37. L.T.C. Rolt, *The Balloonists: The History of the First Aeronauts*, Stroud: Sutton, 2006, p. 163.

38. Montgolfier cited in Rolt, *The Balloonists*, p. 29.

39. John Christopher, *Balloons at War: Gasbags, Flying Bombs, and Cold War Secrets*, Stroud: Tempus, 2004, p. 12.

40. Hallion, *Taking Flight*, p. 56.

41. Christopher, *Balloons at War*, p. 15.

42. For a fuller account of this fascinating chapter in military aerostatic operations, see Peter Mead, *The Eye in the Sky: History of Air Observation and Reconnaissance for the Army, 1785–1945*, London: Her Majesty's Stationery Office, 1983, pp. 13–15.

43. Coutelle cited in Turner, *Astra Castra*, p. 283.

44. Money cited in Turner, *Astra Castra*, p. 285.

45. Despite dedicated enthusiasts such as Major-General Money, the British did not establish a balloon corps until some decades later. See Money cited in Turner, *Astra Castra*, p. 285, also Christopher, *Balloons at War*, p. 53.

46. Aviation historian Richard Hallion writes that there is 'no evidence that he ever had more than a passing interest in military ballooning'. Hallion, *Taking Flight*, p. 65. John Christopher argues more forcefully that Napoleon's deep ties to the cavalry and other experiences turned his relative disinterest in balloons to a 'downright hatred'; Christopher, *Balloons at War*, p. 27. Also in an invested vein, L.T.C Rolt writes: 'Thus was a promising experiment brought to a premature end by one vain and short-sighted man. It had been said, probably with truth, that Napoleon spurned the tactical aid of the new arm as unbecoming a man who, in his own opinion, was an inspired military genius …'; Rolt, *The Balloonists*, pp. 164–5. The history of airpower is often fashioned as a tale of official disinterest, political infighting and even career or technical sabotage (not to mention nationalist posturing and competition between countries). Napoleon's lack of serious support, apocryphal or not, is certainly an urtext.

47. Mead notes that although the French seemed to drop their mercurial advance in military aerostatic experimentation at a surprisingly early stage, other countries that had plenty of dedicated aerial enthusiasts among their citizenry never attempted as much, notably England. See Mead, *The Eye in the Sky*, p. 15.

48. Martin van Creveld, *Technology and War: From 2,000 B.C. to the Present*, New York: The Free Press, 1991, p. 21.

49. Ibid. See also Turner, *Astra Castra*, pp. 284–5.

50. Quentin Hughes, *Military Architecture*, London: Hugh Evelyn, 1974, p. 8.

51. Van Creveld, *Technology and War*, p. 82; Paul Hirst, *War and Power in the 21st Century*, Cambridge: Polity, 2001, p. 17.

52. On the pervasive influence of military culture on urban design in sixteenth- and seventeenth-century Italy, see Martha D. Pollack, *Urban Design, Military*

*Culture, and the Creation of the Absolutist Capital*, Chicago: University of Chicago Press, 1991.

53. Hughes, *Military Architecture*, p. 8.

54. Van Creveld, *Technology and War*, p. 115.

55. Renzo Dubbini, *Geography of the Gaze: Urban and Rural Vision in Early Modern Europe*, Chicago: University of Chicago Press, 2002, p. 30.

56. Charles M. Evans, *War of the Aeronauts: A History of Ballooning in the Civil War*, Mechanicsburg: Stackpole Books, 2002; Stephen Poleskie, *The Balloonist: The Story of T.S.C. Lowe—Inventor, Scientist, Magician, and Father of the U.S. Air Force*, Savannah: Frederic C. Beil, 2007; Captain F. Beaumont, 'Balloon Reconnaissance', in N. Duke and E. Lanchbery (eds), *The Saga of Flight*, New York: Avon, 1964, pp. 44–6.

57. These strategic elements include: (1) psychological (morale); (2) military force (size, organisation, etc.); (3) geometry (relative position, movement of forces in relation to objectives or obstacles, etc.); (4) terrain (topography); and (5) supply (logistics). Louis C. Peltier and G. Etzel Pearcy, *Military Geography*, Princeton: D. Van Nostrand, 1966, pp. 45–6; and Carl von Clausewitz, *Principles of War*, Mineola: Dover, 2003, pp. 45–59.

58. Although balloons proved to be extraordinarily helpful during the siege of Paris in the Franco-Prussian War over the autumn and winter of 1870–1, they were used exclusively to lift key personnel and messages out of the city and over enemy lines. The balloonists faced considerable peril due to the difficulties of aerostatic navigation; several ended up in the freezing ocean far to the west or were blown by storms into Germany or even further north. John Fisher, *Airlift 1870: The Balloon and Pigeon Post in the Siege of Paris*, London: Max Parrish, 1965. For a mid-nineteenth-century perspective on the debates surrounding militarised aerostatic flight, see Chapter VIII, 'War-Balloons', in Turner, *Astra Castra*, pp. 279–98.

59. Michael Dillon and Julian Reid, *The Liberal Way of War: Killing to Make Life Live*, London: Routledge, 2009, p. 15.

60. Newhall, *Airborne Camera*, p. 11.

61. Jean-Marc Besse, 'Aerial Geography', in A.S. MacLean et al. (eds), *Designs on the Land: Exploring America from the Air*, London: Thames and Hudson, 2003, p. 339.

62. Hallion, *Taking Flight*, p. 58.

63. Newhall, *Airborne Camera*, p. 9. I discuss the history of photography and aerial views in more depth in the longer work of which this chapter is a part, *The View from Above: The Visual Culture of Militarization*, forthcoming, Duke University Press.

64. Newhall, *Airborne Camera*, pp. 11–12.

65. Hallion, *Taking Flight*, p. 58.

66. 'Review of *Airopaidia*', *The English Review; or, an Abstract of English and Foreign Literature*, 10, (1787), p. 429.

67. Bernard Comment, *The Panorama*, London: Reaktion Books, 1999, p. 142.

68. Bruno, *Atlas of Emotion*, pp. 60–1.

69. Edward Said, *Orientalism*, New York: Vintage Books, 1979; Derek Gregory, *The Colonial Present*, Malden: Blackwell, 2004; A. Godlewska and N. Smith (eds), *Geography and Empire*, Cambridge: Blackwell, 1994; Inderpal Grewal, *Home and Harem: Gender, Empire, and the Cultures of Travel*, Durham: Duke University Press, 1996.

70. Richard Hamblyn, *The Invention of Clouds: How an Amateur Meteorologist Forged the Language of the Skies*, New York: Picador, 2001.

71. McCormack, 'Aerostatic Spacing', pp. 26–7.

72. Stafford, *Voyage into Substance*, p. 432.

73. Dubbini, *Geography of the Gaze*, p. 51.

74. See Grewal, *Home and Harem*, pp. 23–56, for a discussion of the colonial precepts of British Romanticism and attitudes towards landscape see also Jens Jager, 'Picturing Nations: Landscape, Photography and National Identity in Britain and Germany in the Mid-Nineteenth Century', in J.M. Schwartz and J.R. Ryan (eds), *Picturing Place: Photography and the Geographical Imagination*, London: I.B. Tauris, 2003, pp. 117–40. For the operation of nationalist and colonialist aerial viewing in a later period see M. Christine Boyer, 'Aviation and the Aerial View: Le Corbusier's Spatial Transformations in the 1930s and 1940s', *Diacritics*, 33, 3/4 (2003), pp. 93–116.

75. M. Christine Boyer, *The City of Collective Memory: Its Historical Imagery and Architectural Entertainments*, Cambridge: MIT Press, 1996, p. 247.

76. Clive Adams, 'Between the Angels and the Beasts', in E. Anderson (ed.), *The Impossible View*, Salford Quays: The Lowry Press, 2003, p. 18; Louis Marin, *Utopics: The Semiological Play of Textual Spaces*, New York: Humanity Books, 1984, pp. 208–9.

77. Besse, 'Aerial Geography', p. 343.

78. Ibid.

79. Newhall, *Airborne Camera*, p. 11.

80. Bruno, *Atlas of Emotion*, p. 177.

81. Ibid.

82. Lunardi, who was secretary to the Neapolitan ambassador to England, is almost always recognised as the first person to have ascended in a balloon in Britain, but Hallion cites the flight of Scottish aeronaut James Tytler in 1784, some months after the first French ascents and one month before Lunardi, as the 'first Briton' in the air. See Hallion, *Taking Flight*, p. 59.

83. Thomas Baldwin, *Airopaidia*, London: J. Fletcher, 1786. J.E. Hodgson writes of Baldwin: 'nothing is known beyond his printed account'; see Hodgson, *The*

*History of Aeronautics in Great Britain*, Oxford: Oxford University Press, 1924, p. 131. Baldwin is not mentioned at all in Hallion's *Taking Flight* or Duke and Lanchbery's *The Saga of Flight* (to cite two popular histories of early flight that include significant chapters on ballooning). Turner discusses Baldwin's flight in *Astra Castra* in his chapter on the year 1785 on pages 91–2 while Stafford refers to *Airopaidia* throughout *Voyage into Substance*, see especially pp. 149 and 222–3. I am indebted to conversations with Jen Southern and to her blog post 'Airopaidia' (viewable at http://sketchagraph.wordpress.com/2010/05/20/airo-paidia/). Baldwin aficionados may also be interested in the short video *Arpeggio* made by Ron Kenley in 2010 in collaboration with Jacques Attali, a work that traces Baldwin's route using contemporary aerial and satellite imagery; viewable at http://vimeo.com/14972328

84. 'Review of *Airopaidia*', *The English Review*, p. 429.
85. Stafford, *Voyage into Substance*, p. 24.
86. Baldwin cited in Stafford, *Voyage into Substance*, p. 405.
87. Mason cited in Turner, *Astra Castra*, p. 356.
88. Kees Boeke, *Cosmic View: The Universe in Forty Jumps*, New York: John Day, 1957; Charles Eames and Ray Eames, *Powers of Ten*, New York: Scientific American Library, 1982. See also the films of Charles and Ray Eames, including *A Rough Sketch for a Proposed Film Dealing with the Powers of Ten and the Relative Size of Things in the Universe* (1968) and *Powers of Ten* (1977).
89. Mason cited in Turner, *Astra Castra*, p. 363.
90. Ibid.
91. 'Review of *Airopaidia*', *The Monthly Review; or, Literary Journal*, 75 (July 1786), p. 219.
92. Stafford, *Voyage into Substance*, p. 434.
93. Mason cited in Turner, *Astra Castra*, p. 364.
94. Glaisher cited in Turner, *Astra Castra*, p. 389.
95. Stafford, *Voyage into Substance*, p. 149.
96. Kim Sichel, *To Fly: Contemporary Aerial Photography*, Seattle: University of Washington Press, 2007, p. 11.
97. Mason cited in Turner, *Astra Castra*, p. 377.
98. Howard Zinn writes: 'We might think that at least those individuals in the U.S. Air Force who dropped bombs on civilian populations were aware of what terror they were inflicting, but as one of those I can testify that this was not so. Bombing from five miles high, I and my fellow crew members could not see what was happening on the ground. We could not hear screams or see blood, could not see torn bodies, crushed limbs. Is it any wonder we see fliers going out on mission after mission, apparently unmoved by thoughts of what they have seen.' Howard Zinn, 'Foreword', in Elin O'Hara Slavick, *Bomb After Bomb: A Violent Cartography*, Milan: Edizioni Charta, 2007, p. 9. See also Sven Lindqvist, *A History of Bombing*, New York: The New Press, 2001.

99. Ernst Jünger, 'War and Photography', *New German Critique*, 59 (1993), pp. 24–6.
100. Davide Dériu, 'Picturing Ruinscapes: The Aerial Photograph as Image of Historical Trauma', in F. Guerin and R. Hallas (eds), *The Image and the Witness: Trauma, Memory and Visual Culture*, London: Wallflower Press, 2007, p. 191. See also Karen Frome, 'A Forced Perspective: Aerial Photography and Fascist Propaganda', *Aperture*, 132 (1993), p. 77.
101. See Allan Sekula's important essay, 'The Instrumental Image: Steichen at War', in A. Sekula, *Photography against the Grain: Essays and Photo Works, 1973–1983*, Halifax: The Press of the Nova Scotia College of Art and Design, 1984, pp. 33–52.
102. McCormack, 'Aerostatic Spacing', p. 37.
103. Ibid.
104. Southern, 'Airopaidia'.
105. Ibid.
106. Boyer argues that travel is 'both a vehicle for pleasure and escape into unknown lands and imaginary worlds, in addition to being a metaphor for exploration and economic conquest, for the collection of information and goods that implies mastery and control'. Although representations of travel operate as 'instruments of control', they 'cannot suppress or eradicate the emotional or imaginary effects of wondrous landscapes …' Boyer, *The City of Collective Memory*, p. 205.
107. Stafford, *Voyage into Substance*, p. 435.
108. Ibid.
109. The author thanks Minoo Moallem, Jennifer Terry, Inderpal Grewal, Marisol de la Cadena, Deborah Cowen, Liz Montegary, Toby Beauchamp and Abigail Boggs for their inspiring work and many conversations on militarisation, security states and the politics of representation. In addition to the editors of this volume, the author thanks Steven Groening and Mimi Sheller for invitations to speak at Brown and Drexel universities, respectively, at symposia on the 'Aesthetics of Transport' and 'Mobilities in Motion' where early versions of this essay benefited from insightful comments and suggestions. Eric Smoodin provided kind intellectual and material support on the homefront.

## 2. LINES OF DESCENT

1. Daniel Swift, 'Bomb Proof', *Financial Times*, 4 Sep. 2010. Bombing has a longer history, of course, and pre-dates even the air raids against towns and cities in the First World War. The first (largely unsuccessful) attempts at dropping bombs from aircraft were made over what is now Libya during the Italian–Turkish war of 1911–12. The prospect of bombing by remote control had also been mooted

before the First World War. In May 1910 Thomas Raymond Phillips used a 20-foot model Zeppelin to demonstrate his wireless-controlled 'aerial torpedo' before entranced crowds at the London Hippodrome. 'I can sit in an armchair in London,' he boasted, 'and drop bombs in Manchester, Paris or Berlin.' Given the first city on his target list it is scarcely surprising that he should have been questioned about navigation. He replied that he would work with a large-scale map or a 'telephotographic lens'. I discuss this and other similar experiments at http://geographicalimaginations.com/2012/10/22/over-the-top

2. Mark Clodfelter, *Beneficial Bombing: The Progressive Foundations of American Air Power, 1917–1945*, Lincoln: University of Nebraska Press, 2010. That said, Swift, *Bomber County: The Poetry of a Lost Pilot's War*, London and New York: Hamish Hamilton, 2010, p. 38, claims that 'bombing to the Second World War was what the trenches were to the First: a shocking and new form of warfare, wretched and unexpected, and carried out at a terrible scale of loss'. In fact it was neither new nor unexpected, but the unprecedented scale and scope of strategic bombing confirmed the predictions of the so-called 'prophets' who believed that airpower would create a 'new battlefield' limited only by the boundaries of the belligerent states—the distinction between combatant and civilian would be lost forever—and, crucially, that 'command of the air' would be of decisive importance to the outcome of the war.

3. A.P. de Seversky, *Victory Through Air Power*, New York, 1942, pp. 138–9.

4. Donald Miller, *Masters of the Air*, New York: Simon & Schuster, 2006, pp. 299–303, 415; Ronald Shaffer, 'American Military Ethics in World War II: The Bombing of German Civilians', *Journal of American History*, 67 (1980), pp. 318–34, see pp. 326–7; Michael Sherry, *The Rise of American Air Power: The Creation of Armageddon*, New Haven: Yale University Press, 1987, p. 187.

5. Derek Gregory, 'Doors into Nowhere: Dead Cities and the Natural History of Destruction', in Peter Meusburger, Michael Heffernan and Edgar Wunder (eds), *Cultural Memories*, Dordrecht: Springer, 2011, pp. 249–82.

6. Len Deighton, *Bomber*, London: Pan Books, 1970, p. 383.

7. Graham Swift, *Out of this World*, New York: Poseidon Press, 1993 [1988], p. 47.

8. James Taylor and Martin Davidson, *Bomber Crew*, London: Hodder & Stoughton, 2004, pp. 282–4.

9. Don Charlwood, *No Moon Tonight*, Manchester: Creécy, 2000 [1956], p. 131.

10. Jack Currie, *Lancaster Target*, Manchester: Goodall, 1997, p. 79.

11. Taylor and Davidson, *Bomber Crew*, pp. 283, 447.

12. Sherry, *Rise of American Air Power*, pp. 209–10.

13. 'The American Way of Bombing', *Harper's Magazine*, June 1972, pp. 55–8; Raphael Littauer and Norman Uphoff (eds), *The Air War in Indochina*, Boston: Beacon Press, 1972, pp. 159, 163.

14. This argument is one of the (several) towering achievements of Sherry, *Rise of American Air Power*.

15. Richard G. Davis, *Carl A. Spaatz and the Air War in Europe*, Washington, DC: Center for Air Force History, 1993, p. 435; Tami Davis Biddle, *Rhetoric and Reality in Air Warfare: The Evolution of British and American Ideas about Strategic Bombing, 1914–1945*, Princeton NJ: Princeton University Press, 2002; Randall Hansen, *Fire and Fury: The Allied Bombing of Germany 1942–1945*, Toronto: Doubleday, 2008.

16. Robert Pape, *Bombing to Win: Air Power and Coercion in War*, Ithaca, NY: Cornell University Press, 1996, p. 174; W. Hays Parks, 'Rolling Thunder and the Law of War', *Air University Review*, 33 (1982), pp. 2–23. The 'sticks' were sticks of bombs; the 'carrots' a series of bombing pauses.

17. When the US Air Force was separated from the US Army after the Second World War it established its own Directorate of Targets responsible for the compilation of the 'Bombing Encyclopedia of the World' (later renamed the Basic Encyclopedia). Work started in Jan. 1946 on potential targets in the Soviet Union and in six months IBM cards were punched for 5,594 targets; the database quickly became global and by 1960 it contained 80,000 entries.

18. Mark Clodfelter, *The Limits of Air Power: The American Bombing of North Vietnam*, 2nd edn, Lincoln: University of Nebraska Press, 2006, pp. 76–7; Charles Kamps, 'The JCS 94-Target List', *Aerospace Power Journal*, 15, 1 (2001), pp. 67–80.

19. Clodfelter, *Limits*, pp. 84–8; John Smith, *Rolling Thunder: The Strategic Bombing Campaign against North Vietnam 1964–1968*, Walton-on-Thames UK: Air Research Publications, 1994, p. 58.

20. David Humphreys, 'On the Tuesday Lunch at the Johnson White House: A Preliminary Assessment', *Diplomatic History*, 8 (1984), pp. 81–101.

21. Parks, 'Rolling Thunder'; Pape, *Bombing to Win*, p. 86 insists that civilian planners accepted the logic of interdiction and that during the second phase of Rolling Thunder Johnson 'permitted complete freedom for armed reconnaissance and re-strikes of previously released targets throughout all of North Vietnam, except for small areas around Hanoi, Haiphong and the Chinese border. By the late summer and fall of 1967, nearly all infiltration targets were subject to air attack.'

22. Newly arrived pilots were not cleared to fly combat until they had passed a test on the Rules of Engagement (ROE), but Rasimus claims that this was so heavily annotated 'only a blind man could fail to distinguish the right answers for the 25 multiple-choice questions' and 'no one would be a Rules-of-Engagement lawyer after taking the test, but someone could always say that all of the pilots were trained in the ROE'. Ed Rasimus, *When Thunder Rolled: An F-105 pilot over North Vietnam*, New York: Random House, 2003, pp. 65–6.

23. Peter Braestrup, 'US Morale High at Thai Base, but Raid Curbs are Resented', *New York Times*, 13 Mar. 1967; Harrison E. Salisbury, 'No Military Targets, Namdinh Insists', *New York Times*, 31 Dec. 1966.

24. John Correll, 'The Emergence of Smart Bombs', *Air Force Magazine*, 93, 2010, pp. 60–4; David Koplow, *Death by Moderation: The US Military's Quest for Useable Weapons*, Cambridge: Cambridge University Press, 2010.

25. Perry Lamy, *Barrel Roll, 1968–1973*, Maxwell Air Force Base, AL: Air University Press, 1995, p. 65.

26. Joint Chiefs of Staff, *History of Vietnam War (II) 1965–67*, Vietnam Center and Archive, Texas Tech University, pp. 24–3.

27. John Schlight, *The War in South Vietnam: The Years of the Offensive, 1965–1968 [The United States Air Force in Southeast Asia]*, Honolulu, HI: University Press of the Pacific, 2002; original 1998, p. 50; Donald Mrozek, *Air Power and the Ground War in Vietnam*, Maxwell Air Force Base, AL: Air University Press, 1988, pp. 139–40; William Head, *War From Above the Clouds: B-52 Operations during the Second Indochina War*, Maxwell Air Force Base, AL: Air University Press, 2002, pp. 17–32.

28. Head, *Above the Clouds*, p. 24.

29. Stephen Green, *Meditations in Green*, New York: Bantam Books, 1984, pp. 39–40.

30. Schlight, *War*, p. 52.

31. Joseph Treaster, 'Aboard B-52 Bomber High over Vietnam a Crew Takes Part in an Impersonal War', *New York Times*, 13 Oct. 1972.

32. Joint Chiefs of Staff, *History*, pp. 24–8; MACV Directive 525–3: 'Combat Operations: Minimizing Non-Combatant Casualties'; Schlight, *War*, p. 83.

33. Schlight, *War*, p. 135.

34. Rasimus, *Thunder*, p. 130.

35. Matthew Kocher, Thomas Pepinsky and Stathis Kalyvas, 'Aerial Bombing and Counterinsurgency in the Vietnam War', *American Journal of Political Science*, 55 (2011), pp. 201–18.

36. Bell, *100 Missions*, p. 177.

37. Marshall Harrison, *A Lonely Kind of War: Forward Air Controller, Vietnam*, New York: Xlibris, 2011; original 1997, p. 3.

38. Thomas Ehrhard, *Air Force UAVs: The Secret History*, Arlington, VA: Mitchell Institute Press, 2010, pp. 23–9; Ronald Frankum, *Like Rolling Thunder: The Air War in Vietnam 1964–1975*, Lanham, MD: Rowman & Littlefield, 2005, pp. 96–7; Robert Barkan, 'The Robot Airforce is about to Take Off', *New scientist*, 10 Aug. 1972.

39. Jonathan Schell, *The Real War*, New York: Da Capo Press, 2000, p. 276; Schell's reports were originally published in *The New Yorker* in 1968.

40. Schlight, *War*, p. 74.

41. Harrison, *Lonely*, p. 152. The colonial-frontier figuration of that last sentence is by no means unusual; for a discussion of the role of frontier imagery in the Vietnam War see Richard Slotkin, *Gunfighter Nation: The Myth of the Frontier in Twentieth-Century America*, New York: Atheneum, 1992, Part V.

42. Mike Jackson, *Naked in Da Nang: A Forward Air Controller in Vietnam*, St Paul, MN: Zenith Press, 2004, p. 24.

43. Schell, *War*, p. 325; John Flanagan, *Vietnam above the Treetops: A Forward Air Controller Reports*, New York: Praeger, 1992, p. 24.

44. Seymour Deitchman, 'The "Electronic Battlefield" in the Vietnam War', *Journal of Military History*, 72 (2008), pp. 869–87; Mark, *Aerial Interdiction*, pp. 327–63; Anthony Tambini, *Wiring Vietnam: The Electronic Wall*, Lanham, MD: Scarecrow Press, 2007.

45. George Weiss, 'The Air Force's Secret Electronic War', *Indochina Chronicle*, 15 Oct. 1971; Weiss, 'Battle for Control of the Ho Chi Minh Trail', *Armed Forces Journal* (15 Feb. 1972), pp. 19–22; Herman Gilster, *The Air War in Southeast Asia*, Maxwell Air Force Base, AL: Air University Press, 1993, pp. 16–56; Bernard Nalty, *The War against Trucks: Aerial Interdiction in Southern Laos, 1968–1972*, Washington, DC: Air Force History and Museums Program, 2005.

46. Gibson, *Perfect War*, pp. 396–7.

47. Weiss, 'Secret'; Paul Dickson, *Electronic Battlefield*, Bloomington: Indiana University Press, 1976, p. 88.

48. Dickson, *Electronic Battlefield*, pp. 86–7; William Rosenau, *Special Operations Forces and Elusive Enemy Ground Targets*, Santa Monica: RAND, 2001, p. 13.

49. Paul Dickson and John Rothchild, 'The Electronic Battlefield: Wiring Down the War', *Washington Monthly*, May 1971.

50. Since I completed the original version of this chapter Alfred McCoy has published an essay in which he traces three American 'information revolutions' that map a similar path. In particular, he identifies a 'second information revolution' in Indochina—in which computerised databases and electronic sensors played a leading role—which provided the foundations for what he calls 'an expanded information infrastructure with electronic surveillance, biometric identification and aerial force projection' that he sees being mobilised during the first phases of the 'war on terror'. McCoy argues that the real significance of the drone 'lies in its future role as the lowest tier in an emerging global network of sky/space surveillance' that is presently inaugurating a third 'information revolution'. See his 'Imperial Illusions: Information Infrastructure and the Future of U.S. Global Power', in Alfred McCoy, Josep Fradera and Stephen Jacobson (eds), *Endless Empire: Spain's Retreat, Europe's Eclipse, America's Decline*, Madison: University of Wisconsin Press, 2012, pp. 360–86.

51. My discussion is confined to Afghanistan and Pakistan, but the United States has also deployed remotely piloted aircraft in Iraq, Libya, Somalia and Yemen, and has a ramifying network of bases around the world.

52. Where remotely piloted aircraft are used by Joint Special Operations Command (JSOC), the network includes commanders and image analysts at USAF Special Operations Command at Okaloosa in Florida.

53. Westmoreland was addressing the US Army Association in Washington, DC, on 14 Oct. 1969: Dickson, *Electronic Battlefield*, pp. 220–1.

54. Scott Horton, 'The Trouble with Drones', *Harper's Magazine*, May 2010. The time from finding to engaging emergent targets is now 30–45 minutes; the USAF currently aims to reduce this to less than two minutes, and some commentators think it can be compressed to seconds by 2025: Adam Herbert, 'Compressing the Kill Chain', *Air Force Magazine*, 86, 3 (2003), pp. 50–4.

55. Littauer and Uphoff, *Air War*, p. 160.

56. Gravel was speaking in Oct. 1971 and was quoting Noam Chomsky, *At War with Asia: Essays on Indochina*, Oakland: AK Press, 2004 [1970], p. 67.

57. Dickson and Rothchild, 'The Electronic Battlefield: Wiring Down the War'.

58. Peter Finn, 'A Future for Drones: Automated Killing', *Washington Post*, 19 Sep. 2011.

59. Deitchman, 'Electronic Battlefield', p. 887.

60. McGovern's speech was delivered on 14 Dec. 1971 and is excerpted in 'Automated Warfare' (Jan. 1972) p. 2, Folder 01, Box 02, Douglas Pike Collection: Unit 03—Technology, The Vietnam Center and Archive, Texas Tech University.

61. Schlight, *War*, p. 258.

62. Although there are continuing experiments in detecting voice signatures and chemical signatures (emitted by IED factories), the sensors used in today's remotely piloted aircraft are not primarily acoustic or seismic but optical; the advocates of these platforms make much of their capacity to make the battle space 'transparent', and it is this political technology of vision that I focus on here.

63. The Gorgon Stare and the ARGUS-IS are designed for the Reaper, but the sensor system designed for the RQ-4 Global Hawk, a high-altitude, unarmed remotely piloted aircraft, can cover more than 100,000 square kilometres in a day.

64. Michael Flynn, Rich Juergens and Thomas Cantrell, 'Employing ISR: SOF Best Practices', *Joint Forces Quarterly*, 50, 3 (2008), p. 58.

65. Pat Biltgen and Robert Tomes, 'Rebalancing ISR', *Geospatial Intelligence Forum*, 8, 6 (2010), pp. 14–16; Paul Richfield, 'Intell Video Moves to a Netflix Model', Government Computer News, 6 Apr. 2011 (http://www.geotime.com).

66. Arnie Heller, 'From Video to Knowledge', *Science and Technology Review* (Apr./May 2011), p. 6.

67. Avery Gordon, *Ghostly Matters: Haunting and the Sociological Imagination*, Minneapolis: University of Minnesota Press, 1999, p. 16.

68. Schlight, *War*, p. 37; Rebecca Grant, 'Armed Overwatch', *Air Force Magazine*, 91, 12 (2008), pp. 40–4.

69. Col. James Garrett, 'Necessity and Proportionality in the Operation Enduring Freedom VII Campaign [February 2006–February 2007]', US Army War College Strategy Research Project, 20 Mar. 2008, pp. 19–25; Nick Davies, 'Afghanistan War Logs: Task Force 373—Special Forces Hunting Top Taliban', *Guardian*, 25 July 2010. The most recent counterinsurgency doctrine calls this 'personality targeting' (FM 3–24, pp. 5–30).

70. Flynn, Juergens and Cantrell, 'Employing ISR', p. 59; Anna Mulrine, 'Warheads on Foreheads', *Air Force Magazine*, 91, 10 (2008), pp. 44–7. For discussions of JSOC operations in Afghanistan, see Dana Priest and William Arkin, *Top Secret America: The Rise of the New American Security State*, New York: Little, Brown and Co., 2011, pp. 221–55; they claim that JSOC—which includes USAF components—'flies ten times more drones than the CIA'.

71. Sean Murphy, 'The International Legality of US Military Cross-Border Operations from Afghanistan into Pakistan', George Washington University Law School, Public Law and Legal Theory Working Paper, 451/Legal Studies Research Paper 451 (2009); Mary Ellen O'Connell, 'Unlawful Killing with Combat Drones: A Case Study of Pakistan, 2004–2009', Notre Dame Law School, Legal Studies Research Paper, 09–43 (2009); Jordan Paust, 'Self-Defense Targetings of non-State Actors and Permissibility of US Use of Drones in Pakistan', University of Houston Public Law and Legal Theory Series 2009-A-36; published in *Journal of Transnational Law and Policy*, 19, 2 (2010).

72. Eyal Weizman, 'Thanato-Tactics', in Adi Ophir, Michal Givoni and Sari Hanafi (eds), *The Power of Inclusive Exclusion: Anatomy of Israeli Rule in the Occupied Palestinian Territories*, New York: Zone Books, 2009, pp. 543–73.

73. Brian Glyn Williams, 'The CIA's Covert Predator Drone War in Pakistan, 2004–2010: The History of an Assassination Campaign', *Studies in Conflict and Terrorism*, 33 (2010), pp. 871–92.

74. Tara McKelvey, 'Inside the Killing Machine', *Newsweek*, 13 Feb. 2011; Mark Hosenball, 'Secret Panel can out Americans on "Kill List"', Reuters, 5 Oct. 2011; Partha Chatterjee, 'How Lawyers Sign Off On Drone Attacks', *Guardian*, 15 June 2011; Priest and Arkin, *Top Secret America*, pp. 202–11.

75. I have taken the lower estimates from the Long War Journal at http://www.longwarjournal.org/pakistan-strikes.php, and the higher estimates from the Bureau of Investigative Journalism at http://www.thebureauinvestigates.com

76. Jeremy Scahill, 'The Secret US War in Pakistan', *The Nation*, 23 Nov. 2009; Scahill, 'The Expanding US War in Pakistan', *The Nation*, 4 Feb. 2010.

77. Adam Entous, Julian Barnes and Siobhan Gorman, 'CIA Escalates in Pakistan', *Wall Street Journal*, 2 Oct. 2011; Priest and Arkin, *Top Secret America*, pp. 202–3; Greg Miller and Julie Tate, 'CIA Shifts Focus to Killing Targets', *Washington Post*, 1 Sep. 2011; Adam Entous, Siobhan Gorman and Julian Barnes, 'US Tightens Drone Rules', *Wall Street Journal*, 4 Nov. 2011.

78. Michael Hirsh, 'Slow Dance: Obama's Romance with the CIA', *National Journal*, 11 May 2011; Chris Woods, 'The CIA Drone Strike that Rewrote the Rules' (http://www.thebureauinvestigates.com/2011/07/18/the-cia-drone-strike-that-rewrote-the-rules).

79. The CEP assumes that ordnance will be normally distributed around the target, so that with a CEP of n meters, 50 per cent will be within n meters of the aiming point, a further 43 per cent within the range n–2n, and less than 7 per cent within n–3n. This is an assumption—it is not always met—and neither are the controlled conditions under which the CEP is determined realised in practice: Carl Conetta, *Disappearing the Dead: Iraq, Afghanistan and the Idea of a 'New Warfare'*, Cambridge, MA: Project on Defense Alternatives, Commonwealth Institute, 2004, p. 25.

80. *Wall Street Journal*, 9 Jan. 2010; cf. Maja Zehfuss, 'Targeting: Precision and the Production of Ethics', *European Journal of International Relations*, 17 (2010), pp. 543–66.

81. The aircraft are stationed and maintained at bases in Afghanistan—until recently some CIA-controlled Predators were also based in Pakistan—but the distance from the United States imposes a 1.8 second delay in control inputs ('latency') that makes it impossible for pilots in Nevada to perform take-offs and landings. This delay reinforces the importance of laser-guided weapons.

82. Megan McCloskey, 'Two Worlds of a Drone Pilot', *Stars & Stripes*, 27 Oct. 2009.

83. Tyler Wall and Torin Monahan, 'Surveillance and Violence from Afar: The Politics of Drones and Liminal Security-Scapes', *Theoretical Criminology*, 15 (2011), p. 240; cf. Allen Feldman, 'On the Actuarial Gaze: From 9/11 to Abu Ghraib', *Cultural Studies*, 19 (2005), pp. 203–26, who emphasises 'the hierarchical distance [of the actuarial gaze] from everyday life structures', p. 206.

84. I provide a fuller discussion in Derek Gregory, 'From a View to a Kill: Drones and Late Modern Warfare', *Theory, Culture and Society*, 28, 7–8 (2011), pp. 188–215; for the redacted transcript released under a FOIA request, see David Cloud, 'Combat by Camera: Anatomy of an Afghan War Tragedy', *Los Angeles Times*, 10 Apr. 2011.

85. Gregory, 'From a View to a Kill', pp. 188–215.

86. *Frontline* (PBS), 'Kill/Capture', 9 May 2011 (written and produced by Stephen Grey and Dan Edge).

87. Kate Clarke, 'The Takhar Attack: Targeted Killings and the Parallel Worlds of US Intelligence and Afghanistan', Afghan Analysts Network Thematic Report 5, 2011.

88. David Cloud, 'CIA Drones have Broader List of Targets', *Los Angeles Times*, 5 May 2010; Adam Entous and Siobhan Gorman, 'CIA Strikes Strain Ties with Pakistan Further', *Wall Street Journal*, 29 Aug. 2011; Charlie Savage, 'White House Weighs Limits of Terror Fight', *New York Times*, 15 Sep. 2011.

89. Clarke, 'Takhar Attack', p. 5.

90. Terrie Gent, 'The Role of Judge Advocates in a Joint Air Operations Center', *Air Power Journal*, 13, 1 (1999), pp. 40–55. Discussions of the legality of the air war were retrospective and confined to Rolling Thunder and Linebacker.

91. Joint Targeting Cycle and Collateral Damage Estimation methodology, Joint Chiefs of Staff General Counsel, 10 Nov. 2009; this briefing was made public following a FOIA request from the American Civil Liberties Union.

92. Garrett, 'Necessity and Proportionality'; Chatterjee, 'Lawyers'; Priest and Arkin, *Top Secret America*, p. 215; Jack Beard, 'Law and War in the Virtual Era', *American Journal of International Law*, 103 (2009), pp. 409–45.; Col. James Bitzes, 'Role of an Air Operations Center Legal Advisor in Targeting', presentation to 'Drones, Targeting and the Promise of Law', New America Foundation, Washington, DC, 24 Feb. 2011.

93. Michael Schmitt, 'Drone Attacks under the *Jus ad bello* and the *Jus in bello*: Clearing the "Fog of Law"', *Yearbook of International Humanitarian Law*, 13 (2010), pp. 311–26, see p. 320; Jack Beard, 'Law and War'.

94. Col. Gary Brown, quoted in Charles Dunlap, 'Come the Revolution: A Legal Perspective on Air Operations in Iraq since 2003,' in Raul Pedrozo (ed.), *The War in Iraq: A Legal Analysis*, Newport, RI: Naval War College, 2010, p. 141.

95. Amitai Etzioni, 'The "Secret" Matrix', *The World Today*, 66, 7 (2010), pp. 11–14, see p. 14.

96. Beard, 'Law and War', p. 438; Patricia Owens, 'Accidents Don't Just Happen: The Liberal Politics of High-Technology "Humanitarian" War', *Millennium: Journal of International Studies*, 32 (2003), pp. 595–616.

97. Tom Engelhardt, 'America's Permanent Robot War', *Guardian*, 4 Oct. 2011; Transcript, Department of Defense Bloggers Roundtable with Col. Gary Brown, 27 May 2009.

98. Beard, 'Law and War', p. 419.

99. Anne Orford, 'The Passions of Protection: Sovereign Authority and Humanitarian War', in Didier Fassin and Mariella Pandolfi (eds), *Contemporary States of Emergency: The Politics of Military and Humanitarian Interventions*, New York: Zone Books, 2010, p. 339.

100. David Kilcullen and Andrew Exum, 'Death from Above, Outrage Down Below', *New York Times*, 16 May 2009.

101. Giulio Douhet, *The Command of the Air*, trans. Dino Ferrari, Tuscaloosa, AL: University of Alabama Press, 1998, first published in Italian, 1921, pp. 9–10.

102. I gratefully acknowledge support from the Social Science and Humanities Research Council. I owe a particular debt to Dan Clayton for his comments on a first draft, and I am grateful to Oliver Belcher, Keith Feldman, Craig Jones, Joanne Sharp and Marilyn Young for helpful discussions and to Peter Adey for his encouragement and patience.

## 3. AERIAL SURVEYING, GEOPOLITICAL COMPETITION AND THE FALKLAND ISLANDS AND DEPENDENCIES AERIAL SURVEY EXPEDITION (FIDASE 1955–7)

1. K. Dodds, 'Assault on the Unknown: Geopolitics, Antarctic Science and the International Geophysical Year (1957–8)', in S. Naylor and J. Ryan (eds), *New Spaces of Exploration: Geographies of Discovery in the Twentieth Century*, London: I.B. Tauris, 2009, pp. 148–72.

2. Ibid., pp. 148–72.

3. S. Turchetti, S. Naylor, K. Dean and M. Siegert, 'Accidents and Opportunities: A History of the Radio Echo Sounding of Antarctica', *British Journal for the History of Science*, 41 (2008), pp. 417–44.

4. For an overview of the state of knowledge about the Antarctic, see G. Fogg, *A History of Antarctic Science*, Cambridge: Cambridge University Press, 1992.

5. The Falkland Islands Dependencies Survey (FIDS) was a civilian organisation, managed by the Colonial Office, which operated in the Antarctic region between 1945 and 1961. It owed its origins to a wartime naval operation called Tabarin and was later transformed into British Antarctic Survey in the 1960s. For one account of its historical origins written by a central figure in the organisation's history, see Vivian Fuchs, *Of Ice and Men*, Oswestry, Shropshire: Anthony Nelson, 1982. A contemporary account by one FIDS employee in the 1940s is provided by David James, *That Frozen Land*, London: Falcon Press, 1948.

6. The Falkland Islands Dependencies were listed in the 1908 Letters Patent as 'the groups of islands known as South Georgia, the South Orkneys, the South Shetlands, and the Sandwich Islands, and the territory known as Graham's Land, situated in the South Atlantic Ocean to the south of the 50th parallel of south latitude, and lying between the 20th and the 80th degrees of west longitude'.

7. Cited in M.B. McHugo, *Topographical Survey and Mapping of British Antarctic Territory, South Georgia, South Sandwich Islands 1944–1986*, Cambridge: British Antarctic Survey, 1986, pp. 1 and 3.

8. For a general overview with reference to British Antarctic policy interests, see P. Beck, *International Politics of Antarctica*, London: Croom Helm, 1986.

9. M. Clifford 'Broadcast Address by His Excellency the Governor', in *Sir Miles Clifford Papers*, Oxford: Rhodes House Library, 1948.

10. For one engagement, see R. Drayton, *Nature's Government: Science, Imperial Britain and the 'Improvement' of the World*, New Haven: Yale University Press, 2000.

11. Norwegian whaling activities in the Southern Ocean ceased in the early 1960s.

12. For a contemporary overview of Argentina's long-term engagement with the Antarctic, see S. Rigoz and H. Pujato, *El Conquistador del Desierto Blanco*, Buenos Aires: Editorial Maria Ghirlanda, 2002. It is worth bearing in mind that

Argentine scientists were the first to occupy the Antarctic permanently in 1904 with their base at Laurie Island.

13. A. Howkins, 'Science, Environment and Authority: The International Geophysical Year in the Antarctic Peninsula Region', in R. Launius, J. Fleming and D. Devorkin (eds), *Globalizing Polar Science: Reconsidering the International Polar and Geophysical Years*, London: Macmillan, 2010, pp. 245–64.

14. Denis Cosgrove, *Apollo's Eye: A Cartographic Genealogy of the Earth in the Western Imagination*, London: Johns Hopkins University Press, 2001.

15. K. Hannam, M. Sheller and J. Urry, 'Editorial: Mobilities, Immobilities and Moorings', *Mobilities*, 1 (2006), pp. 1–22.

16. British Antarctic Survey Archives AD3/1/AS/183/6(1) Letter from the Director and General Manager of Hunting Aerosurveys Ltd to the Colonial Office dated 1 Apr. 1955.

17. For example, those operating in the field frequently noted the role of blizzards and low cloud cover in frustrating surveying. A poignant account is provided by Alfred Stephenson, 'Surveying in the Falkland Islands Dependencies', *Polar Record*, 6 (1951), pp. 28–44.

18. Peter Adey, 'Aeromobilities: Geographies, Subjects and Visions', *Geography Compass*, 2 (2008), pp. 1318–36.

19. J. Roscoe, 'Antarctic Photography', in A. Carey, L. Gould, E. Hulbert, H. Odishaw and W. Smith (eds), *Antarctica in the IGY*, Washington, DC: National Academy of Sciences, 1960, pp. 18–21.

20. John Law, 'On the Methods of Long Distance Control: Vessels, Navigation, and the Portuguese Route to India', in J. Law (ed.), *Power, Action and Belief: A New Sociology of Knowledge?* London: Routledge, 1986, pp. 234–63.

21. Ibid., p. 14.

22. This notion of territorial disengagement was seriously raised, however, in 1958 and 1964 on the back of substantial defence spending reductions, and a general re-appraisal of Britain's role in the world following reviews undertaken by the Macmillan government.

23. McHugo, *Topographical Survey and Mapping of British Antarctic Territory*, pp. 5–6.

24. The gendered implications of British polar survey and mapping in the post-war period is discussed in K. Dodds, 'Setting and Unsettling Antarctica', *Signs*, 34 (2009), pp. 505–9.

25. The so-called 'Audit of Empire' undertaken by Prime Minister Harold Macmillan concluded in 1958 that, 'Our withdrawal from Antarctica would mean a loss of UK prestige and influence, especially in scientific circles. It might also involve the loss of strategic minerals, but this would be easier to evaluate when the results have been assessed of the work done during the International Geophysical Year. Argentina and/or Chile, which have claims (partly competing)

to the Dependencies, would probably step in if the United Kingdom withdrew.' The Cabinet Office, *Future Constitutional Developments in the Colonies*, London: The National Archives, 1958.

26. For one account of the institutional and training background to FIDS see M. Smith, *Sir James Wordie, Polar Crusader: Exploring the Arctic and Antarctic*, Edinburgh: Birlinn, 2004.

27. For example, see John Rymill, 'British Graham Land Expedition, 1934–7', *Geographical Journal*, XCI, 4 (1938), pp. 297–312 and 424–38.

28. On this period, in the interwar period and British imperial dilemmas, see Peder Roberts, *The European Antarctic: Science and Strategy in Scandinavia and the British Empire*, London: Palgrave, 2011.

29. On the changing political economy of whaling in the 1920s and 1930s see Peter Beck, *International Politics of Antarctica*, London: Croom Helm, 1986.

30. Simon Nasht, *The Last Explorer: Hubert Wikins, Hero of the Great Age of Polar Exploration*, New York: Arcade Publishing, 2006.

31. A critical factor in this anxiety was the polar over-flights by the American explorer, Richard Byrd. In 1928, Byrd flew several airplanes over the continental interior and most notably in Nov. 1929 flew from his base 'Little America' to the South Pole and back again in eighteen hours and forty-one minutes. His achievement, widely publicised around the world, caused immediate imperial anxieties as Australians, Britons and New Zealanders charged with administering polar territories pondered the implications of this aerial development.

32. The US Navy-led Operation High Jump in 1946–7 was indicative of this militarisation of flight in the Antarctic. Naval planes, supported by a flotilla of ship including an aircraft, were dispatched to explore, to photograph and to test polar equipment.

33. Klaus Dodds, *Pink Ice: Britain and the South Atlantic Empire*, London: I.B. Tauris, 2002, p. 16.

34. James Scott, *Seeing like a State*, London: Yale University Press, 1998.

35. E. Bingham, 'The Falkland Islands Dependencies Survey 1946–7', *Polar Record*, 6 (1947), pp. 27–39 and V. Fuchs 'The Falkland Islands Dependencies Survey 1947–50', *Polar Record*, 6 (1951), pp. 7–27.

36. British Antarctic Survey Archives AD3/1/A5/183/6 (1) Falkland Islands and Dependencies Aerial Survey Expedition Report on First Season's Work 1955–6.

37. Dodds, *Pink Ice*, p. 19.

38. David Mason, 'The Falkland Islands Dependencies Survey: Explorations of 1947–8', *Geographical Journal*, CXV (1950), pp. 145–60.

39. Dodds, *Pink Ice*, p. 26.

40. Peter Mott, *Wings over Ice: The Falkland Islands and Dependencies Aerial Survey Expedition*, London: Miles Apart, 1986, p. 60.

41. Dodds, *Pink Ice*, p. 60.

42. Peter Adey, *Aerial Life: Spaces, Mobilities and Affects*, London: Routledge, 2009, pp. 86–90.

43. The actual aerial photography carried out by the BGLE was modest. The Williamson Eagle III camera had to be hand-held because the mount was considered to take up too much of the aircraft load.

44. Trimetrogon photography is usually described as a method of aerial photography for rapid topographic mapping, in which one vertical and two oblique photographs are taken simultaneously.

45. See, for example, M.B. McHugo, 'Mapping the Falkland Islands Dependencies and British Antarctic Territory', *Polar Record*, 12 (1964), pp. 395–402 and Barbara McHugo, *Topographical Survey and Mapping of British Antarctic Territory, South Georgia and the South Sandwich Islands 1944–1986*, Cambridge: British Antarctic Survey, 2003, p. 47.

46. Oblique photography was taken on reconnaissance flights and was not usually intended for mapping purposes. It was really intended to help with overland journeys and fieldwork planning.

47. Vertical photography contributed to the science of photogrammetry, which was using aerial photographs to derive maps.

48. For example, as noted in Peter Mott, 'Air survey of the Falkland Islands Dependencies 1955–56', *Polar Record*, 8 (1956), pp. 237–45 but also in the BAS archives including the original proposals for the 'aerial survey of Grahamland' submitted in Apr. 1955.

49. Peter Mott, 'The Falkland Islands and Dependencies Aerial Survey Expedition', *Photogrammetric Record*, 2 (1958), pp. 309–29. Quote taken from p. 327.

50. Peter Mott, *Wings over Ice*, p. 60.

51. British Antarctic Survey Archives AD3/1/A5/183/6 (1) Copy of Diary of John Saffrey, Flying Manager, during the FIDAS Expedition 1955/6.

52. Ibid.

53. Ibid.

54. Peter Mott, 'The Falkland Islands and Dependencies Aerial Survey Expedition', *Photogrammetric Review*, II (1960), pp. 309–29. Quote taken from p. 314.

55. British Antarctic Survey Archives AD3/1/A5/183/6(1) Falkland Islands and Dependencies Aerial Survey Expedition Report on the Second Season's Operations, 1956–7.

56. Ibid.

57. Mott, 'The Falkland Islands and Dependencies Aerial Survey Expedition', quote taken from p. 315.

58. In the aftermath of FIDASE, map production continued hand in hand with other practices designed to consolidate imperial control. In mapping terms, the FIDASE contributed to a new generation of 1: 200,000 Scale mapping of the

British Antarctic Territory (as it became in 1962 and thus replacing the Falkland Island Territories) and each sheet produced covered 1 degree of latitude and 4 degrees of longitude south of 75 degrees south, while the coverage north of that latitude addressed 1 degree of latitude and 2 degrees of latitude.

59. The Antarctic Treaty, 1 Dec. 1959, 402 UNTS 71. Article IV of the Antarctic Treaty effectively states that all claims to territory are considered suspended for the duration of the treaty in order to promote international cooperation and scientific collaboration in particular.

60. The definitive source on British place names and the accelerating naming trend is Geoffrey Hattersley-Smith, *The History of Place-Names in the British Antarctic Territory*, Cambridge: British Antarctic Survey, 1991.

## 4. NETWORKS, NODES AND DE-TERRITORIALISED BATTLESPACE: THE SCOPIC REGIME OF RAPID DOMINANCE

1. The author would like to thank those who have commented on this chapter in its various incarnations. Luis Lobo-Guerrero, Emily Jackson, Alison Williams and Chris Zebrowski offered helpful comments. I am particularly grateful to Peter Adey and Tarak Barkawi: the former for the initial invitation to contribute and subsequent important contributions, the latter for a very thorough set of thought-provoking comments at a crucial time in the writing of this chapter. All errors remain, of course, my own.

2. Harlan K. Ullman and James P. Wade, *Shock and Awe: Achieving Rapid Dominance*, Washington, DC: National Defense University, 1996. In what follows I will capitalise Shock and Awe and thus treat it as a proper noun, referring to a singular event with distinctive, temporally and spatially limited, characteristics. This is to indicate that, rather than the affective states of shock or awe, Shock and Awe refers to a singular event.

3. US Department of Defense, 'DoD News Briefing—ASD PA Clarke and Maj. Gen. McChrystal, March 22, 2003 4:00 PM EDT', 2003 (http://www.defense.gov/Transcripts/Transcript.aspx?TranscriptID=2077; accessed Jan. 2011)

4. Ubiquitous footage can be found on YouTube, but for illustrative purposes see the Sky News picture gallery: http://news.sky.com/home/world-news/media-gallery/1084644 (accessed 31 July 2011).

5. Martin Shaw, *The New Western Way of War: Risk Transfer War and its Crisis in Iraq*, Cambridge: Polity, 2005.

6. Jean Baudrillard, *Simulations*, New York: Semiotext(e), 1983.

7. Allen Feldman, *Formations of Violence*, Chicago: University of Chicago Press, 1991.

8. Jean Baudrillard, *The Spirit of Terrorism and Other Essays*, London: Verso, 2003, p. 72. Or, perhaps more accurately, a display masquerading as a token of the real,

geographically distant source of power that is itself more real than that hypothetical power. An appearance that both invokes *and* performs its essence.

9. Debbie Lisle, 'Sublime Lessons: Education and Ambivalence in War Exhibitions', *Millennium: Journal of International Studies*, 34, 3 (2006), pp. 841–62.

10. Martin Coward, 'Network-centric Violence, Critical Infrastructure and the
Urbanisation of Security', *Security Dialogue*, 40, 4–5 (2009), pp. 399–418.

11. E. Markusen and D. Kopf, *The Holocaust and Strategic Bombing: Genocide and
Total War in the Twentieth Century*, Boulder, CO: Westview Press, 1995.

12. Ronnie D. Lipschutz, 'Imperial Warfare in the Naked City: Sociality as Critical Infrastructure', *International Political Sociology*, 2, 3 (2008), pp. 204–18.

13. Edward A. Smith, *Effects Based Operations: Applying Network Centric Warfare
in Peace, Crisis, and War*, Washington, DC: US Department of Defense, Command and Control Research Program, 2003; Arthur K. Cebrowski and John J.
Garstka, 'Network-Centric Warfare: Its Origin and Future', *Proceedings of the
U.S. Naval Institute*, Jan. 1998.

14. John A. Warden III, 'The Enemy as a System', *Airpower Journal* (Spring 1995);
John Arquilla and David Ronfeldt, 'The Advent of Netwar (revisited)', in J.
Arquilla and D. Ronfeldt (eds), *Networks and Netwars: The Future of Terror,
Crime, and Militancy*, Santa Monica, CA: RAND, 2001.

15. Effects Based Operations (EBO) refers to a wide and loose grouping of tactics
and strategies designed to multiply effects by striking key vulnerabilities. This
looseness has made the core of EBO hard to locate and has led to it being
rejected as doctrine by the US Army. For a summary of the problems inherent
in this loose gathering of ideas (and a refutation of the use of EBO for US Army
warfighting doctrine) see James N. Mattis, 'USJFCOM Commander's Guidance for Effects-Based Operations', *Parameters*, Autumn 2008, pp. 18–25.

16. On 'logistical life' see Julian Reid, *The Biopolitics of the War on Terror: Life Struggles, Liberal Modernity and the Defence of Logistical Societies*, Manchester: Manchester University Press, 2009.

17. Stephen Graham, 'Switching Cities Off: Urban Infrastructure and US Air
Power', *City*, 9, 2 (2005), pp. 169–93.

18. This is echoed in warden's formula '(Physical) x (Morale) = Outcome' where
'physical refers to the material assets of any enemy'. As Warden notes 'strategic
entities, be they an industrial state or a guerrilla organization, are heavily dependent on physical means. If the physical side of the equation can be driven close
to zero, the best morale in the world is not going to produce a high number on
the outcome side of the equation.' Warden, 'The Enemy as a System'.

19. Azar Gat, *A History of Military Thought: From the Enlightenment to the Cold
War*, Oxford: Oxford University Press, 2001.

20. Andrew Latham, 'Re-Imagining Warfare: The "Revolution in Military Affairs"',
in Craig A. Snyder (ed.), *Contemporary Security and Strategy*, London: Macmillan, 1999.

21. It is, of course, worth subjecting this public perception to a certain amount of critical scrutiny: the engagement of land forces was asymmetrical, pitching the technologically advanced professionalised US war machine against the technologically underdeveloped conscript army of Iraq after a month of air bombardment.

22. United States General Accounting Office, 'Operation Desert Storm: Evaluation of the Air War', 1996, p. 4 (available online: http://www.gao.gov/archive/1996/pe96010.pdf; accessed Feb. 2011); Thomas A. Keaney and Eliot A. Cohen, *Gulf War Air Power Survey: Summary Report*, Washington, DC: US Government Printing Office, 1993

23. The Gulf War Air Power Survey notes the significance and novelty of the high proportion (40 per cent) of 'Beyond Visual Range (BVR) kills'. These represent instances in which targets were destroyed beyond the visual rage of those firing munitions and are a consequence of being able to en-vision the enemy across distance greater than visual range (Keaney and Cohen, *Gulf War Air Power Survey*, pp. 302–3).

24. Michael Russell Rip and James M. Hasik, *The Precision Revolution: GPS and the Future of Aerial Warfare*, Annapolis, MD: Naval Institute Press, 2002, pp. 143–90.

25. Latham, 'Re-Imagining Warfare', p. 226.

26. Ibid., pp. 224–6.

27. It is for this reason that the scopic regime of the RMA can be considered a novel recasting of Virilio's 'logistics of perception' (Paul Virilio, *Cinema and War: The Logistics Of Perception*, London: Verso, 1989). While 'en-visioning' suggests that the RMA is merely an evolution of the trajectory mapped by Virilio in *Cinema and War*, the centrality of data to the scopic regime of the RMA suggests something subtly different is in play. Where the logistics of perception evolve from the line of sight and are largely ocular in orientation, data constitutes a mechanism for mediating between that which cannot be perceived and that which can be understood. As such this is less perception than cognition, though the primary form in which it is rendered is often visual (though not simply image-based since much of the rendering is in the form of text). 'En-visioning' thus refers to the process by which something is brought within our purview and we have a sense—though not necessarily sight—of its location.

28. Martin Jay, 'Scopic Regimes of Modernity', in Hal Foster (ed.), *Vision and Visuality*, San Francisco: Bay Press, 1988.

29. On the manner in which data is mined to en-vision hidden dangers, see: Louise Amoore, 'Vigilant Visualities: The Watchful Politics of the War on Terror', *Security Dialogue*, 38, 2 (2007), pp. 215–32; Louise Amoore, 'Lines of Sight: On the Visualization of Unknown Futures', *Citizenship Studies*, 13, 1 (2009), pp. 17–30.

30. US Marine Corps, *Marine Corps Operations: MCDP 1–0*, Washington, DC: United States Marine Corps, 2001, pp. 4–3 (available online: http://www. marines.mil/news/publications/Documents/MCDP%201–0%20Marine%20 Corps%20Operations.pdf; accessed Feb. 2011).

31. US Army, *Field Manual No. 100–12: Army Theater Missile Defense Operations*, Washington, DC: Department of the Army, 2000: Appendix A (available online: http://www.globalsecurity.org/space/library/policy/army/fm/100–12/index. html; accessed Feb. 2011).

32. Jane Bennett, 'The Agency of Assemblages and the North American Blackout', *Public Culture*, 17, 3 (2005), p. 445.

33. William E. Conolly, *The Ethos of Pluralization*, Minneapolis: University of Minnesota Press, 1995, p. 1.

34. Jan Aart Scholte, *Globalization: A Critical Introduction*, London: Macmillan, 2000, p. 55.

35. Martin Heidegger, *Being and Time*, trans. John Macquarie and Edward Robinson, Oxford: Blackwell, 1962, p. 141.

36. Globe-spanning bombing flights are a legacy of Cold War refinements of the strategic bombing doctrine—a consequence of plans for the destruction of Soviet cities by nuclear munitions. Nuclear strategic bombing focused on massive destruction, targeting cities as the natural centre of gravity of the population of the enemy state. As such this form of bombing belongs firmly in the same trajectory as the strategic bombing of Dresden, lacking the so-called precision munitions that have characterised the globe spanning bombing raids of the 2003 war on Iraq. Thus it is not globality per se or, as I will suggest below, targeting the city that is unique to Shock and Awe. Rather it is the manner in which a certain scopic regime underpins the latter leading to a focus on nodal targets rather than entire urban zones.

37. The perception that urban space comprised an impassable quagmire led military commentators to discuss 'indirect' approaches to urban centres: establishing a military cordon around cities in order to avoid having to enter their 'complex' space and risk the casualties and difficulties that would entail (see, Maj. Robert F. Whittle Jr, *Can the U.S. Avoid Urban Combat in Baghdad?* Fort Leavenworth, Kansas: US Army School of Advanced Military Studies United States Army Command and General Staff College, 2003. (available online: http://www.dtic.mil/cgibin/GetTRDoc?Location=U2&doc=GetTRDoc. pdf&AD=ADA420053; accessed 2 Aug. 2011). It is important to note that attitudes toward the civilian nature of the city have been ambivalent. On the one hand the laws of war have attempted to safeguard the city from 'wanton destruction'. On the other Cold War targeting gravitated to the destruction of the city as the site where the enemy population was most numerous.

38. On re-enchantment in/of warfare see: Christopher Coker, *The Future of War:*

*The Re-Enchantment of War in The Twenty-First Century*, Oxford: Blackwell, 2004. In this case, re-enchantment should not be taken to imply the city was wholly disenchanted as a target or arena of warfare prior to Shock and Awe: the citizens of Hue, Sarajevo, Grozny (to name but a few of the urban areas targeted in the twentieth century) can attest to this not being the case. Rather this re-enchantment is a distinctive re-inflection of the relation between organised violence and urban areas in the military imaginaries of advanced industrial states.

39. Stephen Graham, *Cities Under Siege: The New Military Urbanism*, London: Verso, 2010.

40. James N. Mattis, 'USJFCOM Commander's Guidance for Effects-Based Operations', *Parameters*, Autumn 2008, p. 23.

41. Human Rights Watch, *Off Target: The Conduct of the War and Civilian Casualties in Iraq*, New York: Human Rights Watch, 2003, p. 6.

42. Ibid., pp. 24–39.

43. Madelyn Hsiao-Rei Hicks, Hamit Dardagan, Gabriela Guerrero Serdán, Peter M. Bagnall, John A. Sloboda and Michael Spagat, 'Violent Deaths of Iraqi Civilians, 2003–2008: Analysis by Perpetrator, Weapon, Time, and Location', *PLoS Medicine*, 8, 2 (2011).

44. Associated Press, 'NATO Air Strikes Can No Longer Target Houses in Afghanistan, Says Karzai', *The Guardian*, Tuesday 31 May 2011 (available online: http://www.guardian.co.uk/world/2011/may/31/nato-target-houses-afghanistan-karazai?INTCMP=SRCH; accessed 4 Aug. 2011); Declan Walsh, 'Pakistan Orders US Out of Drone Base', *The Guardian*, Thursday 30 June 2011 (available online: http://www.guardian.co.uk/world/2011/jun/30/pakistan-orders-us-out-drone-base?INTCMP=SRCH; accessed 4 Aug. 2011).

45. Peter Berger and Katherine Tiedemann, 'The Year of the Drone: An Analysis of U.S. Drone Strikes in Pakistan, 2004–2010', New America Foundation Counterterrorism Strategy Initiative Policy Paper, 24 Feb. 2010 (available online: http://counterterrorism.newamerica.net/sites/newamerica.net/files/policydocs/bergentiedemann2.pdf; accessed 3 Aug. 2011).

46. Network-oriented forms of warfighting have, however, remained an integral part of US counterinsurgency, especially in Afghanistan where Special Forces raids increased to close to 3,000 between Apr. and July 2011 (cf. Peter Beaumont, 'Afghanistan Civilian Death Toll has Risen Sharply, Says United Nations', *The Guardian*, Tuesday 19 July 2011 (available online: http://www.guardian.co.uk/world/2011/jul/19/afghanistan-civilian-deaths-rise-un; accessed 4 Aug. 2011).

47. Dexter Filkins, 'U.S. Tightens Airstrike Policy in Afghanistan', *New York Times*, 21 June 2009 (available online: http://www.nytimes.com/2009/06/22/world/asia/22airstrikes.html; accessed 4 Aug. 2011).

48. Mattis, 'USJFCOM Commander's Guidance for Effects-Based Operations'.

49. The White House, 'Remarks by the President on a New Strategy for Afghanistan and Pakistan', 27 Mar. 2009 (available online: http://www.whitehouse.gov/the-press-office/remarks-president-a-new-strategy-afghanistan-and-pakistan; accessed 4 Aug. 2011).

50. Mattis, 'USJFCOM Commander's Guidance for Effects-Based Operations', p. 23.

51. Graham, *Cities Under Siege*; Coward, 'Network Centric Violence'.


## 5. PHOTOMOSAICS: MAPPING THE FRONT, MAPPING THE CITY

1. On aerial-reconnaissance-aided archaeology, or 'phytoarchaeology', see esp. the writings of O.G.S. Crawford; Kitty Hauser, *Bloody Old Britain: O.G.S. Crawford and the Archaeology of Modern Life*, London: Granta, 2008.

2. Paul K. Saint-Amour, 'Modernist Reconnaissance', *Modernism/modernity*, 10 (2003), pp. 349–80.

3. Roland Barthes, 'The Eiffel Tower', in *The Eiffel Tower and Other Mythologies*, trans. Richard Howard, New York: Hill and Wang, 1979, p. 9.

4. W.T. Lee, *The Face of the Earth as Seen from the Air: A Study in the Application of Airplane Photography to Geography*, New York: American Geographical Society, 1922, p. 22.

5. Michel de Certeau, *The Practice of Everyday Life*, trans. Steven Rendall, Berkeley: University of California Press, 1984, p. 92.

6. Barthes, 'Eiffel', p. 10.

7. De Certeau, *Practice*, p. 108.

8. Henri Lefebvre, *The Production of Space*, trans. Donald Nicholson-Smith, Oxford, and Cambridge, MA: Blackwell, 1991, p. 52.

9. In making this case, I contest what used to be (though no longer is) a dominant scholarly narrative about modernism. 'Western' literary modernism, in particular, was influentially read as subordinating the local, the partial and the differential to totalising structures such as myth, epic, encyclopedia, long poem, *Gesamtkunstwerk*. One locus classicus for this reading was T.S. Eliot's characterisation of Joyce's *Ulysses* as employing a 'mythical method'—one that 'manipulat[es] a continuous parallel between contemporaneity and antiquity'—as a way of 'controlling, of ordering, of giving a shape and a significance to the immense panorama of futility and anarchy which is contemporary history'. See T.S. Eliot, '*Ulysses*, Order, and Myth', in F. Kermode (ed.), *Selected Prose of T.S. Eliot*, New York: Harcourt Brace Jovanovich. 1975, p. 177. The critical reception of Anglophone literary modernism, at least, might have looked rather different had Eliot entertained the possibility that *Ulysses* deploys Homer not to impose an ancient or ahistorical order on the chaotic present but to insist that the drive to impose order in such a manner is itself a historically, geographically, politically situated

one—that the total is, again, a special case or tendentious longing of the partial. Despite its brittle monologism, however, Eliot's formulation does contain an insight pertinent to the present discussion: that the 'immense panorama', far from being an intrinsically ordered or intentional vantage, offers anarchy and futility to the eye. Power, in other words, has to work to distil order or produce control from the Icarian view.

10. Our most influential analyst of the kill-chain is Paul Virilio, with whose work my own is clearly in dialogue. Part of the polemical and formulative vigour of Virilio's work stems from its willingness to extrapolate general, even total, maxims from an extreme example. For instance, Virilio has repeatedly invoked as 'perfectly express[ing] the new geostrategic situation and partially explain[ing] the current round of disarmament' a statement by former US Undersecretary of State for Defense W.J. Perry: 'Once you can see the target, you can expect to destroy it.' 'If *what is perceived is already lost*', writes Virilio, 'it becomes necessary to invest in concealment what used to be invested in simple exploitation of one's available forces.' See Paul Virilio, *War and Cinema: The Logistics of Perception*, trans. P. Camiller, London: Verso, 1989, p. 4, original emphasis. Virilio's compression of Perry in paraphrasing him—from 'expect to destroy it' to 'already lost'—exemplifies a widespread tendency in Virilio's work to harden and shorten the kill-chain to the point where a whole critical vision is based on a model of unerring, instantaneous targeting that continues to be belied by the inaccuracy of actual military targeting. By exhibiting this faith in the most extreme military accuracy-claims, Virilio oddly forgoes the chance to engage with the terrible consequences of their failure—with the disparity between, say, the ideal of a precision target 'already lost' and the realities of 'collateral damage'. This aspect of Virilio's work provides a sharp example of how one finds abstract space in the very theoretical sites where one might most expect to find differential space. For a more extended discussion of Virilio in relation to cognate work by Samuel Weber, Rey Chow and Caren Kaplan, see Paul K. Saint-Amour, 'War, Optics, Fiction', *NOVEL: A Forum on Fiction*, 43 (2010), pp. 93–9; see also Armitage earlier in this volume.

11. Harold E. Porter marvels that 'From only 4,000 feet ... with a 10-inch lens and a 4x5 plate, the longer side of the plate covers about 2,000 feet of ground, and the shorter side covers about 1,600 feet, with a total area of 450,000 square yards, or 175 acres. At 7,000 feet [an altitude safely out of artillery range], the area shown is over 1,000,000 square yards.' Harold E. Porter, *Aerial Observation: The Airplane Observer, the Balloon Observer, and the Army Air Corps Pilot*, New York: Harper & Brothers, 1921, pp. 161–2.

12. Allan Sekula, 'The Instrumental Image: Steichen at War', *Artforum*, 14 (1975), p. 28.

13. My description of the reconnaissance flights and subsequent photomosaic production is largely a paraphrase of Porter's. Porter, *Aerial Observation*, p. 164.

14. Sekula, 'Instrumental', pp. 27–8.

15. John Welchman, for instance, finds the RAF atlas's citation of cubism and futur5sm 'quite remarkabl[e], given its origin in the military establishment of a country many of whose few "advanced" cultural commentators were still fulminating against the esthetic degeneracy of the continental avant-garde'. John Welchman, 'Here There & Otherwise', *Artforum*, 27 (1988), p. 18.

16. Herbert E. Ives, *Airplane Photography*, Philadelphia: J.B. Lippincott, 1920, p. 353.

17. I am indebted to Smith for the notion that controlled mosaics effect a truce or 'compromise' (in Smith's term) between perspectival and planimetric or orthographic projections. Harold Theodore Uhr Smith, *Aerial Photographs and Their Applications*, New York: D. Appleton-Century Co., 1943, p. 52.

18. Talbert Abrams, *Essentials of Aerial Surveying and Photo Interpretation*, New York: McGraw-Hill, 1944, pp. 228–9.

19. For an excellent reading of the sovereign and biopolitical dimensions of RAF colonial air surveys between the world wars, see Peter Adey, *Aerial Life: Spaces, Mobilities, Affects*, Oxford: Wiley-Blackwell, 2010.

20. George W. Goddard, *Overview: A Life-long Adventure in Aerial Photography*, Garden City, NY: Doubleday, 1969, p. 101, n. 5. One Captain Bouché of the French Service reportedly said, 'Aerial photographs are most valuable to show the present use of land, the density of population and the relation of open space to built over space. In the devastated regions they have been of the greatest help in making surveys, making it possible to study ensemble improvements, to reparcel property, to verify the lines of plans and to make bird's-eye perspectives of proposed improvements.' Guy Wilfrid Hayler, 'The Aeroplane and City Planning: The Advantages of Viewing Cities from Above', *American City*, 23 (1920), pp. 575–6.

21. Nelson P. Lewis, 'A New Aid in City Planning', *American City*, 26 (1922), p. 212.

22. Arthur Brock Jr, and L.J.R. Holst, 'The Future of Airplane Photography', *Aviation* (1 Jan. 1919), p. 707. The war, according to Brock and Holst's fascinatingly resentful piece, had distracted photogrammetry from its true concerns with mechanics and optics—with camera automation, regular coverage, better shutters and lenses. In the meantime, they claim, 'Our actual experimental work in photographic surveying and in the design of cameras to obtain these results has been carried far beyond experimental work in this line by any individuals or by any of the Governments who have participated in the European War.' Brock and Holst offer themselves, in effect, as an isolationist counterfactual to the dominant narrative, which made peacetime aerial surveying a happy byproduct of wartime research and development; in contrast, they represent the advances that aerial photogrammetry would have made had it not gone to war. Their article presses the counterfactual point farther, asking whether the cur-

rent motion picture industry would have reached its 'present stage of [mechanical] perfection' if it 'had had a military use and military development'. With the exception of developing negatives and printing from them, airplane photography, they conclude, 'is a matter which should be entirely in the hands of civil and mechanical engineers'.

23. Unsigned, 'Mosaic Maps of Cities', *American City*, 27 (1922), pp. 253–5; Unsigned, 'Practical Aerial Photography', *Photographic Journal of America*, 59 (1922), pp. 113–9; Unsigned, 'New York Mapped by Sky Camera in 69 Minutes!' *Popular Science Monthly*, 100 (1922) p. 46; Unsigned, 'Air Map of Greater City', *New York Times*, 30 Sep. 1924, 37, p. 3.

24. Sherman M. Fairchild, 'Aerial Mapping of New York City', *American City*, 30 (1924), p. 74.

25. Lowell Thomas, *European Skyways: The Story of a Tour of Europe by Aeroplane*, London: W. Heinemann, 1928, p. 1678. For a discussion of Thomas's book in the context of 1920s aerial tourism, see Davide Deriu, 'The Ascent of the Modern *Planeur*: Aerial Images and the Urban Imaginary in the 1920s', in Christian Emden, Catherine Keen and David Midgley (eds), *Imagining the City, Vol. 1: The Art of Urban Living*, Bern: Peter Lang, 2006, pp. 189–212.

26. Barthes, 'Eiffel', p. 9.

27. Fairchild in Jennifer S. Light, *From Warfare to Welfare: Defense Intellectuals and Urban Problems in Cold War America*, Baltimore: Johns Hopkins University Press, 2003, p. 132. Light offers a comprehensive account of Fairchild and other interwar aerial surveyors as an early incarnation of what she calls 'moon-shot management for American cities'—a phrase that indexes the intersections between the Cold War military–industrial–academic complex and urban planning during the same period.

28. Granted, the later map covered not just Manhattan but all five boroughs (Fairchild, 'Aerial Mapping of New York City', p. 74).

29. [Sherman M. Fairchild], 'The Making of Greater New York's Air Map', *Aviation*, 16 (1924), p. 16. Whereas Fairchild's signed Jan. 1924 article in *American City* omits mention of the many obstacles to 'suitable' photographic mapping days, this longer, unsigned *Aviation* piece from the same month goes into great detail on this count. For the urban planning readership of *American City*, the emphasis is on the clarity and instantaneity of the photogrammetric signal, as it were; for the airminded readership of *Aviation*, the emphasis falls on the aviator-photographer's heroic agon with noise. My account of this noise is largely derived from the longer, unsigned piece.

30. The aerial coverage of the event allegedly beat the ground photos into distribution; see Unsigned, 'Practical', pp. 114, 117.

31. On Le Corbusier and the aerial view, see Anthony Vidler, 'Photourbanism: Planning the City from Above and Below', in Gary Bridge and Sophie Watson

(eds), *A Companion to the City*, Oxford: Blackwell, 2000, pp. 35–45; and Adnan Morshed, 'The Cultural Politics of Aerial Vision: Le Corbusier in Brazil (1929)', *Journal of Architectural Education*, 55 (2002), pp. 201–10.

32. I borrow the juxtaposition of *flâneur* and *planeur* from Deriu ('The Ascent of the Modern *Planeur*'), in which the latter term is resignified (*planeur* being French for 'glider') to denote urban planners of the Le Corbusier type.

33. That horizontality can accommodate menacingly penetrating and centralised scopic regimes has recently been demonstrated by the rash of complaints, lawsuits and protests against Google Street View (GSV), whose latter-day *flânerie* has proven at least as controversial as the planimetric Google Earth. I would suggest, additionally, that the wide array of ludic and oppositional gestures GSV has attracted—the websites where self-described 'GSV voyeurs' post screenshots of funny, glitchy, baffling, shocking, transgressive, illicit and otherwise anomalous events serendipitously captured by the camera car; the performance art pieces staged for the sake of the camera car by local residents to whom its shooting schedule was leaked in advance—arise not because the horizontal is innately the plane of local resistance or (after de Certeau) pedestrian enunciations but because horizontality is emerging as the site or mode par excellence of penetrative seeing—of what Google itself calls 'geoimmersive data production'—and therefore attracts the most urgent and spectacular counter-gestures.

34. The Prague image has since been surpassed in information-size by over a dozen panoramic photographs, including a 26-gigapixel image of Paris, a 45-gigapixel image of Dubai and a 272-gigapixel image of Shanghai—all of them replete with examples of the ghosting and doubling I discuss here. Martin's 80-gigapixel image of London is currently the largest spherical panoramic image in the world.

## 6. 'CONCEALING THE CRUDE': AIRMINDEDNESS AND THE CAMOUFLAGING OF BRITAIN'S OIL INSTALLATIONS, 1936–9

1. Peter Adey, Lucy Budd and Phil Hubbard, 'Flying Lessons: Exploring the Social and Cultural Geographies of Global Air Travel', *Progress in Human Geography*, 31, 6 (2006), p. 775.

2. Joseph Corn, *The Winged Gospel: America's Romance with Aviation*, Oxford: Oxford University Press, 1983, p. xi.

3. Mark Gottdiener, *Life in the Air: Surviving the New Culture of Air Travel*, Lanham, MD: Rowman & Littlefield, 2000; Brian Graham, *Geography and Air Transport*, Chichester: Wiley, 1995.

4. Peter Adey, 'Aeromobilities: Geographies, Subjects and Vision', *Geography Compass*, 2, 5 (2008) p. 1319; Peter Adey, '"May I have your attention": Airport Geographies of Spectatorship, Position and (Im)mobility', *Environment and Planning D*, 25, 3 (2007), pp. 515–36; Saulo Cwerner, Sven Kesslering and John Urry (eds), *Aeromobilities*, London: Routledge, 2009.

5. Peter Fritzsche, *A Nation of Fliers: German Aviation and the Popular Imagination*, Cambridge, MA: Harvard University Press, 1992, Robert Wohl, *A Passion for Wings: Aviation and the Western Imagination, 1908–1918*, London: Yale University Press, 1996.

6. Corn, *The Winged Gospel*, p. 12.

7. L. Edmunds, 'How Australians Were Made Airminded', *Continuum: The Australian Journal of Media and Culture*, 7, 1 (1993), pp. 183–206.

8. Ibid., pp. 184–5.

9. Peter Adey, 'Airports and Air-Mindedness: Spacing, Timing and Using the Liverpool Airport, 1929–1939', *Social and Cultural Geography*, 7, 3 (2006), p. 346.

10. The National Archives (hereafter: TNA) HO191/8: 'Summary Report No. 3: Camouflage Research, Part I: The Camouflage Problem', undated.

11. Lucy Budd, 'The View from the Air: The Cultural Geographies of Flight', in P. Vannini (ed.), *The Cultures of Alternative Mobilities: Routes Less Travelled*, Aldershot: Ashgate, p. 71.

12. Kitty Hauser, *Shadow Sites: Photography, Archaeology, and the British Landscape 1927–1955*, Oxford: Oxford University Press, 2007.

13. M. Schwarzer, *Zoomscape: Architecture in Motion and Media*, New York: Princeton Architectural Press, p. 123.

14. D. Pascoe, *Airspaces*, London: Reaktion, 2001, pp. 37–9.

15. D. Edgerton, *England and the Aeroplane: An Essay on a Militant and Technological Nation*, London: Macmillan, 1991, p. xiv.

16. P. Fearon, 'The Growth of Aviation in Britain', *Journal of Contemporary History*, 20, 1 (1985), pp. 21–40; J. Myerscough, 'Airport Provision in the Inter-War Years', *Journal of Contemporary History*, 20, 1 (1985), pp. 41–70.

17. L.E.O Charlton, *War from the Air*, London: T. Nelson and Sons, 1935.

18. Ibid., p. 5.

19. Ibid., p. 35.

20. L.E.O. Charlton, *The Menace of the Clouds*, London: W. Hodge and Co., 1937, pp. 53–4.

21. Fritzsche, *A Nation of Fliers*, p. 205.

22. Charlton, *War from the Air*, p. 60.

23. Robert Wohl, *The Spectacle of Flight: Aviation and the Western Imagination, 1920–1950*, London: Yale University Press, 2005, p. 217.

24. 'Mr Baldwin on Aerial Warfare—A Fear for the Future', *The Times*, 11 Nov. 1932.

25. Tim H. O'Brien, *Civil Defence*, London: Longmans, Green and Co., 1955, pp. 56–7.

26. TNA, CAB16/170: 'Minutes of the Third meeting of the Camouflage Sub-Committee of the C.I.D. on the 13th May 1937', p. 3.

27. TNA, CAB16/170: 'C.A.M. 1—Composition and Terms of Reference', dated 23 Oct. 1936.

28. TNA, HO186/14: 'C.I.D. Report 1516-B', dated 9 Feb. 1939, p. 3.

29. Ibid., p. 3.

30. Guy Hartcup, *Camouflage: A History of Concealment and Deception in War*, 2nd edn, Barnsley: Pen and Sword Military, 2008 [1979].

31. TNA, HO191/8: 'Summary Report No. 3: Camouflage Research, Part I: The Camouflage Problem', undated.

32. TNA, HO191/8: 'Summary Report No. 3: Camouflage Research, Part I: The Camouflage Problem', undated.

33. TNA, CAB16/170: 'Minutes of the Third Meeting of the Camouflage Sub-Committee of the C.I.D. on 13th May 1937', p. 6.

34. TNA, CAB16/170: 'Minutes of the First Meeting of the Camouflage Sub-Committee of the C.I.D. on 30th Oct 1936', p. 6.

35. TNA, CAB16/170: 'C.A.M.16—Memorandum by the Air Raid Precautions Department on the Protection of Important Undertakings against Aimed Bombing', dated 18 Feb. 1937, p. 1.

36. TNA, CAB16/170: 'Minutes of the Third Meeting of the Camouflage Sub-Committee of the C.I.D. on 13th May 1937', p. 14.

37. TNA, CAB16/170: 'Minutes of the Second Meeting of the Camouflage Sub-Committee of the C.I.D. on 9th Dec 1936', p. 12.

38. TNA, HO186/14: 'C.I.D. Report 1301-B', dated 3 Feb. 1937, p. 4.

39. TNA, CAB16/170: 'Minutes of the Second Meeting of the Camouflage Sub-Committee of the C.I.D. on 9th Dec 1936', p. 10.

40. TNA, CAB16/170: 'C.A.M.18—Correspondence', Col. J. Turner, to Major Dykes, Secretary of the Sub-Committee, dated 1 Mar. 1937.

41. TNA, HO186/390: 'Correspondence', A. Johnston to Lt-Col. R.A. Thomas, dated 20 Nov. 1937.

42. TNA, CAB16/170: 'C.A.M.4—Examination of Booklet "Methods of Concealment of Oil Fuel Tanks by Using Colour Only"', p. 2.

43. Ibid., p. 8.

44. TNA, CAB16/170: 'Extract from the Minutes of the 288th Meeting of the Committee of Imperial Defence, held 11th Feb 1937', p. 1.

45. TNA, HO186/14: 'C.I.D. Report 1516-B', dated 31 Jan. 1939, p. 4.

46. TNA, CAB16/170: 'C.A.M.39—Second Progress Report of the Camouflage Department', dated 12 Nov. 1938, p. 1.

47. TNA, CAB16/170: 'C.A.M.30—First Progress Report of the Camouflage Section', dated 7 Apr. 1938, p. 7.

48. TNA, CAB16/170: 'C.A.M.39—Second Progress Report of the Camouflage Department', dated 12 Nov. 1938, p. 1.

49. Ibid., p. 1.

50. Ibid., p. 2.

51. The author would like to thank Dr Peter Merriman and Dr Mark Whitehead

for reading through earlier forms of this chapter and Lucy Jackson and Dr Sarah Mills for further helpful comments in the writing of this chapter. Thanks also to Pete, Alison and Mark for the invitation to contribute to this collection.

## 7. FLYING INTO THE UNKNOWN: CINEMATIC CULTURES OF WAR AND THE AESTHETICS OF DISAPPEARANCE

1. Paul Virilio, *The Aesthetics of Disappearance*, New York: Semiotex(e), 2009 [1980], p. 209.
2. Paul Virilio with John Armitage 'In the Cities of the Beyond: An Interview with Paul Virilio' in *OPEN No. 18: 2030: War Zone Amsterdam, Cahier on Art and the Public Domain*, Amsterdam, 2009, pp. 100–11; Virilio, *Aesthetics*.
3. Joy Garnett is an artist living and working in New York. She has utilized representations of the 'techno-military sublime' in her paintings since the late 1990s. The paintings included here are part of her 'Predator' series (2011–ongoing), comprised of target or grid-like compositions that re-purpose the night vision and heat-seeking images she gathers as source material. Based on screen-grabs from declassified or leaked US military videos of secret operations, the Predator series takes technical images of machine vision and translates them as painterly objects for contemplation.
4. Paul Virilio, *War and Cinema: The Logistics of Perception*, London: Verso, 1989; Paul Virilio, *Strategy of Deception*, London: Verso, 2000; Paul Virilio, *Desert Screen: War at the Speed of Light*, London: Continuum, 2002. I do not discuss Baudrillard's analysis of the Iraq War of 2003 in this chapter because his examination of that war in *The Intelligence of Evil Or the Lucidity Pact* (Oxford and New York: Berg, 2005) does not incorporate the aesthetics of disappearance.
5. Jean Baudrillard, 'The Evil Demon of Images', in Clive Cazeaux (ed.), *The Continental Aesthetics Reader*, London: Routledge, 2000, pp. 444–52.
6. Virilio, *Aesthetics*, p. 102.
7. Virilio, *War*, p. 8.
8. Ibid., original emphasis.
9. Ibid.
10. Ibid.
11. Ibid.
12. Susan Tegel, *Nazis and the Cinema*, London: Hambledon Continuum, 2007.
13. Ibid., pp. 8–9.
14. Ibid., p. 8.
15. Ibid., p. 9.
16. Roel Vande Winkel and David Welch (eds), *Cinema and the Swastika: The International Expansion of Third Reich Cinema*, London: Palgrave Macmillan, 2010.
17. Virilio, *War*, p. 9.

18. Ibid., pp. 9–10.
19. Ibid., p. 10
20. Ibid., pp. 9–10; Virilio, *Aesthetics*, p. 60.
21. Virilio, *War*, p. 10.
22. Ibid.
23. Virilio, *War*, p. 11.
24. Ibid., pp. 11–30.
25. Ibid., p. 11.
26. Ibid.
27. Ibid., pp. 11–13; Virilio, *Desert*, pp. 87–8.
28. Virilio, *War*, pp. 11–13, original emphasis.
29. Virilio, *Aesthetics*, pp. 102–3.
30. Ibid., pp. 85–107.
31. Virilio, *War*, p. 14.
32. Ibid., p. 15.
33. Ibid.
34. Ibid.
35. Ibid.
36. Ibid., p. 16.
37. Ibid.
38. Ibid.
39. Ibid.
40. Ibid., p. 17.
41. Ibid., original emphasis.
42. Ibid.
43. Gerry Coulter, 'Jean Baudrillard and Cinema: The Problems of Technology, Realism and History', *Film-Philosophy*, 14, 2 (2010), pp. 6–20.
44. Mike Gane, *Jean Baudrillard: In Radical Uncertainty*, London: Pluto, 2000.
45. Jean Baudrillard, 'The Evil Demon of Images', in Clive Cazeaux (ed.), *The Continental Aesthetics Reader*, London: Routledge, 2000, p. 450.
46. Baudrillard, 'The Evil Demon', p. 450.
47. Ibid., pp. 450–1.
48. Ibid., p. 451.
49. Ibid.
50. Ibid.
51. Jean Baudrillard, *The Gulf War Did Not Take Place*, Bloomington: Indiana University Press, 1995; Baudrillard, *The Intelligence of Evil Or the Lucidity Pact*.
52. Baudrillard, *Gulf*, p. 62.
53. Ian Almond, 'Baudrillard's Gulf War: Saddam the Carpet-Seller', *International Journal of Baudrillard Studies*, 6, 2 (2009), pp. 1–9.
54. Baudrillard, *Gulf*, p. 63.

55. Kim Toffoletti and Victoria Grace, 'Terminal Indifference: The Hollywood War Film Post-September 11', *Film-Philosophy*, 14, 2 (2010), pp. 62–83.
56. Virilio, *War*, p. 19.
57. Virilio, *Desert*, p. 45.
58. Virilio, *War*, p. 20.
59. Ibid.; Virilio, *Desert*, p. 54.
60. Virilio, *War*, p. 20.
61. Virilio, *Strategy*, p. 224.
62. Virilio, *War*, p. 20.
63. Ibid.
64. Ibid.
65. Ibid.
66. Ibid., pp. 20–1
67. Ibid., p. 21, my emphasis.
68. Ibid.
69. Virilio, *Desert*, p. 107.
70. Virilio, *War*, p. 24.
71. Ibid.
72. Ibid.
73. Ibid.
74. Ibid.
75. Erkki Huhtamo, 'The Sky is (not) the Limit: Envisioning the Ultimate Media Display', *Journal of Visual Culture*, 8, 3 (2010), pp. 329–48.
76. Ibid., p. 329.
77. Virilio, *War*, p. 49.

## 8. PROJECT TRANSPARENT EARTH AND THE AUTOSCOPY OF AERIAL TARGETING: THE VISUAL GEOPOLITICS OF THE UNDERGROUND

1. Jean-Luc Nancy, *The Ground of the Image*, New York: Fordham University Press, 2005, p. 5.
2. Jean Baudrillard, *Impossible Exchange*, New York: Verso, 2001, p. 11.
3. Jonathan Swift, *Gulliver's Travels*, London, 1726.
4. I would like to thank the editors of this volume for their comments as well as thank John Armitage, John Beck and Sean Cubitt, who all read an earlier draft of this chapter and provided wonderful suggestions for it.
5. Swift, *Gulliver's*, p. 165.
6. The classic text on the relationship between visualising machines and the war machine is Paul Virilio's *War and Cinema*, though he also has pursued these links in several other works. Similarly many other theorists have explored this rela-

tionship, particularly Peter Adey, Louise Amoore, Ben Anderson, John Armitage, Benjamin Bratton, David Campbell, Karen Caplan, Jordan Crandall, Rey
Chow, James Der Derian, Manuel DeLanda, Gilles Deleuze and Felix Guat-
tari, Steve Graham, David Gregory, Martin Heidegger, Sven Lindquist, Paul
Saint-Amour and Eyal Weizman, just to name a few. Also, my own work has
been involved with these issues for some time, both writing done alone or with
Greg Clancey and/or John Phillips. I am indebted to all of these scholars and
our many discussions and collaborations.

7.  Peter Adey, *Aerial Life: Spaces, Mobilities, Affects*, London: Wiley-Blackwell,
    2010, p. 3.
8.  Tom Vanderbilt, *Survival City*, Princeton: Princeton Architectural Press, 2002,
    p. 129.
9.  For an extended discussion of the relationship between sensory manipulation
    in art, literature and music and the way in which it pertains to developments
    in military technology, see Ryan Bishop and John Phillips, *Modernist Avant-
    Garde Aesthetics and Contemporary Military Technology*, Edinburgh: Edinburgh
    University Press, 2010.
10. Jack Goldstein, *Aphorisms*, 'Jack Goldstein, Films, Records, Performances and
    Aphorisms 1971–1984', Koln, and Verlag der Buchhandlung Walther: Konig,
    Koln, p. 22.
11. Jean Baudrillard, *The Perfect Crime*, London: Verso, 1996, p. 71.
12. Ovid, *Metamorphoses*, Oxford: Oxford University Press, 1986, pp. 22–30.
13. John Beck, *Dirty Wars: Landscape, Power and Waste in Western American Liter-
    ature*, Lincoln and London: University of Nebraska Press, 2009, p. 179.
14. Ryan Bishop and John Phillips, *Modernist Avant-Garde Aesthetics and Contem-
    porary Military Technology: Technicities of Perception*, Edinburgh: Edinburgh
    University Press, 2010, pp. 63–6.
15. Jean-Luc Nancy, *The Ground of the Image*, New York: Fordham University Press,
    2005, p. 2.
16. See Rosalind Williams's evocative book on the underground in relation to the
    literary and technological imagination, Rosalind Williams, *Notes on the Under-
    ground: An Essay on Technology, Society, and the Imagination*, Cambridge, MA,
    and London: MIT Press, 1990.
17. Mark Smith of the Geospatial Corporation, a solutions provider for the under-
    ground infrastructure industry. 'Geospatial Corporation Maps the World under
    the Earth's Crust', 10 Mar. 2010 (http://homelandsecuritynewswire.com/geo-
    spatial-corporation-maps-world-under-earths-crust).
18. Akira Mizuta Lippit, *Atomic Light (Shadow Optics)*, Minneapolis: University of
    Minnesota Press, 2005, p. 31
19. Ovid, *Metamorphoses*, pp. 253–9.
20. 'The Last Frontier: DARPA Wants to Make the Earth's Crust Transparent', 10

Mar. 2010 (http://homelandsecuritynewswire.com/last-frontier-darpa-wants-make-earths-crust-transparent).

21. 'The Last Frontier'; Keith Button, 'Tunnel Vision: U.S. Intel Community Seeks New Ways to Peer into Underground Sites', 1 Aug. 2009 (http://www.c4isr-journal.com/story.php?F=4151031).

22. Lewis Page, 'Underground Mole-Satnavs to Work Off Lightning Strikes: "Sferic" Zap-Sniff Tech for Future Subterranean Warriors', The Register, 10 Mar. 2010 (http://www.theregister.co.uk/2010/03/10/darpa_s_bug/).

23. 'HAARP Detection and Imaging of Underground Structures Using ELF/VLF Radio Waves', Federation of American Scientists (http://www.fas.org/irp/program/collect/haarp.htm).

24. Michel Foucault, *The Birth of the Clinic*, New York: Vintage, 1994.

25. Charles Lyell, *Principles of Geology*, 1830.

26. *Fortune Magazine*, 1963, p. 125.

27. Plato, 'The Allegory of the Cave', from *The Republic*, New York: Dover Editions, 2000.

## 9. AFP-731 OR THE OTHER NIGHT SKY: AN ALLEGORY

1. Much of this chapter develops ideas from Trevor Paglen, *Blank Spots on the Map*, New York: Dutton, 2009.

2. 'Secret Mission', *Aviation Week and Space Technology*, 132, 4 (2010), p. 23.

3. For the National Reconnaissance Office, see Jeffrey T. Richelson, 'The NRO Declassified', National Security Archive Electronic Briefing Book No. 35, 27 Sep. 2000 (http://www.gwu.edu/~nsarchiv/NSAEBB/NSAEBB35/; accessed 25 Mar. 2011); for the NRO budget, see Commission on the Roles and Capabilities of the United States Intelligence Community, 'The Cost of Intelligence', Chapter 13 in 'Preparing for the 21st Century: An Appraisal of U.S. Intelligence', website dated 13 Feb. 1996, accessed 25 Mar. 2011, http://www.fas.org/irp/offdocs/report.html

4. Dwayne A. Day, 'The Spooks and the Turkey: Intelligence Community Involvement in the Decision to Build the Space Shuttle', The Space Review, 20 Nov. 2006 (http://www.thespacereview.com/article/748/1; accessed 25 Mar. 2011).

5. Ibid.

6. William E. Burrows, *Deep Black*, New York: Random House, 1986, p. 104.

7. Ibid., pp. 84, 90–1.

8. Ibid., p. 107.

9. For John H. Casper's NASA biography, see http://www.jsc.nasa.gov/Bios/html-bios/casper.html (accessed 25 Mar. 2011).

10. For the 'Red Hats' and history of purloined MiGs, see Curtis Peebles, *Dark Eagles*, Novato, CA: Presidio, 1995, pp. 217–44; for patches, see Trevor Paglen,

*I Could Tell You But Then You Would Have to Be Destroyed by Me*, Brooklyn, NY: Melville House, 2008.

11. Jeffrey T. Richelson, *America's Secret Eyes in Space*, New York: Harper and Row, 1990, p. 231.
12. 'Secret Mission', p. 23.
13. For launch details on STS-36, see NASA's web page about the mission (http://science.ksc.nasa.gov/shuttle/missions/sts-36/mission-sts-36.html; accessed 25 Mar. 2011).
14. Warren Leary, 'Problems Are Reported with New Spy Satellite', *The New York Times*, 18 Mar. 1990, part 1, p. 20.
15. All Molczan quotes come from two days of interviews in Toronto, Canada, in July 2008, and from a year's worth of e-mail correspondence with the author.
16. For amateur astronomy and professional science, see Timothy Ferris, *Seeing in the Dark*, New York: Simon and Schuster, 2002.
17. See Anthony Eccles, 'UFOs and the NOSS Problem', The Anomalist (http://www.anomalist.com/features/Noss.html; accessed 35 Mar. 2011); 'Recent Australian UFOs Were Just U.S. Navy Satellites', Listserv post under Reader Feedback (http://www.ufoinfo.com/roundup/v08/rnd0804.shtml; accessed 25 Mar. 2011).
18. W. Patrick McCray, *Keep Watching the Skies! The Story of Operation Moonwatch and the Dawn of the Space Age*, Princeton: Princeton University Press, 2008, pp. 108–9.
19. Quoted in Patrick Radden Keefe, 'I-Spy: Former Intelligence Officials Take to the Novel to Tell Their Stories', *The Boston Globe*, 4 May 2006 (http://www.boston.com/news/globe/ideas/articles/2006/05/14/i_spy/; accessed 25 Mar. 2011).
20. For the full story of the USA 144Deb object, see Trevor Paglen, *Blank Spots on the Map*.
21. For some of the classic critiques of absolute space, see Henri Lefebvre, *The Production of Space*, trans. Donald Nicholson, London: Blackwell, 1990; Neil Smith and Cindi Katz, 'Grounding Metaphor: Towards a Spatialized Politics', in Michael Keith and Steve Pile (eds), *Place and the Politics of Identity*, London: Routledge, 1993, pp. 67–83.
22. The literature on these questions is of course far too vast to address here in any substantive way.
23. For example, Michel Foucault, *The Foucault Reader*, ed. Paul Rabinow, New York: Vintage, 1984, pp. 59–60.
24. Winston Smith was the protagonist in George Orwell's 1949 classic novel, *1984*.
25. US Patent No. 5345238 Eldridge et al., 'Satellite Signature Suppression Shield', filed 14 Mar. 1990, granted 6 Sep. 1994 (http://www.gwu.edu/~nsarchiv/NSAEBB/NSAEBB143/nph-Parser.htm; accessed 25 Mar. 2011).

26. 'Memorandum for Deputy for Technology/CSA, Subject: A Covert Reconnaissance Satellite', 17 Apr. 1963, declassified 26 Nov. 1997.

27. US Patent No. 5345238.

28. Warren Leary, 'Space Shuttle Lifts Off with Secret Military Cargo', *New York Times*, 16 Nov. 1990.

## 10. THE PAIN OF LOVE: THE INVENTION OF AERIAL SURVEILLANCE IN BRITISH IRAQ

1. Besides wartime uses, in 1919 airpower was used to put down unrest in Egypt, Punjab, Somaliland, Afghanistan and the North West Frontier (see David Omissi, *Air Power and Colonial Control*, Manchester: Manchester University Press, 1990, p. 11; Sven Lindqvist, *A History of Bombing*, New York: Granta, 2000 pp. 42–3). It was also used against the Red Army in South Russia. These were 'spasmodic, almost casual affairs'; it was in Iraq that British military history was transformed (John Laffin, *Swifter Than Eagles: The Biography of Marshal of the Royal Air Force Sir John Maitland Salmond*, Edinburgh: W. Blackwood, 1964, p. 192). On pre-war imaginings of the uses of airpower, see Michael Paris, *Winged Warfare: The Literature and Theory of Aerial Warfare in Britain, 1859–1917*, Manchester: Manchester University Press, 1992; Robert Wohl, *A Passion for Wings: Aviation and the Western Imagination, 1908–1918*, New Haven: Yale University Press, 1994. The war sections of these books focus on the western front.

2. The Anglo-Ottoman accord of 1901 committed both parties to maintaining the status quo in the Ottoman Empire. This in itself made intelligence more important, as British officials remained puzzled as to what the status quo was. None of this, however, prevented the British from pursuing (secretly) their private arrangements with Gulf potentates—which only stoked Ottoman paranoia about British commitment to the status quo. Hence, the Ottomans banned British travel in the region just when Britons were becoming most desperate to venture there, forcing intelligence into a semi-covert and thus semi-autonomous sphere made up of civilians and the off-the-record activities of various local representatives.

3. Douglas Carruthers, *Arabian Adventure: To the Great Nafud in Quest of the Oryx*, London, 1935, p. 68; F.R. Maunsell, 'The Hejaz Railway', *Geographical Journal*, 32 (1908), p. 570.

4. See for instance Capt. Shakespear, 2. Feb. 1910, tour diary 1910, to Cox, 23 Mar. 1910, 44/29805/10, FO 371/1013, PRO; Bell to her family, 19 Feb. 1914, 5 Mar. 1911, and 17 Feb. 1911, in *The Letters of Gertrude Bell*, ed. Lady Bell, 2 vols, London, 1927, vol. 1, pp. 276, 288, 339; Douglas Carruthers, 'Journey in North-Western Arabia', *Geographical Journal*, 35 (1910), pp. 234–40; J.G. Lorimer (ed.), *Gazetteer of the Persian Gulf, Oman, and Central Arabia*, vol. 2, *Geographical and Statistical*, Calcutta, 1908, pp. 199, 759n., 767, IOR: L/PS/20/

C91/4, BL. For further support of this and other claims in this chapter, see Satia, *Spies in Arabia*, Oxford: Oxford University Press, 2008. Geography had always been the 'material underpinning for knowledge about the Orient', Edward Said, *Orientalism*, New York: Patheon, 1978, p. 216, and was central to intelligence work anywhere, but acquired a special importance in Ottoman Arabia, then an infamous 'white spot' on British maps. Theories of environmental determinism also suggested agents could learn much about Arabs simply by studying their landscape, in any case the primary factor in a region of 'small wars', which were 'in the main campaigns against nature', Col. C.E. Callwell, *Small Wars: Their Principles and Practice*, 3rd edn, London: Harrson and Sons, 1906, p. 44. The urgent need to obtain details of the German-backed Baghdad Railway made map-making even more central, while agents found in geography a suitably scholarly cover for other kinds of intelligence work in the region. See for instance Capt. Fraser Hunter to Assistant Surveyor-General, 6 Apr. 1910, 3263, India Office Records (IOR): L/PS/10/259, British Library (BL); Bury to Sir Richmond Ritchie, IO, 15 Nov. 1909, 3216, IOR: L/PS/10/135, BL.

5. G.E. Leachman, quoted in N.N.E. Bray, *A Paladin of Arabia: The Biography of Brevet Lieut.-Colonel G. E. Leachman, C. I. E., D. S. O., of the Royal Sussex Regiment*, London, 1936, p. 171. See also Callwell, *Small Wars*, pp. 49–50; Aubrey Herbert, *Ben Kendim: A Record of Eastern Travel*, ed. Desmond MacCarthy, 2nd edn, New York: Hutchison, 1925, p. 67.

6. Louisa Jebb, *By Desert Ways to Baghdad*, Boston: T.F. Unwin, 1909, pp. 224–5; David Hogarth, comment on Butler, 'Baghdad to Damascus', p. 533. Meredith Townsend recognised early on that most Englishmen, 'filled ... with the "idea" of Arabia', tended to exaggerate the region's aridity (*Asia and Europe: Studies Presenting the Conclusions Formed by the Author in a Long Life Devoted to the Subject of the Relations between Asia and Europe*, 2nd edn, New York: GP Putnam and Sons, 1904, p. 161.

7. G. Wyman Bury, *Land of Uz*, London, 1911, p. xxi. Also see Peter Brent, *Far Arabia: Explorers of the Myth*, London: Weidenfield and Nicholson, 1977, p. 26, 133; Jeremy Wilson, *Lawrence of Arabia: The Authorised Biography of T.E. Lawrence*, London: Heinemann, 1989, p. 95; Kathryn Tidrick, *Heart-Beguiling Araby*, Cambridge: Tauris Park, 1981, pp. 124, 166–9, 178. Aubrey Herbert was unanimously described as a 'knight'; Bray titled his biography of Leachman, '*A Paladin of Arabia*'. Lawrence was famously obsessed with medieval warfare; his first steps in the region were taken to research his thesis on the influence of the Crusades on European military architecture. The Romantics had also looked to Eastern philosophy and culture for alternatives to Occidental materialism and mechanism, but unlike these forbears, Edwardians did not so much urge Europe to copy Arabia as seek individual escape into it, combining fulfilment of this wish with intelligence work. Nor, once there, did they recoil, as the Romantics

had, from the 'real' Orient; they nurtured a need for aesthetic experience so desperate that they did not even see mundane Arabia when they got there, only their prefiguring vision of it. See for instance David Fraser, *The Short Cut to India: The Record of a Journey along the Route of the Baghdad Railway*, Edinburgh: W. Blackwood and Sons, 1909, p. 234. On the Romantics, see Said, *Orientalism*, pp. 100–15.

8. Mark Sykes, *The Caliph's Last Heritage: A Short History of the Turkish Empire*, London: Macmillan, 1915, pp. 5, 118.

9. D.G. Hogarth, 'Problems in Exploration I: Western Asia', *Geographical Journal*, 32 (1908), pp. 549–50. See also Ali Behdad, *Belated Travelers: Orientalism in the Age of Colonial Dissolution*, Durham, NC: Alpha Academic, 1994, p. 93; Felix Driver, *Geography Militant: Cultures of Exploration and Empire*, Oxford: Blackwell, 2001, p. 199. The Amazon and Tibet also fascinated Edwardians, but significant progress had been made on mapping Tibet by the time of the 1907 convention with the Russians, a fact that Hogarth marshalled to demonstrate just how unknown Arabia was. Tibet emerged as a utopic archive-state only in the British imagination to the 1930s; Thomas Richards, *The Imperial Archive: Knowledge and the Fantasy of Empire*, London: Verso, 1993, p. 32. The North and South Poles were also being explored for the first time, but Arabia was the last inhabited area where Europe could still dream of shining the light of civilisation and progress. Agents who had travelled in other vast spaces (the veldt, the Sahara, the Gobi, the Mexican plains, etc.) singled out Arabia as the most sublime and mysterious among these. See, for instance, Carruthers, *Arabian Adventure*, pp. 122–3.

10. Ferdinand Tuohy, *The Secret Corps: A Tale of 'Intelligence' on All Fronts*, London: A.J. Murray, 1920, p. 172.

11. On this see also Christopher Andrew, *Her Majesty's Secret Service: The Making of the British Intelligence Community*, 1985; reprinted edn, New York: Viking, 1986, pp. 34–58.

12. Jon Thompson, *Fiction, Crime, and Empire: Clues to Modernity and Postmodernism*, Urbana, IL: University of Illinois Press, 1993, p. 87. Also see Richards, *The Imperial Archive*, p. 26. Almost everything in the novel relating to intelligence work in India was a product of Kipling's imagination (Gerald Morgan, 'Myth and Reality in the Great Game', *Asian Affairs*, 60 (1973), p. 197; Michael Silvestri, 'The Thrill of "Simply Dressing Up": The Indian Police, Disguise, and Intelligence Work in Colonial India', *Journal of Colonialism and Colonial History*, 2 (2001), paragraph 13, http://muse.jhu.edu/journals/journal_of_colonialism_and_colonial_history/v002/2.2silvestri.html; C. A. Bayly, '"Knowing the Country": Empire and Information in India', *Modern Asian Studies*, 27 (1993), p. 36), but that *Kim* influenced *real* intelligence afterwards is evident in 'Intelligence Methods in Peace Time', 1909, KV 1/4, PRO. Arendt also sees

*Kim* as the foundational legend of the British secret agent (Hannah Arendt, *Imperialism: Part Two of 'The Origins of Totalitarianism'*, 1951; repr. edn, New York: Harcourt, Brace & World, 1968, p. 97).

13. Townsend, *Asia and Europe*, pp. 305–6.

14. Bell, 1928, quoted in Rosemary O'Brien (ed.), *Gertrude Bell: The Arabian Diaries, 1913–1914*, Syracuse, NY: Syracuse University Press, 2000, pp. 9–10.

15. Jebb, *By Desert Ways*, pp. 264–5. Emphasis added.

16. Archaeology was not merely 'cover' for secret service work, nor was it entirely innocent. Lawrence worked for David Hogarth, who served on the PEF and had close social and professional ties to the intelligence network. Archaeology was also a prestige issue impinging on geopolitical considerations. Lawrence and Leonard Woolley definitely were involved in a military intelligence mission in Sinai in 1913 which led to further work with Stewart Newcombe on the Baghdad railway. These prewar experiences laid the foundation for the induction of Hogarth, Lawrence and Woolley into wartime intelligence.

17. Sykes, *The Caliph's Last Heritage*, p. 57. These travellers did not seek the sensual indulgence that the Orient is generally supposed to have offered Europeans but escape from what they saw as the moral decadence of their *own* society.

18. Townsend, *Asia and Europe*, p. 167; Jebb, *By Desert Ways*, pp. 16–17.

19. Said, *Orientalism*, pp. 120–2, 230–1. On the links between primitivism and modern occultism see Patrick Brantlinger, *Rule of Darkness: British Literature and Imperialism, 1830–1914*, Ithaca, NY: Cornell University Press, 1988, p. 232.

20. Frederic Lees, introduction to Philip Baldensperger, *The Immovable East: Studies of the People and Customs of Palestine*, Boston: Pitman, 1913, p. vii; N.N.E. Bray, *Shifting Sands*, London: Unicorn Press, 1934, p. 14. Emphasis added. This argument differs from Kathryn Tidrick's argument in *Heart-Beguiling Araby* that the British felt endowed, by their intrinsic similarity, with a miraculous insight into Arab affairs. Tidrick's book does not take into account significant changes in the Edwardian era, when intelligence efforts were stepped up markedly and British agents grew convinced that the seemingly noble Arab and his landscape lied almost compulsively. Intuition became the centrepiece of their epistemology in a new way: if Arabia was unintelligible, a rational British mind could have no insight into its workings, regardless of cultural familiarity. Instead, the aspiring agent would have to learn to think irrationally, like an Oriental.

21. See for instance Lorimer, Tour Journal no. 1 of 1910, Voyage to Salman Pak, 11 Feb. 1910, 44/11909/10, FO 371/1008, PRO; Knox, report on the outlook in Nejd, 26 Sep. 1906, 44/40123/06, FO 371/156, PRO; Lorimer, *Gazetteer*, 2:iii, 1352; Shakespear, Notes on tour via as-Safa, al-Batin and back to Kuwait, to Cox, 4 Apr. 1910, 44/29805/10, FO 371/1013, PRO; Sykes, notes on Kurdistan, submitted to Grey, 14 Aug. 1907, 44/28053/07, FO 371/354, PRO; J.W.A. Young, 'A Little to the East', p. 141, memoir, *c*.1940s, Papers of

J.W.A. Young, MEC; A fellow officer, obituary for Leachman, *Daily Telegraph*, 21 Aug. 1920.

22. AIR 1/426/15/260/3: Air Staff, 'On the Power of the Air Force and the Application of that Power to Hold and Police Mesopotamia', Mar. 1920. For a general narrative of the RFC's work in the Middle East, see H.A. Jones, *The War in the Air*, vol. 5, Oxford, 1935, Chapters 3–5, and vol. 6, 1937, Chapters 5–6.

23. Dickson to Gwenlian Greene, 7 Feb. 1915, 1st booklet, Papers of H.R.P. Dickson, MEC. The first two airplanes finally arrived in Aug. 1915, the germ of Squadron 30, but they were in poor condition and the force still lacked repairmen, photographic equipment, spare parts, sufficient pilots or sufficient knowledge about flying in 'tropical' climates. Only after the Kut debacle and transfer of the command to the War Office did new planes and the latest photographic equipment arrive. Near the end of 1916, the RFC in Mesopotamia was made part of the new Middle East Brigade, which included units in Egypt, Salonika, East Africa and, later, Palestine.

24. Philby, 'Mesopotage', Chapter 7; *Arabian Days*, p. 129.

25. AIR 1/140/15/40/306: General, Force 'D' to WO, 5 Feb. 1916; AIR 2/940: Tennant, Resume of operations, 30 Mar. 1917; Wauchope, Chapter 7, 70.

26. Tennant, *In the Clouds*, pp. 35, 38–9, 141; FO 686/6/Pt. 1: Joyce to Wilson, 24 Mar. 1917.

27. AIR 10/1001: RAF, Preface of 'Notes on Aerial Photography, Part II, The Interpretation of Aeroplane Photographs in Mesopotamia', 1918; Occleshaw, *Armour Against Fate*, pp. 61–2. On the agents' development of aerial photography in Egypt, see J. Wilson, *Lawrence of Arabia*, pp. 189, 198 n. 77.

28. Tennant, *In the Clouds*, p. 38, pp. 60–1.

29. WO 158/626: Salmond to CGS EEF, 12 Nov. 1916 and Note in Egyptforce to Wingate, 14 Nov. 1916; Lawrence to GOC, Egypt and DMI, 17 Nov. 1916 in Arabian Report no. 18; Leith-Ross, 'Tactical Side', pp. 8–9; Tennant, *In the Clouds*, p. 163.

30. Bray, 8 Nov. 1916, in Arabian Report no. 18.

31. AIR 1/2399/280/1: Squadron Leader L.G.S. Payne, 'The Use of Aircraft in Connection with Espionage', 7 Nov. 1924.

32. Lawrence to C.E. Wilson, 6 Dec. 1916, in *Letters of T.E. Lawrence*, pp. 211–12; Marlowe, *Late Victorian*, p. 129; BL: Add. MSS: Sir A.T. Wilson Papers: 52455C: Wilson to Hirtzel, 22 May 1918.

33. This fairly large conventional war lasted several months and involved much of the country, including Kurdistan. Roughly 1,000 British and Indian troops were killed, and another 1,000 wounded. Roughly 10,000 Iraqis were killed. For an Iraqi scholar's perspective, see Ghassan Attiyah, *Iraq, 1908–1921: A Socio-Political Study*, Beirut: Arab Institute for Research and Publication, 1973, Chapters 7 and 8, in which the rising figures as an independence movement.

34. See Churchill to Shuckburgh, 11 Jan. 1922, in *Churchill*, 4/3: 1723; Omissi, *Air Power*, pp. 28–9, 44–5; Charles Townshend, *Britain's Civil Wars: Counterinsurgency in the Twentieth Century*, Boston: Faber and Faber, 1986, pp. 99–113. The squadrons based in Iraq, Palestine and Transjordan came under the Middle East Command at Cairo. In 1928, the RAF took over the Sudan and Aden, while also striving to maintain Ibn Saud's dependence on their assistance. RAF squadrons could also be found in India, Malta and Singapore. Air control eventually began to 'substitute' for traditional forces in other parts of the empire. See David Killingray, '"A Swift Agent of Government": Air Power in British Colonial Africa, 1916–1939', *Journal of African History*, 25 (1984), pp. 429–44; Omissi, *Air Power*, pp. 39–59.

35. Lawrence to Liddell Hart, 1933, in *The Letters of T.E. Lawrence*, p. 323; emphasis in original.

36. See, for instance, Memorandum on the scheme for the employment of the forces of the crown in Mesopotamia, n.d., AIR 20/526, PRO; CAS, Scheme for the Control of Mesopotamia by the Royal Air Force, 12 Mar.[?] 1921, AIR 5/476, PRO; Salmond to Air Ministry, 22 May 1923, AIR 1/2132/207/136/2, PRO; Air Marshal Sir J.M. Salmond, Report on Command from 1 Oct. 1922 to 7 Apr. 1924, n.d. [ca. Apr. 1924], AIR 23/542, PRO; John Glubb, *War in the Desert: An R.A.F. Frontier Campaign*, London: Norton, 1960, p. 69; A.T. Wilson, memo on RAF Scheme, to IO, [early 1921], quoted in Marlowe, *Late Victorian*, p. 201; A.T. Wilson, *Mesopotamia, 1917–1920: A Clash of Loyalties—A Personal and Historical Record*, London: Oxford University Press, 1931, pp. 238–9; A.T. Wilson, memo, 16 Feb. 1921, quoted in Martin Gilbert, *Winston S. Churchill*, vol. 4: *1916–1922: The Stricken World*, Boston: Houghton Mifflin, 1975, p. 532; Air Staff, 'On the Power of the Air Force and the Application of That Power to Hold and Police Mesopotamia'; Sir A. Haldane, GOC-in-C, Mesopotamia Expeditionary Force, report on RAF work in Mesopotamia, to WO, 25 Nov. 1920, Cabinet Paper, Feb. 1921, AIR 5/1253, PRO; 'Notes on the value of the air route between Cairo and Baghdad for strategic and other purposes', n.d., AIR 9/14, PRO; Edmonds, 22 Nov. 1924, in Diary 1915 to 1924, Papers of Cecil J. Edmonds, MEC; Churchill, memo, 1 May 1920, quoted in Gilbert, *Churchill*, 4: 481; Churchill, Cabinet Memorandum, 'Policy and Finance in Mesopotamia, 1922–23', 4 Aug. 1921, in *Churchill, Companion, Part 3: Documents April 1921—November 1922*, ed. Martin Gilbert, London: Houghton Mifflin, 1977, pp. 1576–81; Churchill to Trenchard, 29 Feb. 1920, quoted in Gilbert, *Churchill*, 4, p. 217.

37. Robert Brooke-Popham, 'Aeroplanes in Tropical Countries', lecture, in Proceedings of Meeting of the Royal Aeronautical Society on 6 Oct. 1921, *Aeronautical Journal*, 25, 1 Mar. 1922, Brooke-Popham Papers, Liddell Hart Centre for Military Archives, King's College, London (LHCMA).

38. Major General T. Fraser, Commanding British Forces in Iraq, to WO, 3 Aug. 1922, AIR 5/202, PRO. See also John Glubb, *The Changing Scenes of Life: An Autobiography*, London: Quartet Books, 1983, p. 60; *Arabian Adventures: Ten Years of Joyful Service*, London: Cassell, 1978, p. 135; Report regarding the value of aeroplanes as main weapon of an Administration in the maintenance of law and order, compiled from reports submitted by military commanders and political officers in experience gained during the 1920 insurrection in Mesopotamia, n.d., AIR 23/800, PRO; C.H. Keith, 29 Oct. 1926, Hinaidi, in *Flying Years*, Aviation Book Club ed. London, 1937, p. 16; SSO Basrah, Memo on operations against outlaws of Albu Khalifah, to GHQ, 19 July 1921, CO 730/4, PRO.

39. See for instance 18th Division, Intelligence report, 15 June 1921, AIR 1/432/15/260/23 (A-B), PRO; Dr W.A. Wigram, 'Problems of Northern Iraq', *Journal of the Central Asian Society*, 15 (1928), p. 331.

40. Sir A. Haldane, report on RAF work in Mesopotamia; Air Staff comments in column parallel with Haldane's report, Jan. 1921; Glubb, 'Conduct of the operations: 1928–29 Year of Sibilla', draft, Glubb Papers, Iraq Southern Desert (3), 1928–30, MEC; Keith Jeffery, *The British Army and the Crisis of Empire 1918–22*, Manchester: Manchester University Press, 1984, p. 153.

41. E.A.S., minute, 30 Mar. 1922, on a phone conversation with Wilson, CO 730/20, PRO. See also Bullard, minute, 29 Mar. 1922, on Cox to S/S CO, 25 Mar. 1922.

42. Salmond, Air Ministry, Iraq Command Report for Oct. 1922–Apr. 1924, Nov. 1924, AIR 5/1253, PRO.

43. See Commanding Officer of 17th Division, report, 26 June 1921, AIR 1/432/15/260/23 (A-B), PRO; Thomas, Memorandum, to PO Muntafik, 13 July 1920, E11758/2719/44/1920, FO 371/5230, PRO; [Hall?], minute, 11 Aug. 1921, on Cox to CO, 30 June 1921, 39645, CO 730/2, PRO; SSO Basrah, Memo on operations against outlaws of Albu Khalifah; Cox to Churchill, 6 Oct. 1921, on operations at Batas and elsewhere, CO 730/7, PRO; Omissi, *Air Power*, p. 174. Townshend confirms that the 'gentle vision' of air blockade was a self-deception, if not a conscious fraud ('Civilization and 'Frightfulness', p. 153). Many published sources provide descriptions of exemplary episodes; see, for instance, Peter Sluglett, *The British in Iraq, 1914–1932*, London: Tauris, 1976, pp. 262–70 (available online at http://globalpolicy.igc.org/security/issues/iraq/history/1976sluglett.htm; accessed 20 Nov. 2005); Dodge, *Inventing Iraq*, 150–4; Corum, 'The Myth of Air Control', p. 67. It is difficult to say how many Iraqis were killed in these operations, given the official refusal to gather data on results.

44. Worthington-Evans, quoted in Townshend, 'Civilization and "Frightfulness"', p. 147. On these criticisms see also Omissi, *Air Power*, pp. 151–83.

45. See Sluglett, *The British in Iraq*, pp. 262–70; Satia, *Spies in Arabia*, Chapter 9.
46. Air Staff, Memorandum, in Air Ministry to CID, 26 Nov. 1921, CO 730/18, PRO. See also Brooke-Popham, 'Some Notes on Aeroplanes, with Special Reference to the Air Route from Cairo to Bagdad', *Journal of the Central Asian Society*, 9 (1922), p. 139.
47. Chairman [Lord Peel?], comment on Right Hon. Lord Thomson (S/S Air), 'My Impressions of a Tour in Iraq', *Journal of the Central Asian Society*, 12 (1925), p. 211.
48. Basil Liddell Hart, *Paris, or the Future of War*, New York: Garland, 1925, p. 44. See also Fussell, *The Great War*, p. 190.
49. Others also find this argument specious but have not explained exactly what is wrong with it. See Geoffrey Best, *Humanity in Warfare*, New York: Columbia University Press, 1980, pp. 274–5; Townshend, 'Civilization and "Frightfulness"', pp. 147–8. Several air control historians, on the other hand, find the argument compelling. See Omissi, *Air Power*, p. 169; Meilinger, 'Trenchard and "Morale Bombing"', p. 259; James S. Corum and Wray R. Johnson, *Airpower in Small Wars: Fighting Insurgents and Terrorists*, Lawrence, KA: University Press of Kansas, 2003, p. 59.
50. Thomson, 'My Impressions of a Tour in Iraq', p. 211; An officer in Iraq, quoted in 'With the RAF in Iraq', *Basrah Times*, 3 May 1924, in AIR 5/1298, PRO. Jon Lawrence argues that the myth of peaceableness triumphed in post-war culture and that by 1921 militarism had become marginalised as the preserve of an ultra-right wing rump ('Forging a Peaceable Kingdom: War, Violence, and Fear of Brutalization in Post-First World War Britain', *Journal of Modern History*, 75 (2003), p. 558), but my effort here is to make sense of how the militarism perpetuated in many parts of the empire—as Lawrence acknowledges—was made acceptable to those who otherwise deemed Britain a uniquely peaceable kingdom. On the aeroplane's centrality to interwar militarism, see David Edgerton, *England and the Aeroplane: An Essay on a Militant and Technological Nation*, Basingstoke: Macmillan, 1991.
51. Glubb, Note on the Southern Desert Force [*c.*1930s], Glubb Papers, Box I, File: Iraq S. Desert (1), 1927–1928; Keith, 29 Oct. 1926, Hinaidi, in *Flying Years*, p. 18.
52. John Glubb, *Story of the Arab Legion*, London: Hodder and Stoughton, 1948, p. 149; *Arabian Adventures*, p. 148. The tribal principle of communal responsibility was also held to recommend indiscriminate punishment as a mark of cultural respect. See for instance Sir John Slessor, *The Central Blue: The Autobiography of Sir John Slessor, Marshal of the RAF*, New York: Praeger, 1957, pp. 54–5. Also see Omissi, *Air Power*, pp. 167–8 on this point.
53. Wilson to the Chief of the General Staff, Mesopotamia, 4 Mar. 1920, in Air Staff, Memo, n.d., 58212, CO 730/18, PRO; Trenchard, maiden speech to the

House of Lords, 1930, quoted in Townshend, 'Civilization and "Frightfulness"', p. 155.

54. Tidrick, *Heart-Beguiling Araby*, p. 215 makes a similar point about British rule in the Middle East more generally.

55. J.M. Spaight, *Air Power and War Rights*, London: Longmans Green, 1924, pp. 23–4, 102–3. Most of his examples of airpower's chivalric influence are taken from the campaigns in the Middle East pp. 105–6. Lawrence frequently invoked Coleridge's 'The Rime of the Ancient Mariner' to describe man's conquest of the air, 'as lords that are expected' (see for instance Lawrence to Charlotte Shaw, 9 Dec. 1933, quoted in J. Wilson, *Lawrence*, pp. 911–12). On the perceived chivalry of aircraft, also see Edgerton, *England and the Aeroplane*, pp. 43–58. In France too aircraft were known as 'knights of the air'. See Vennesson, 'Institution and Airpower', p. 48.

56. Haldane to Churchill, 26 Nov. 1921, in *Churchill*, 4/3:1676; Lawrence to Liddell Hart, June 1930, quoted in Mack, *A Prince of Our Disorder*, p. 385. Also see Omissi, *Air Power*, pp. 164–5. Mack says Lawrence did not anticipate the monstrous use to which bombing would be put (*A Prince of Our Disorder*, p. 385); indeed, an early letter confirms that he thought airpower affected irregulars only through its 'moral value' (Lawrence to A.P. Wavell, 21 May 1923, in *The Letters of T.E. Lawrence*, p. 423). Nevertheless, the above comments point to a more selective humanitarianism. Even his guerrilla theory had been rooted in a highly qualified horror of bloodshed: he had written afterwards that he was proudest 'that I did not have any of our own blood shed. All our subject provinces to me were not worth one dead Englishman' ('The Suppressed Introductory Chapter for *Seven Pillars of Wisdom*', in *Oriental Assembly*, p. 144). He does seem to have reformed his view of air control later (see Lawrence to Thurtle, 1933, quoted in *A Prince of Our Disorder*, p. 395).

57. See Elleke Boehmer, '"Immeasurable Strangeness" in Imperial Times: Leonard Woolf and W.B. Yeats', in Howard J. Booth and Nigel Rigby (eds), *Modernism and Empire*, Manchester: Manchester University Press, 2000, p. 99.

58. Meinertzhagen, minute, 29 Mar. 1922, on Cox to S/S CO, 25 Mar. 1922, CO 730/20, PRO. The RAF also doctored its reports to make casualties look smaller (Townshend, 'Civilization and "Frightfulness"', p. 147).

59. F.H. Humphreys to Sir John Simon, 15 Dec. 1932, AIR 8/94, PRO. Stanley Baldwin began to find air warfare utterly repugnant after the conference and called for its abolition, but others vehemently protested the importance of air policing to the colonies. Unsurprisingly, the conference achieved little—particularly after Hitler came to power in 1933 and Germany rejected disarmament altogether. See J. Cox, 'Splendid Training Ground', p. 173; Omissi, *Air Power*, pp. 178–81. See Lindqvist, *A History of Bombing*, p. 49 on similar failures to regulate air war at The Hague in 1923.

60. [Lawrence, June 1930], quoted in Basil Liddell Hart, *The British Way in Warfare* (New York, 1933), p. 159.

61. A.T. Wilson, Note on Use of Air Force in Mesopotamia, 26 Feb. 1921, AIR 5/476, PRO; Office of no. 30 Squadron, RAF, MEF, Baghdad, Report on RAF operations in South Persia, to GOC, 8 Apr. 1919, AIR 20/521, PRO.

62. Some disparaged this tactic as 'bluff'—see 'Old notes on "substitution" (dictated as a basis for a talk to the Parliamentary Army and Air Committees on the 21st June, 1932)', AIR 9/12, PRO—to which its defenders replied that 'bluff' was proof of willingness to take risks, use new technology, and rely on 'racial superiority'.

63. Air Staff, 'On the Power of the Air Force'. It differed crucially from Bentham's Panopticon in that there was no provision for public surveillance of the aerial inspectors—ultimately a source of the British public's increasingly deep suspicion of the British administration in Iraq.

64. Philby, Note on the Khurma dispute, *c.* July 1919, IS/19/37, FO 882/XXI. On the effectiveness of this principle, see Knox, Koweit, telegram, 3 Apr. 1924, CO 727/9, PRO; Glubb, 'Conduct of the operations: 1928–29 Year of Sibilla'.

65. Air Staff, 'On the Power of the Air Force'; CAS, Memo on Air Force Scheme of Control in Mesopotamia, 5 Aug. 1921, AIR 5/476, PRO; Townshend, 'Civilization and "Frightfulness"', pp. 148–9, and pp. 150–1 on a 1922 memo for the Air Staff by the deputy director of operations on 'Forms of Frightfulness'; Townshend, *Britain's Civil Wars*, p. 98. See Lindqvist, *A History of Bombing*, pp. 46–7, on the play of this theory among contemporary military theorists such as J.F.C. Fuller, J.M. Spaight, and Basil Liddell Hart (Lawrence's great admirer and friend). Contemporaries credited this allegedly humane tactic to Lawrence and his guerrilla theory: see, for instance, John Buchan, *Memory Hold the Door*, London: Hodder and Stoughton, 1940, p. 214.

66. Townshend, 'Civilization and "Frightfulness"', pp. 149–50; Wing-Commander, J.A. Chamier, 'The Use of Air Power for Replacing Military Garrisons', *RUSI Journal*, 66 (1921), p. 210, quoted in Corum, 'The Myth of Air Control', p. 66.

67. Glubb, *Arabian Adventures*, p. 148; Glubb, *Story of the Arab Legion*, pp. 149, 159, 161; Genstaff, 'Notes on Modern Arab Warfare Based in the Fighting round Rumaithah and Diwaniyah, July–August 1920', Appendix IX in Haldane, *The Insurrection in Mesopotamia*, p. 333.

68. Bell to her parents, 16 Mar. 1922, quoted in Burgoyne, *Gertrude*, 2:266 (my italics); 'The Akhwan Operations (1928) and the General Service Medal', 6 May 1929, AIR 8/94, PRO (unsigned, but clearly by a high-ranking individual in the RAF in Iraq (other than Salmond)). The argument that Wahhabis were somehow beyond the pale is mutely accepted in Omissi, *Air Power*, p. 170. In fact, not all the victims of bombings in Wahhabi raiding areas were Wahhabis, and bombing was often used there for tax collection.

69. Cf. Townshend, 'Civilization and "Frightfulness"', p. 159; Omissi, *Air Power*, p. 109; Lindqvist, *A History of Bombing*; Dodge, *Inventing Iraq*, p. 64. Many, including Salmond, argued that people were the same everywhere and would respond in the same way to the bomber (through stages of panic, indifference, weariness and longing for peace; See 'Civilization and "Frightfulness"', pp. 149–50; *Air Power*, p. 110). Certainty of this universal pattern underlay the theory of air control. True, when critics appealed to the British memory of being bombarded, the RAF replied that it was 'fantastic to suggest that the psychology of the tribesmen, who spend half their lives shooting each other, is similar to that of an English villager' (1936, quoted in 'Civilization and "Frightfulness"', p. 158), but the particularism here applies not to bombardment's power to enforce submission but to the tribes' ability to cope with it.

70. Leachman, 20 July 1920, quoted in Bray, *A Paladin of Arabia*, p. 391; *A Paladin of Arabia*, p. 406. My italics.

71. Capt. Hon. W. Ormsby Gore, MP, 'The Organization of British Responsibilities in the Middle East', *Journal of the Central Asian Society*, 7 (1920), pp. 95–6.

72. Leith-Ross, 'Tactical Side', pp. 3, 7, 8.

73. See Trenchard, personal letter to Sir W. Tyrrell, 8 May 1928, printed as CP 160 (28) 'Some Problems of Air Power Illustrated by the Recent Operations in North and South Arabia', and Trenchard to Lord Reading, 13 Apr. 1931, in Private Papers of Rufus Daniel Isaacs (Reading), IOR: Eur Mss F118/86, BL; Glubb, *War in the Desert*, pp. 69–70; *Arabian Adventures*, p. 32. This relationship between airmen and agents on the ground was formally enshrined in Draft Chapters of RAF manual, enclosed in Air Ministry to AOC RAF Uxbridge, 5 July 1933, AIR 5/1203, PRO. On the ground, agents continued to experience difficulties observing the lie of the land and often asked the RAF to fly them over their area to better learn the terrain.

74. Glubb, *Arabian Adventures*, p. 125.

75. E.B. Howell, Deputy Civil Commissioner of Mesopotamia, personal letter re: Arab amir for Iraq, 4 Dec. 1918, Baghdad, IOR: L/PS/10/755, BL.

76. [Flight Lieut.? AI5], Future Intelligence Organisation in Iraq, 21 July 1930, AIR 2/1196, PRO; [document on air intelligence in Iraq], n.d., AIR 2/1196, PRO. In 1921, a Secret Service Committee determined that the SIS should be exclusively responsible for espionage on an inter-Service basis. It would be under Foreign Office control but would retain its military title, MI6. See F.H. Hinsley, *British Intelligence in the Second World War: Its Influence on Strategy and Operations*, 5 vols, New York: Cambridge University Press, 1979, vol. 1, p. 17. The SIS was keenly interested in expanding into the Middle East by the late twenties. This was the last big gap in the SIS organisation—a mark of the region's perceived uniqueness as an object of surveillance. See Minutes of 4th Meeting of sub-committee on SIS in Arabia on 5 Oct. 1926, AIR 4/485, PRO.

77. As for instance in the case of Glubb and Major W.J. Bovill. See J.F.A. Higgins, AVM, AOC, British Forces in Iraq to Air Ministry, 19 Nov. 1926, AIR 5/1254, PRO; Administrative Inspector, Muntafiq Liwa, Nasiriya, Arabia Report, to Advisor to the Ministry of the Interior, Baghdad, 3 Jan. 1925, AIR 23/3, PRO. On the importance of wireless to this work, see Draft Chapters of RAF manual.

78. Bullard and Meinertzhagen, minutes, Sep. 1921, on Cox, telegram, 24 Sep. 1921, 48218, CO 730/5, PRO.

79. CAS to Sir R. Maconachie, 10 Jan. 1933. Tidrick argues that a confident belief that they 'knew' the Arabs helped the British, psychologically, to maintain a commanding position in an area in which international public opinion made it *difficult* to use force (*Heart-Beguiling Araby*, p. 208), but after the war this same confidence underwrote the use of intense force and helped deflect public criticism of it.

80. Slessor, *The Central Blue*, p. 57; Glubb, *The Changing Scenes of Life*, p. 105; Towle, *Pilots and Rebels*, p. 54. Towle is cited repeatedly in much of this secondary literature as well as in Capt. David Willard Parsons, USAF, 'British Air Control: A Model for the Application of Air Power in Low-Intensity Conflict', *Airpower Journal*, Summer 1994, http://www.airpower.maxwell.af.mil/airchronicles/apj/apj94/parsons.html, which extends the ideas of Iraq's peculiar suitability to air operations up to the Persian Gulf War and then explores the uses of air control in Bosnia. See Corum, 'The Myth of Air Control', pp. 62 n.2, 73 n. 76, 73 n. 79, 73 n. 84 for other recent American works looking to the RAF in Iraq as a model. Corum cautions against swallowing this 'Myth of Air Control.'

81. Glubb, *War in the Desert*, pp. 66–7, 92, 94–5, 104, 179, 215, 255; *Arabian Adventures*, pp. 63, 93, 94, 147; Report on the defensive operations against the Akhwan, Winter 1924–5. Also see Codrington, 'Gathering Moss', pp. 195, 211, 215, 332; Mann to Major Cumberbatch, 31 Aug. 1919, Najaf, and to his mother, 11 Nov. 1919, in *Administrator in the Making: James Saumarez Mann, 1893–1920*, ed. his father (London, 1921), pp. 149, 168; Edmonds to 'Chick', 8 Dec. 1926, Edmonds Papers, Box 12, File 2; 12 Mar. 1923, in diary, Edmonds Papers.

82. Brooke-Popham, 'Aeroplanes in Tropical Countries'; Glubb, 'Monotony of the Desert', in Appendix I: Flying in the Desert, in Report on the defensive operations against the Akhwan, Winter 1924–5; Keith, 30 Apr. 1929, Mosul, in *Flying Years*, pp. 240–1; Hill, *To Know the Sky*, pp. 96–7.

83. In explaining the Amritsar massacre, Derek Sayer similarly argues that the British viewed India as a different moral universe; massacre was viewed as a 'moral education' for a subject population configured in this case as children ('British Reaction to the Amritsar Massacre, 1919–1920', *Past and Present*, 131 (1991), pp. 130–64). The British critique of this kind of moral relativism is also virtually as old as the Eastern empire—in response to Warren Hasting's defence that

the arbitrary exercise of power was the norm in India, Burke insisted, 'the laws of morality are the same everywhere'. It is perhaps a measure of the historically complex play of moral relativism and universalism in British liberal thought that Burke found Thomas Paine's defence of the *Rights of Man* equally objectionable.

84. Quoted in Townshend, 'Civilization and "Frightfulness"', p. 159.

85. Quoted in Omissi, *Air Power*, p. 154. The draft of the Air Staff's 'Notes on the Method of Employment of the Air Arm in Iraq', presented to Parliament in Aug. 1924, carried this sentence almost verbatim. Later drafts omitted it and stressed air control's humaneness.

86. Churchill, quoted in Gerard J. De Groot, 'Why Did They Do It?' *Times Higher Education Supplement*, 16 Oct. 1992, 18; W.G. Sebald, *On the Natural History of Destruction*, trans. Anthea Bell (1999; trans., New York, 2003), 19–24. Omissi argues that air policing had little influence on the development of the RAF, partly because he, like others, is refuting a historiography that blames the RAF's inadequate state of preparation for operations against an industrial power in 1939 on its imperial preoccupations (*Air Power*, pp. 147–9, 134–5, 210; Meilinger, 'Trenchard and "Morale Bombing"', p. 244; J. Cox, 'A Splendid Training Ground', p. 176; Clayton, '"Deceptive Might"', p. 286). But Harris's presence in Iraq is not merely a fact made trivial by the dissonance of interwar RAF views on strategic bombing; air control did train the RAF in bombardment, and it was the only significant British experience of bombing before the Second World War. Two and a half times as many British pilots served in Iraq as elsewhere ('A Splendid Training Ground', p. 176). Harris himself testifies that his faith in the heavy bomber as the only possible salvation against the Germans was grounded in his past experience in the Middle East (*Bomber Offensive*, pp. 9–33). It was in Iraq that Harris, bored with his work in a troop-carrying squadron, first laid the foundation of the long-range heavy bomber—by crudely but effectively converting a transport plane—and developed night bombing strategies as a means of terrorising Arabs into thinking aeroplanes could see them even in the dark. At his side were Fl. Lt RHMS Saundby and Hon. R.A. Cochrane, later his right-hand men in Bomber Command. The difference in British and American understandings of strategic bombing— the former favouring general area bombing and the latter 'precision' (see Biddle, 'British and American Approaches', pp. 115–16; W. Hays Parks, '"Precision" and "Area" Bombing: Who Did Which, and When?' in Gooch (ed.), *Air Power*, pp. 145–74)—can also be traced to British experts' feeling, based on the Iraqi experiment, that accuracy was of less account than moral effect. The RAF itself thought it was getting good training for its future role. See Extracts from First Lord of the Admiralty, letter, 26 Dec. 1922, AIR 8/57, PRO; Graves, *Lawrence and the Arabs*, p. 395. To be sure, this is not the whole story of how bombard-

ment became part of this world—it cannot explain the Japanese attacks on China, the Spanish Civil War, the Italians in Ethiopia, the Blitz, Hiroshima and the rest—but it is striking that Erich Ludendorff had British colonial bombings in mind when he wrote *Der Totale Krieg* (1935) (Lindqvist, *A History of Bombing*, p. 68).

87. Peck, 'Aircraft in Small Wars', p. 545.

88. Townshend, 'Civilization and "Frightfulness"', p. 158.

89. A view most famously propounded in Hannah Arendt, *Imperialism: Part Two of 'The Origins of Totalitarianism'*, 1951; repr. edn, New York, 1968, p. 101.

90. Lt Col. David J. Dean, USAF, *The Air Force Role in Low-Intensity Conflict*, Maxwell Air Force Base, Alabama 1986, Chapter 2: 'Air Power in Small Wars: The British Air Control Experience'.

91. Lindqvist, *History of Bombing*, p. 68. See also Malcolm Smith, *British Air Strategy between the Wars*, Oxford: Oxford University Press, 1984, p. 29.

92. Quoted in Dexter Filkins, 'Tough New Tactics by U.S. Tighten Grip on Iraq Towns', *New York Times*, 7 Dec. 2003, p. 13.

93. Consultant quoted in Seymour Hersh, 'Up in the Air: Where is the Iraq War Headed Next?' *New Yorker*, 5 Dec. 2005, available at http://www.newyorker.com/fact/content/articles/051205fa_fact

94. Air Policy with Regard to Iraq, n.d. [Oct.–Nov. 1929], AIR 2/830, PRO; Air Staff, Note on the Status of the RAF in Iraq when that country becomes a member of the League of Nations, 7 Sep. 1929, AIR 2/830, PRO.

95. This was explicitly stated in Salmond, Report on Command.

96. Air Policy with Regard to Iraq, n.d. [Oct.–Nov. 1929], AIR 2/830, PRO; Air Staff, Note on the Status of the RAF in Iraq when that country becomes a member of the League of Nations, 7 Sep. 1929, AIR 2/830, PRO.

97. Draft memo based on Flood's draft letter for Cabinet discussion, 7 Oct. 1932, E4224, FO 371/16041, PRO; Ludlow-Hewitt to Air Ministry, 22 May 1931, AIR 2/1196, PRO; Oswald Scott, Baghdad to Rendel, FO, 13 Aug. 1937, E5095, FO 371/20794, PRO; Air Ministry to G.W. Rendel, FO, 4 Dec. 1933, AIR 2/1196, PRO; Barnes, minute to Rendel, 1 Dec. 1933, E7529, FO 371/16925, PRO.

## 11. TARGETING AFFECTIVE LIFE FROM ABOVE: MORALE AND AIRPOWER

1. Harlan Ullman and James P. Wade, *Shock and Awe: Achieving Rapid Dominance*, Washington, DC: National Defense University, 1996, pp. 20, 61, 62.

2. Manuel Delanda, *War in the Age of Intelligent Machines*, New York: Zone Books, 2003.

3. On the constitutive role of hope see Ben Anderson, 'Hope for Nanotechnology:

Anticipatory Knowledge and the Governance of Affect', *Area*, 19 (2007), pp. 156–65.

4. Sven Lindqvist, *A History of Bombing*, London: Granta, 2001. For a summary of this work see: Peter Adey, 'Aeromobilities: Geographies, Subjects, and Vision', *Geography Compass*, 2, 5 (2008), pp. 1318–36.

5. See for example; Philip Meilinger, 'Ten Propositions Regarding Airpower', *Airpower Journal*, 10, 1 (1996), pp. 52–72; Daniel Liddell, 'Operational Art and the Influence of Will', *Marine Corps Gazette*, 82, 2 (1998), pp. 50–5; Karl Mueller, 'The Essence of Coercive Air Power: A Primer for Military Strategists', *Air and Space Power Journal—Chronicles*, online journal, 2001 (available at: http://www.airpower.maxwell.af.mil/airchronicles/cc/mueller.html; accessed 15 Feb. 2008); Ellwood Hinman, *The Politics of Coercion: Toward a Theory of Coercive Airpower for post-Cold War Conflict*, Maxwell: Air University Press, 2002; Kenneth Rizer, 'Bombing Dual-Use Targets: Legal, Ethical, and Doctrinal Perspectives', *Air and Space Power Journal*, 2001 (available at: www.airpower.maxwell.af.mil/airchronicles/cc/Rizer.html; last accessed 7 Mar. 2007).

6. On naming as evocative and referential see: Denise Riley, *Impersonal Passion: Language as Affect*, London: Duke University Press, 2005; Alexander Galloway and Eugene Thacker, *The Exploit: A Theory of Networks*, London: University of Minnesota Press, 2007.

7. The analysis is based on texts of three types: current US military doctrine; a literature on effects-based airpower generated in the intersections between think tanks, universities and the military before and after the occupations of Iraq and Afghanistan; and texts in the 'classic' literature on airpower. All material is publically available.

8. See Riley, 'Impersonal Passion'.

9. For citation of this special issue in the context of counterinsurgency discussion see Charles Dunlap Jr, 'Air-Minded Considerations for Joint Counterinsurgency Doctrine', *Air and Space Power Journal*, Winter 2007, (http://www.airpower.maxwell.af.mil/airchronicles/apj/apj07/win07/dunlap.html; last accessed 23 May 2008).

10. Eri Ash, 'Terror Targeting: The Morale of the Story', *Aerospace Power Journal*, 13, 4 (1999), pp. 33–47.

11. Richard Ek, 'A Revolution in Military Geopolitics', *Political Geography*, 19 (2000), pp. 841–74; A. Latham, 'Warfare Transformed: A Braudelian Perspective on the "Revolution in Military Affairs"', *European Journal of International Relations*, 8, 2 (2002), pp. 231–66. It is worth noting that 'effects-based' targeting extends back to the targeting of 'vital nodes' in World War Two; Philip Meilinger, 'A History of Effects-Based Operations', *The Journal of Military History*, 71 (2007), pp. 139–68.

12. US Joint Forces Command, J9 Joint Futures Lab, 'A Concept for Rapid Deci-

sive Operations', RDO White Paper 2.0. Norfolk, VA; US Joint Forces Command.

13. On different network forms see John Arquilla and David Ronfeldt, *The Advent of Netwar*, Santa Monica, CA: RAND, MR-789-OS, 1996.

14. Mick Dillon. 'Poststructuralism, Complexity and Poetics', *Theory, Culture and Society*, 17, 1 (2000), pp. 1–26.

15. Samuel Weber, *Targets of Opportunity: On the Militarization of Thinking*, New York: Fordham University Press, 2005.

16. On network movement see Arquilla and Ronfeldt, *Networks and Netwars*; John Arquilla and David Ronfeldt, 'Swarming—The Next Face of Battle?' 2003 (available at: http://www.rand.org/commentary/2003/09/29/AWST.html; last accessed 23 July 2007).

17. For debates on the 'centre of gravity' see Milan Vego, *Operational Warfare*, Newport, RI: Naval War College, 2002; Antulio Echevarria, *Clausewitz's Center of Gravity: Changing our Warfighting Doctrine—Again!* Carlisle, PA: US Army War College, Strategic Studies Institute, 2002.

18. Hannah Arendt, *The Human Condition*, Chicago: University of Chicago Press, 1958, p. 220.

19. Ibid., p. 245

20. Dunlap, 'Air-Minded Considerations'.

21. Especially: USSBS, *Effects of Bombing* on *German Morale*, 2 vols, European Survey Report no. 64B Morale Division, Washington, DC, 1947; USSBS, *The Effects* of *Strategic Bombing* on *Japanese Morale*, Pacific Survey Report no. 14 Morale Division, Washington, DC, 1947.

22. On morale in the context of 'total war' see, for example: James Ulio, 'Military Morale', *The American Journal of Sociology*, 47, 3 (1941), pp. 321–30; Robert Park, 'Morale and the News', *The American Journal of Sociology*, 47, 3 (1941), pp. 360–77.

23. Ash, 'Terror Targeting', p. 33.

24. JWARS is a 'campaign level' computer simulation war gaming tool that includes 'soft factors' such as the interplay of morale and cohesion as it affects 'will to fight'. 'Morale and cohesion' is given a numerical value on the basis of a set of sub-elements including: the amount of time soldiers and leaders serve together, unit pride, unit discipline, leader-to-subordinate loyalty, soldier-to-soldier loyalty and unit morale. 'Perception assessment matrices' are used in counterinsurgency operations to establish a surface of contact between psychological processes and counterinsurgency operations.

25. For discussion of morale/will in the context of the role of 'cognitive' or 'psychological' effects see Donald Chisholm, 'The Risk of Optimism in the Conduct of War', *Parameters*, Winter 2003, pp. 114–31; Grant Hammond, *The Mind of War: John Boyd and American Security*, Washington: Smithsonian Insti-

tution Press, 2001; Stephen Carey and Robyn Read, 'Five Propositions Regarding Air Power', *Air & Space Power Journal*, 20, 1 (Spring 2006), pp. 63–74; David Deptula, 'Effects-Based Operations', *Air & Space Power Journal*, 20, 1 (Spring 2006), pp. 4–5.

26. Mckenzie Wark, *Gamer Theory*, Cambridge, MA: Harvard University Press, 2007.

27. John Warden, 'The Enemy as a System', *Airpower Journal*, 9, 1 (1996), pp. 41–55.

28. Philip Meilinger, 'Air Strategy: Targeting for Effect', *Aerospace Power Journal*, 13, 4 (1999), pp. 48–61.

29. Ibid., p. 50.

30. Ibid.

31. Ibid.

32. Weber, 'The Militarisation of Thinking'.

33. Levy cited in Samuel Weber, 'Special Effects and Theatricality', *Emergences*, 10, 1 (2000), pp. 119–26. See also Pierre Levy, *Becoming Virtual: Reality in the Virtual Age*, New York: Plenum Trade, 1999.

34. Even though unlike in classic airpower theory civilians are not targeted 'directly'—compare with Gulio Douhet, *Command of the Air*, New York: Arno Press, 1972 [1921].

35. Estimates suggest a total of around 80 million leaflets were dropped and a total of 610 hours of TV and radio were broadcast in the first phase of 'Operation Iraqi Freedom'. The stated aim was to stop Iraqi troops fighting for the regime; Steven Collins, 'Mind Games', *Nato Review*, Summer 2003 (available at: http://www.nato.int/docu/review/2003/issue2/english/art4.html; last accessed 12 July 2008).

36. See US Army/Marine Corps, *Counterinsurgency Fieldmanual*, Chicago and London; University of Chicago Press, 2007. The consensus in an emerging literature on counterinsurgency and airpower is that airpower has a logistical role in support of operations, a central role in 'precision' killing and interdiction of insurgent-held territory, and information superiority, and various 'non-kinetic' actions such as perception management. See: James Corum and Wray Johnson, *Airpower in Small Wars: Fighting Insurgents and Terrorists*, Kansas: University Press of Kansas, 2003; see special issue of *Air and Space Power Journal* on 'irregular warfare' and airpower, Winter 2007 (available at: http://www.airpower.maxwell.af.mil/airchronicles/apj/apj07/win07.html; last accessed 10 Feb. 2008); Ronald Stuewe, 'One Step Back, Two Steps Forward: An Analytical Framework for Airpower in Small Wars', *Air and Space Power Journal*, Spring 2006 (available at: http://www.airpower.maxwell.af.mil/airchronicles/apj/apj06/spr06/stuewe.html; last accessed 10 Feb. 2008).

37. For an analysis of the biopolitical logic of focusing on 'popular support' in counterinsurgency see Ben Anderson, 'Population and Affective Perception: Biopol-

itics and Anticipatory Action in US Counterinsurgency Doctrine', *Antipode*, 43 (2011), pp. 205–36.

38. Ullman and Wade, 'Achieving Rapid Dominance', p. 64.
39. COIN FM: 5–27.
40. For discussions of effects-based airpower see US Joint Warfighting Centre, *Operational Implications of Effects-Based Operations (EBO)*, Washington: US Government, 2004; S. Smith, *Complexity, Networking and Effects-Based Approaches to Operations*, Washington: Center for Advanced Concepts and Technology, 2006; Vego, *Operational Warfare*; Robert Freniere, John Dickmann and Jeffrey Cares, 'Complexity Based Targeting: New Sciences Provide Effects', *Air and Space Power Journal*, 2003 (available at: http://www.airpower.maxwell.af.mil/airchronicles/apj/apj03/spr03/freniere.html; last accessed 10 July 2006).
41. RAND, *Effects-Based Operations: A Grand Challenge for the Analytical Community*, Santa Monica: RAND, 2001, p. 7.
42. Both 'full spectrum' counterinsurgency and action over the 'total situation' in rapid dominance involve simultaneous action over all dimensions of life without limit. The counterinsurgency manual, in a fairly typically statement, stresses that: 'Counterinsurgency (COIN) operations require synchronized application of military, paramilitary, political, economic, psychological, and civic actions' (COIN FM: 5–1).
43. Weber, 'Special Effects'.
44. Peter Sloterdijk, 'Airquakes', *Environment and Planning D: Society and Space*, 27, 1 (2009), pp. 41–57.
45. Michel Foucault, *The Birth of Biopolitics*, trans. Graham Burchell, London: Palgrave Macmillan, 2008 [2004]. Environmental-type interventions function on a far-from-equilibrium environment by modulating the aleatory 'transversal processes' that make up that environment, p. 261. See also Massumi, 'Future Birth', and Anderson 'Population and Affective Perception' for an expansion of Foucault's brief comments on environmentality.
46. Dunlap 'Air-Minded Considerations', p. 2. See also Major Jon Huss on how the B-52 offers 'extreme psychological effect' due to the noise, intensity and duration of airstrikes that causes what Huss terms 'significant emotional events in the lives of survivors', p. 28; Jon Huss, 'Exploiting the Psychological Effects of Airpower: A Guide for the Operational Commander', *Airpower Journal*, 13, 4 (1999), pp. 23–32.
47. Robyn Read, 'Irregular Warfare and the US Airforce', *Air and Space Power Journal*, 2007 (available at: http://www.airpower.maxwell.af.mil/airchronicles/apj/apj07/win07/read.html; last accessed 10 Mar. 2008).
48. COIN FM: E-5.
49. COIN FM: E-5.
50. COIN FM: E-6

51. See Anderson 'Population and Affective Perception' for an expansion on this point.
52. John Bellflower, 'The Indirect Approach', *Armed Forces Journal*, Jan. 2007 (available at: http://www.armedforcesjournal.com/2007/01/2371536; last accessed 10 May 2008).
53. Ibid., no pagination.
54. Galloway and Thacker, 'The Exploit'.

## 12. ECOLOGIES OF THE WAYWARD DRONE

1. BBC News, 'Mexico Drone Crashes in El Paso in Texas', 17 Dec. 2010, http://www.bbc.co.uk/news/world-us-canada-12024766
2. Alison J. Williams, 'Blurring Boundaries/Sharpening Borders: Analysing the US's Use of Military Aviation Technologies to Secure International Borders, 2001–2008', in D. Wastl-Walter (ed.), *The Ashgate Research Companion to Border Studies*, Aldershot: Ashgate, 2011, pp. 283–300.
3. CNN, 'Drone Crash in El Paso under Investigation', 17 Dec. 2010, http://edition.cnn.com/2010/US/12/17/texas.mexican.drone/index.html
4. Diana Washington Valdez and Daniel Borunda, 'Mexican Drone Crashes in Backyard of El Paso Home', El Paso Times, 17 Dec. 2010, http://www.elpasotimes.com/ci_16875462
5. Aeronautics-sys.com, 'Orbiter—UAV System', 2007, http://www.aeronautics-sys.com/orbiter_mini_uav_muas
6. Bill Yenne, *Attack of the Drones: A History of Unmanned Aerial Combat*, St Paul: Zenith Press, 2004.
7. See Williams, 'Blurring Boundaries'.
8. American Border Patrol, http://www.americanborderpatrol.com/, 2013.
9. Parc Aberporth is the UK's drone training centre, housing a number of drone manufacturers and providing restricted airspace that can be used for aerial testing, see West Wales UAV Centre, http://www.wwuavc.com/index.html, 2011.
10. For more information about these and other drone crashes, see Drone Wars UK, Drone Crash Database, http://dronewarsuk.wordpress.com/drone-crash-database/, 2013.
11. Ibid.
12. Ibid.
13. See, for example, Free Republic, 'Air Crash Near Bushehr, Drones Slam into Reactor Dome', 17 Aug. 2010, http://www.freerepublic.com/focus/f-news/2572492/posts
14. Washington Valdez and Borunda, 'Mexican Drone Crashes in Backyard of El Paso Home'.
15. Globalsecurity.org, 'RQ-4A Global Hawk', http://www.globalsecurity.org/intell/systems/global_hawk-program.htm, 2012.

16. Robert Kaplan, 'Hunting the Taliban in Las Vegas', *Atlantic Monthly*, Sep. 2006, http://theatlantic.com/doc/print/200609/taleban-vegas

17. Bohemia Interactive Solutions, 'Virtual Battlespace 2', http://products.bisimulations.com/products/vbs2/overview, 2011.

18. Tim Blackmore, *War X: Human Extensions in Battlespace*, Toronto: University of Toronto Press. 2005.

19. Ellen Nakashima and Craig Whitlock, 'With Air Forces' Gorgon Drone "We Can See Everything"', *Washington Post*, 2 Jan. 2011, http://www.washingtonpost.com/wp-dyn/content/article/2011/01/01/AR2011010102690.html

## 13. SATELLITE IMAGES, SECURITY AND THE GEOPOLITICAL IMAGINATION

1. Details on the Syrian case come from Peter Hart, 'Coverage of Syria Airstrike', *FAIR Extra!* Dec. 2007; 'Syria had "Covert Nuclear Scheme"', BBC News Online, 25 Apr. 2008; and 'Bush Administration Release Images to Bolster Claim about Syrian Reactor', *New York Times*, 25 Apr. 2008.

2. David Albright and Paul Brannan, *Suspect Reactor Construction Site in Eastern Syria: The Site of the September 6 Israeli Raid?* Washington: The Institute for Science and International Security, 23 Oct. 2007; David Albright, Paul Brannan and Jacqueline Shire, *Syria Update: Suspected Reactor Site Dismantled*, Washington: The Institute for Science and International Security, 25 Oct. 2007; David Albright and Paul Brannan, *The Al Kibar Reactor: Extraordinary Camouflage, Troubling Implications*, Washington: The Institute for Science and International Security, 12 May 2008.

3. 'Colin Powell Addresses Security Council', CNN.com, 5 Feb. 2003.

4. 'Yet Another Photo of Site in Syria, Yet More Questions', *New York Times*, 27 Oct. 2007.

5. This historical overview draws on Jeffrey T. Richelson, 'U.S. Satellite Imagery, 1960–1999', *National Security Archive Briefing Book No. 13*, 14 Apr. 1999; and Laurie J. Schmidt, 'New Tools for Diplomacy: Remote Sensing Use in International Law', *Earth Observatory*, NASA Earth Science Enterprise Data Center, 12 Jan. 2001.

6. Richelson, 'U.S. Satellite Imagery, 1960–1999'.

7. Lisa Parks, *Cultures in Orbit: Satellites and the Televisual*, Durham, NC: Duke University Press, 2005.

8. Chad Harris, 'The Omniscient Eye: Satellite Imagery, "Battlespace Awareness", and the Structures of the Imperial Gaze', *Surveillance and Society*, 4, 1/2 (2006), p. 101.

9. 'Colin Powell Addresses Security Council', CNN.com, 5 Feb. 2003.

10. Harris, 'The Omniscient Eye: Satellite Imagery, "Battlespace Awareness", and

the Structures of the Imperial Gaze', p. 106; Richelson, 'U.S. Satellite Imagery, 1960–1999'.

11. Parks, *Cultures in Orbit*, p. 91.

12. Office of the Director of National intelligence, 'Background Briefing with Senior U.S. Officials on Syria's Covert Nuclear Reactor and North Korea's Involvement', GlobalSecurity.org, 24 Apr. 2008.

13. Schmidt, 'New Tools for Diplomacy'; Richelson, 'U.S. Satellite Imagery, 1960–1999'.

14. Amnesty International, 'Zimbabwe: Satellite Images Provide Shocking Evidence of the Obliteration of a Community', Press Release AFR 46/008/2006, 31 May 2006.

15. Barbara Cochran, 'Fighting the Feds over Shutter Control', Radio Television Digital News Association, Dec. 1999; David Corn, 'Their Spy in the Sky', *The Nation*, 8 Nov. 2001; Jennifer LaFleur, 'Government, Media Focus on Commercial Satellite Images', *The News Media and the Law*, Summer 2003.

# INDEX

Page numbers in *italics* indicate illustrations

363

# INDEX

*Airborne Camera* 30
airmindedness 146–61
*Airopaidia* 35, 36–7
al-Qaeda 58, 59, 63, 64, 271
Al Udeid Air Base 54
Albright, David 290
Allied Central Interpretation Unit 291
Almond, Ian 177
American Association for the
    Advancement of Science 296
American Border Patrol 267–9
*American City Magazine, The* 134
American Civil War 12, 29
American Technicolor 166
Amnesty International 13, 296
Anacostia River, USA *122*
Anderson, Ben 14
Antarctic Place Names Committee 93
Antarctic Treaty 1959 92
Antarctica 4, 6, 71–93
Apollinaire 172
Apollo 8 9
Apollonian gaze 15, 22, 294
Arab Revolt 227–30
Arabia 224–42
archaeology 2, 120
Argentina 72, 73, 76, 79, 80, 82, 83,
    85, 92
ARGUS-IS 56, *57*
*Armed Forces Journal* 259
Armed Services Committee 53
Armitage, John 5, 10
Armstrong, Neil 204
*Arnold*, Henry Harley 42, 43–4
Arquilla, John 102
artificial lighting *188*, 196–7, 198
*Atlantis* space shuttle 203–4, 207–8
*Atlas Eclipticalis* 209
atomic bombings of Hiroshima and
    Nagasaki 44, 168, 183, 190, 198
Auschwitz concentration camp 292
Auster aircraft 84

Australian Air Force 274
autoscopy 201–2
*Avatar* 180
*Aviation Week and Space Technology*
    204, 208, 213
Avro *Lancaster bombers 42*

B-17 Flying Fortress *100*
B-52 Stratofortresses 1, 47–9, 68,
    111, 258
Báky, Josef von 166
Baldwin, Stanley 152, 153
Baldwin, Thomas 38, 40
    'A Balloon Prospect from Above
        the Clouds' *34*, 35
Balla, Giacomo 129
'Balloon Prospect from Above the
    Clouds, A' *34*, 35
balloons 4, 7, 15, 24, 25–7, 29–40,
    *34*, 75, 165, 291
Barbari, Jacopo de' 33
Barthes, Roland 127, 135, 140
'The Eiffel Tower' 121–2, *123*
bats 190–91
Battle of Fleurus 24, 25–6, 31
Baudrillard, Jean 97, 165, 174–8,
    184, 189, 196
    'Evil Demon of Images, The' 175
    *Gulf War Did Not Take Place, The*
        176
    *Impossible Exchange* 185
Beard, Jack 65–6
Beck, John *Dirty Wars* 190–91
Bečvář, Antonín: *Atlas Eclipticalis* 209
Bedouins 229, 231, 233–4, 236–7
*Being and Time* 110
Bell 47D 90
Bell Eagle Eye 275
Bell OH-58 Kiowa 63
Bell, Gertrude 228, 237
Bellaschi, P.L. 198
Benjamin, Walter 171

# INDEX

# INDEX

'East Coast Glacier, The' *86*
'Helicopter Wreck, A' *91*
Mount Weather Emergency Operations Center 199
Museum of Modern Art (MoMA) 30

Nagasaki 44, 168, 183, 190, 198
Nagl, John 63
Nakhon Phanom Royal Thai Air Force Base 52
Nancy, Jean-Luc: *Ground of the Image, The* 185
napalm 42, 53
Napoleon I, king of the French 27
*National Aeronautics and Space Administration* (NASA) 9, 203–19
National Photographic Interpretation Center 295
National Reconnaissance Office (NRO) 204–19, 297
National Security Agency (NSA) 204, 210
National Security Council (NSC) 46, 205
National Transportation Safety Board 264, 272
Naval Ocean Surveillance Satellites (NOSS) 211
Nazism 166–8, 184, 292
Neidrick, Pierre 211
Neocleous, Mark 3
Network Centric Warfare (NCW) 101–2, 251–61
New America Foundation 115
*New Scientist* 50
New York City 122, 134–8, *136*, *138*
*New York Times* 46, 208, 219
Newcombe, Stewart 228
Newhall, Beaumont 32
    *Airborne Camera* 30
Newton, Isaac 215
Nicaragua 259

Nietzsche, Friedrich Wilhelm 216–17
NIMBUS 197, 200
Nixon, Richard Milhous 47, 50
nodal targeting 113–17
NORAD (North American Aerospace Defense Command) 199
North Atlantic Treaty Organization (NATO) 54, 114, 115
North Korea 295–6
Northrop Grumman RQ-4 Global Hawk 266, *267*, 275–6, *276*
Norway 72, 77, 78
Nostovi 208

Obama, Barack 58, 243, 244
Obsreduce 210
*Oluf Sven* 90, 91
ontology 107–12, 278–87
Operation Arc Light 47–9
Operation Desert Storm 104–5
Operation *Igloo White* 52
Operation Linebacker 47, 50
Operation Moonwatch 212
Operation Ranch Hand 51
Operation Rolling Thunder 45–7, 52
Operation Tabarin 77, 80–81, 82
Orbiter Mini UAV 263–9, 271, 273, 274, 286
Orford, Anne 66
Orwell, George *1984* 217
Ottoman Empire 224
Ovid 189, 193

Paglen, Trevor 9, 11
*Pakistan 4*, 41–2, 43–4, 58, 60, 65, 66, 68, 114, 115–16, 199, 243
Palestine 2
panopticon 23–4, 31, 39, 284
Paris 26, *100*, 122–3
Parks, Lisa 292, 295
Pastrone, Giovanni 172–3
Perception Assessment Matrices 253

# INDEX

# INDEX